W0227654

Meteorological Essays and Observations

J. FREDERIC DANIELL

CAMBRIDGE
UNIVERSITY PRESS

CAMBRIDGE UNIVERSITY PRESS

Cambridge, New York, Melbourne, Madrid, Cape Town,
Singapore, São Paolo, Delhi, Mexico City

Published in the United States of America by Cambridge University Press, New York

www.cambridge.org
Information on this title: www.cambridge.org/9781108056571

© in this compilation Cambridge University Press 2013

This edition first published 1827
This digitally printed version 2013

ISBN 978-1-108-05657-1 Paperback

This book reproduces the text of the original edition. The content and language reflect the beliefs, practices and terminology of their time, and have not been updated.

Cambridge University Press wishes to make clear that the book, unless originally published by Cambridge, is not being republished by, in association or collaboration with, or with the endorsement or approval of, the original publisher or its successors in title.

CAMBRIDGE LIBRARY COLLECTION

Books of enduring scholarly value

Physical Sciences

From ancient times, humans have tried to understand the workings of the world around them. The roots of modern physical science go back to the very earliest mechanical devices such as levers and rollers, the mixing of paints and dyes, and the importance of the heavenly bodies in early religious observance and navigation. The physical sciences as we know them today began to emerge as independent academic subjects during the early modern period, in the work of Newton and other 'natural philosophers', and numerous sub-disciplines developed during the centuries that followed. This part of the Cambridge Library Collection is devoted to landmark publications in this area which will be of interest to historians of science concerned with individual scientists, particular discoveries, and advances in scientific method, or with the establishment and development of scientific institutions around the world.

Meteorological Essays and Observations

By the early nineteenth century, meteorologists were equipped with plenty of useful devices: barometers, thermometers, hygrometers, and any number of variations thereon. But the nature of these instruments was not wholly understood. While it was possible to take accurate measurements with a barometer, what physical process made the mercury move? What exactly is atmospheric pressure? And how can one measure sunlight? Ranging from wild theories of gravity-resistant air particles to the latest experiments in altitude, chemist and physicist John Frederic Daniell (1790–1845) presents his answers in this collection of essays. First published in 1823, this enlarged second edition of 1827 includes his work on the climate of London, the effect of atmospheric conditions on human health, and suggested improvements for the design of a new hygrometer. Daniell later became the first professor of chemistry at King's College, London, and foreign secretary of the Royal Society.

Cambridge University Press has long been a pioneer in the reissuing of out-of-print titles from its own backlist, producing digital reprints of books that are still sought after by scholars and students but could not be reprinted economically using traditional technology. The Cambridge Library Collection extends this activity to a wider range of books which are still of importance to researchers and professionals, either for the source material they contain, or as landmarks in the history of their academic discipline.

Drawing from the world-renowned collections in the Cambridge University Library and other partner libraries, and guided by the advice of experts in each subject area, Cambridge University Press is using state-of-the-art scanning machines in its own Printing House to capture the content of each book selected for inclusion. The files are processed to give a consistently clear, crisp image, and the books finished to the high quality standard for which the Press is recognised around the world. The latest print-on-demand technology ensures that the books will remain available indefinitely, and that orders for single or multiple copies can quickly be supplied.

The Cambridge Library Collection brings back to life books of enduring scholarly value (including out-of-copyright works originally issued by other publishers) across a wide range of disciplines in the humanities and social sciences and in science and technology.

METEOROLOGICAL

ESSAYS AND OBSERVATIONS.

BY

J. FREDERIC DANIELL, F.R.S.

SECOND EDITION, ENLARGED AND REVISED

The wind goeth toward the south, and turneth about unto the north; it whirleth about continually, and the wind returneth again according to his circuits.

All the rivers run into the sea, yet the sea is not full; unto the place from whence the rivers come thither they return again.

ECCLESIASTES, Chap. I.

LONDON:

PRINTED FOR THOMAS AND GEORGE UNDERWOOD,

32, FLEET-STREET.

MDCCCXXVII.

LONDON:
Printed by William Clowes,
Stamford-street.

TO

WILLIAM THOMAS BRANDE, Esq.,

FELLOW OF THE ROYAL SOCIETY, AND PROFESSOR OF
CHEMISTRY AT THE ROYAL INSTITUTION
OF GREAT BRITAIN.

THESE PAGES

ARE INSCRIBED AS A TESTIMONY OF REGARD
AND ESTEEM.

PREFACE

TO

THE SECOND EDITION.

The reasons which originally induced me to adopt the form of Essays in the arrangement of my Observations and Speculations upon Meteorology have acquired new force from further reflection. Too many branches of the science are still open to experiment and discussion to allow of their connexion with advantage, on account of the details which their progressive state requires. In this edition I have, therefore, adhered to my original plan, and have availed myself of the facility which it affords to throw all the new matter into a separate form, which will enable the possessors of the former edition to complete the work by the purchase of the Second Part. This mode of proceeding is attended by some inconveniences, such as repetitions, two or three inconsistencies, &c., which might otherwise have been avoided, but none of which, I conceive, counterbalance its manifest advantages.

I have, however, taken the opportunity of correcting all the greater errors which I have either discovered myself, or which have been pointed out to me by others,—the most important of which is the miscalculation of the Table in the essay upon the Hygrometer, for finding the specific gravity of any mixture of atmospheric air and aqueous vapour. This was first indicated to me by an unknown friend in the Dublin Philosophical Journal, and afterwards by Monsieur Gay Lussac, who most obligingly favoured me with the formula from which the present Table is derived. This corrected table, as being of considerable importance, I have also added to the Second Part.

To my other critics, known and unknown, I beg to return my best acknowledgments, and I have endeavoured to profit by many of their hints. In many instances I am afraid that the praise bestowed upon my labours has been beyond their desert; and if, on one occasion, I have been rather roughly roused from the dream of satisfaction which such approbation was calculated to produce, I trust that I do not retain my opinion upon the point of controversy without a due consideration of all the real argument which has been brought against it.

In one particular the present edition will be found to differ from the last, and that is, in the careful avoidance of all allusion to subjects which, however intense the personal interest which they excite at the moment of their discussion, can be but of ephemeral importance. The motives to this forbearance will not, I trust, be misunderstood.

The assistance and advice of my friend, Capt. Sabine, R.A., have been of the utmost importance to me in the progress of my labours; and the interest which he has kindly taken in my work, from its commencement, has excited me to perseverance, under circumstances, at times, of no encouraging nature.

To Capt. Basil Hall, R.N., my special acknowledgments are also due for the great attention which he has given to my speculations, and for the practical illustrations with which his experience has supplied me. His remarks upon the Trade Winds have, of course, afforded me the highest gratification, and will doubtless be read with interest by all those who think upon Meteorology.

To Lieut. H. Foster, R.N., and to many others, I am indebted for the loan of some interesting registers; and to Mr. Galbraith of Edinburgh, for a very accurate Table of the force of Steam at various temperatures.

I greatly regret the failure of some attempts which have been made to produce more unity of action between those who are practically engaged in advancing our knowledge of atmospheric phenomena. Their labour and perseverance lose more than half their value by the want of a well-digested plan of mutual co-operation. I trust that untoward circumstances have only delayed such a union of those who are interested in the science, and that a Meteorological Society, or Committee, may yet be formed under more auspicious circumstances. If ever this should be effected, they cannot better commence their career than by adopting, as a model, the plan of the Meteorological Society of the Palatinate, of whose constitution and labours some account will be found in the following pages.

AN ESSAY

UPON THE

CONSTITUTION OF THE ATMOSPHERE.

MAN may almost with propriety be said to be a meteorologist by nature: he is naturally placed in such a state of dependance upon the elements, that to watch their vicissitudes and anticipate their disturbances, becomes a necessary portion of the labour to which he is born. The daily tasks of the mariner, the shepherd, and the husbandman, are regulated by meteorological observations; and the obligation of constant attention to the changes of the weather, has endued the most illiterate of the species with a certain degree of prescience of some of its most capricious alterations. Nor, in the more artificial forms of society, does the subject lose any of its universality or interest: much of the tact of experience, indeed, is blunted and lost; but artificial means, derived from science, supply, perhaps inadequately, the deficiency; and the general influence is still felt and acknowledged, though not so accurately appreciated. The generality of this interest is, indeed, so absolute, that the common form of salutation amongst many nations is a meteorological wish, and the first introduction between strangers a meteorological observation.

But although the atmospheric phenomena have excited the attention of all classes of men, from the earliest ages of the world; and have probably formed the most ancient and universal theme of conversation and speculation, both with the learned and unlearned; and although they may have been, daily, nay hourly, discussed since the time when the human race were first exposed to their influence; the observations of the vulgar, and the theories of the philosopher, have been alike insufficient for a rational and satisfactory explanation of their general laws. Many and ingenious are the instruments which the science of modern ages has constructed for the accurate appreciation of these perpetual changes; and diligent have been the observers who have dedicated their time to the science of meteorology: but, from the first contrivance of the barometer to the present day, the great and unceasing fluctuations of the vast aërial ocean, denoted by that instrument, are unexplained. "The wind bloweth where it listeth, and we cannot tell whence it cometh nor whither it goeth." The complication of the processes, carried on in the immense laboratory of nature, the wide-extended circle of their agency, and the ever-varying results of their compound influences, appear to have been too much for the mind to comprehend as a whole; and the powers of reason have been bewildered in the inextricable labyrinth of causes and effects.

Being myself deeply interested in these inquiries, and having devoted much of my time and attention to experimenting upon this subject, it occurred to me to consider, that although the science of meteorology, contemplated as a whole, had lately made but little progress towards perfection; yet, that the parts of which it is composed, comprising nearly the whole circle of the natural sciences, had been by no means stationary; but, on the contrary, were making rapid strides towards perfection. The elements of the science, considered as founded upon experiment and observation, have been largely extended and deeply explored; and a rich accumulation of facts have been collected, which only require, perhaps, to be properly adjusted, to enable us to raise the superstructure with security. I reflected that, in the present state of our knowledge, this might probably be done *synthetically* with the greatest advantage; and that by setting out from a few plain and established principles, and by accurately appreciating their mutual influences, there was a probability of ascending to more complicated relations; till at length, by gradual steps, we might possibly accomplish the explanation of those atmospheric phenomena, the analysis of which has hitherto been perplexed with insurmountable difficulties. This idea has been so strongly impressed upon my mind, that I have resolved to institute such a process, and with this clue, to venture in a path in which so many have failed before me.

Before I proceed, however, to attempt the problem which I have contemplated, it may not be improper to prove the necessity of further illustrating a subject which has already exercised the ingenuity of so many and such distinguished philosophers. For this purpose I cannot, I think, do better than refer to the latest hypothesis upon the cause of the rise and fall of the barometric column, which has recently been advanced by Professor Leslie, in the supplement to the Encyclopædia Britannica*. This distinguished philosopher, previous to offering a solution of the difficulty, passes a sentence upon the attempts of all those who have preceded him in the task, with which the scientific world in general will be disposed to agree.

"Philosophers have eagerly sought to explain the fluctuations of the mercurial column. They have tried every principle that might appear to exert any influence in modifying the local weight of the atmosphere; but their very numerous attempts, it must be confessed, have hitherto been singularly unsuccessful. It was requisite to shew that such causes would not only give results of the kind expected, but were, besides, fully adequate to the production of the phenomena. In most instances, however, either none of those effects could have followed, or they would occur in a very inferior degree and disproportionate extent."

The principal of these attempts he then proceeds to examine; and having shewn their fallacy, pro-

* Ency. Brit., Sup., Article METEOROLOGY, p. 329.

pounds the following explanation, as hitherto overlooked, and capable of furnishing a satisfactory solution of a great variety of phenomena.

"It is obvious that a *horizontal* current of air must, from the globular form of the earth, continually deflect from its *rectilineal* course. But such a deflection being, precisely of the same nature as a centrifugal force, must hence diminish the weight or pressure of the fluid. The only question, is, to determine the amount of that disturbing influence. Though it should appear quite inconsiderable, in the interval of a short space, it may yet accumulate to a very notable quantity, through the wide extent over which the same wind is known to travel: suppose a current to begin to flow from A, Fig. I,

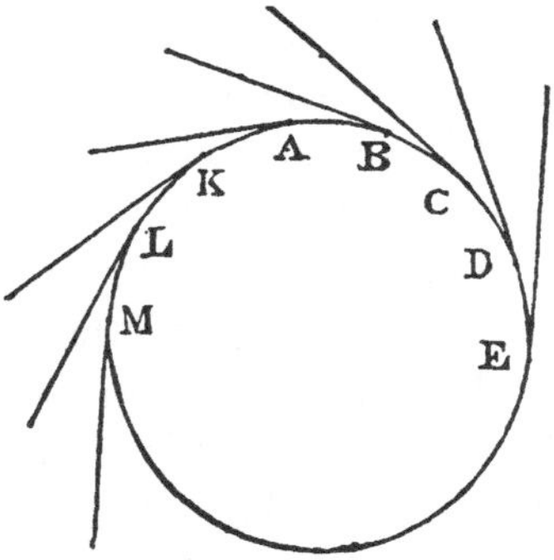

Fig. I.

in the direction of a tangent: it will successively bend from a rectilineal track, at the points B, C, D, &c., on the surface of the earth. The particles of the fluid are, therefore, drawn incessantly from their course by the action of gravity. Their vertical pressure is consequently diminished by the force spent in producing this deflection. Wherefore, during the prevalence of the wind, the atmospheric column will press with inferior weight at B, than at A, at C than at B, at D than at C; thus gradually decreasing through the whole chain. Suppose the intervals A B, B C, C D, D E, &c., to be each of them a mile, and that the current reaches the points B, C, D, E, &c., in successive minutes, a celerity which frequently happens; the deflection at B, owing to the curvature of the earth, would be eight inches, or two thirds of a foot; but the space through which a body would descend in a minute by the action of gravity is, $60 \times 60 \times 16 = 57600$ feet or 86400 times greater than the deviation from the tangent. Wherefore the atmospheric pressure would, on that hypothesis, be diminished by the 86400th part, for each interval of a mile from A to D. In the space of 288 miles, this diminution would consequently be the 300th part of the incumbent weight; and over an extent of 2880 miles, it would amount to the 30th part. If we assume the very probable estimate, that storms involve the whole region of the clouds, or attain an elevation of near three miles, the diminution of pressure, occasioned by a long series of deflections in the stream would affect one half of the atmosphere. Wherefore a wind which has blown over a track of 2880 miles at the rate of 60 miles an hour, might cause the mercurial column to subside half an inch. If the velocity of the wind were doubled, which is probably the limit of the most tremendous hurricane, the fall of the Barometer would be four times greater, and amount to two inches."

Now I conceive, that it will be no very difficult task to shew that the Professor has been as unfortunate as his predecessors, in his proposed solution: and nothing can better illustrate the difficulty of the problem than such a failure. His error, appears to me, to lie in the misapplication of the term *horizontal,* in the first sentence of the above extract: as there applied, it is made to signify *rectilineal,* contradistinguished to parallelism to the surface of the globe. Now what power can be supposed to produce such horizontality as this? Mr. Leslie observes, that deflection from it is "of the same nature as a centrifugal force:" but is it not obvious that it is itself a centrifugal force? And then whence does such an impulse originate? He has not revealed to us the manner in which he supposes the wind, which he employs, to arise, (and this alone is a defect in his theory;) but upon no known principle, I conceive, can its tendency be tangental to the circumference of the earth. But, granting for a moment, the possibility of such a direction, let us suppose a current to begin to flow from A, in the direction of a tangent, and that it is bent from its rectilineal track at the point B by the action of gravity, how is the unknown force to be renewed, so that the wind at B is again to assume a tangental course? But the hypothesis not only supposes this, but further, that it is infinitely renewable; and the effect which is at first scarcely perceptible, is "accumulated by a long series of deflections."

When the foundations of a theory are, as I think, so palpably erroneous, it can scarcely be necessary to remark that, the accordance of the phenomena with it, is not so close as has been supposed; otherwise many inexplicable cases might be adduced to prove its insufficiency. One of the strongest of these is, that wind does not always precede a fall of the mercurial column; but, on the contrary, the greatest depressions of the mercury generally precede a wind. Sometimes also great falls are not attended with wind, and sometimes, the mercury has been depressed to leeward of the storm.

It is the more singular that Professor Leslie should have fallen into this error, as in referring in his treatise to the well-known experiment of Mr. Hauksbee, at the beginning of the last century, he refutes its fallacy in a way which is equally applicable to his own hypothesis; if, indeed, it be not the very same idea, clothed in another dress.

"To explain," says he, "the descent of the Barometer during wind, a very ingenious idea has been proposed, which, being apparently confirmed by experiment, has obtained general reception. It is conceived that a current of air, in sweeping over the surface of the earth, must cease to exert any vertical pressure. But this assumption can hardly be reconciled with any strict principle in science, *for the particles of air will not for a moment cease to gravitate, nor will any horizontal motion of them produce the slightest derangement in a perpendicular direction.*" Now the tangental direction of Mr. Leslie's wind, is nothing less than a *cessation to gravitate*; and its horizontal motion produces *derangement in a perpendicular direction*, in violation of the very law which he has himself so properly and carefully explained.

M. Biot, one of the great masters of science, after adverting to the different hypotheses, which have been framed for the explanation of the motions of the Barometer, thus candidly concludes his review of them all.

"Le parti le plus sage est de considérer ces faits comme des résultats d'observation dont on ne peut jusqu'à présent donner aucune explication satisfaisante. La hauteur du Baromètre éprouve des élévations et des abaissemens qui paraissent ténir aux modifications de l'atmosphère, mais dont la cause est encore inconnue."—*Traité de Physique,* Tome I., p. 95.

The method, which has occurred to me, of viewing this complicated subject, promises to simplify the conditions of the acknowledged problem; and I trust to the candour of the learned, to receive the following attempt at its solution with indulgence.

I pass over the particulars of the chemical composition of the atmosphere, as foreign to my present purpose, and consider it as essentially composed of an homogeneous, permanently-elastic fluid, mixed with varying proportions of condensible elastic vapour. I omit, likewise, its relations to light and electricity, and assume that radiant heat passes through it with little or no interruption.

I shall divide the proposed inquiry, thus restricted, into four parts. In the first, I shall consider the habitudes of an atmosphere of perfectly-dry, permanently-elastic fluid, under certain conditions; in the second, those of an atmosphere of pure, aqueous vapour; in the third, the compound relations of a mixture of the two; and in the fourth, I shall endeavour to apply such principles as may legitimately be deduced from the previous investigation, to some of the observed phenomena of the atmosphere of the earth. Many of the observations which I shall have to make, will appear, at first, trite and uninteresting; but let it be remembered that it is from self-evident axioms that the most complicated problems are solved.

PART I.

On the Habitudes of an Atmosphere of perfectly-dry, permanently-elastic Fluid.

In proposing the following hypothetical cases, my object has been to assimilate the conditions as much as possible to those of the atmosphere of the earth; separating only the phenomena into classes, that, in considering them singly, we may trace, without confusion, the ultimate effects of each simple cause.

I shall, therefore, propose as the first problem, the natural state of an atmosphere, of perfectly-dry permanently-elastic fluid, surrounding a sphere in a state of rest, of uniform temperature in all its parts; to the centre of which it gravitates equally?

Its height, density, and elasticity, would every where be equal, at equal elevations; and the column of mercury, which it would support in the Barometer, would be the same every where at the surface of the sphere. These conclusions rest upon the fundamental laws of Hydrostatics, and need no demonstration here. The first condition, therefore, of its state, must be that of perfect equilibrium and rest.

The second condition is, that its density must decrease in a geometrical progression, in ascending through equal stages to its higher regions: for the density would every where be proportionate to the superincumbent weight.

The calculation of this progression is simple: for in logarithmic curves when the ordinates are the same, the intercepted portions of the abscissæ are proportional to the subtangents, and the process may be conducted in the following manner: We will suppose the length of a column of mercury, at the surface of the sphere, equivalent to the weight of an equal column of the air, to be 30 inches, and its temperature to be 32°; the height of an homogeneous atmosphere, of equal density in all its parts, would therefore be 26250 feet; for the specific gravity of dry air at 32°, and under a pressure equal to 30 inches of mercury, is to that of mercury, as 1 to 10500. From these *data*, the density for any height may be found by the use of Logarithms*; for as the height of the homogeneous atmosphere, or atmospheric subtangent, is to the height proposed, so is the modulus of the common system of Logarithms, or logarithmic subtangent, to the difference of the Logarithms of the densities. Thus, under the conditions just named, let it be required to know the density or elasticity of the atmosphere at the height of 5000 feet;

	Feet.	Feet.	Modulus.	Difference.
then	26250 :	5000 ::	.4342945 :	.0827227.

which, deducted from .4771212 the Logarithim of 30 inches, leaves .3943985, the Logarithm of 24.797 inches, the density required. The following table represents the decrements of Density, for the different heights subjoined.

* See Young's Natural Philosophy. Vol. I. page 272.

TABLE I. *Shewing the Decrease of Density, and Fall of the Barometer at different Heights in an Atmosphere of uniform Temperature throughout.*

Height in Feet.	Barometer Column inches.	Density.	Temp.
0	30.000	1.00000	32
5000	24.797	.82656	
10000	20.499	.68321	
15000	16.941	.56472	
20000	14.000	.46677	
25000	11.575	.38582	
30000	9.567	.31890	

The third condition of the atmosphere must be, that its sensible heat shall decrease progressively from below upwards. Experiment has proved that the specific heat of atmospherical air, relative to its mass, increases as the density diminishes: the absolute quantity, therefore, of heat contained in every part of any vertical column remaining unchanged, this gradation of temperature must naturally flow from the enlarged capacity which the air acquires from rarefaction.

The temperature, due to any given height, may easily be found as follows. Reckoning the density of the air at the surface of the sphere = 1, the difference between the density at any given altitude, and its reciprocal being multiplied by 45, will express the mean diminution.

For example, 30,000 : 24.797 :: 1.000 : .826 which is the density at 5000 feet, the density at the surface being 1. Therefore, .826 : 1.000 :: 1.000 : 1.210, and 1.210 − 826 = 384 × 45 = 17.2, the diminution due to that elevation*. The scale of temperature, appropriate to the preceding heights and densities, is as follows:

TABLE II. *Shewing the Decrease of Density and Temperature due to different Elevations in an Atmosphere of permanently-elastic Fluid.*

Height in feet.	Density.	Temp.
0	1.00000	32.
5000	.82656	14.8
10000	.68321	− 3.1
15000	.56472	−22.4
20000	.46677	−43.6
25000	.38582	−67.5
30000	.31890	−95.1

We have here another cause developed, which, as well as gravity, affects the constitution of permanently-elastic fluids: namely, an alteration of temperature. A difference of 1 degree, upon Fahrenheit's scale, causes a contraction or expansion of $\frac{1}{480}$th part of their volume; which, under equal pressure, proportionally increases or diminishes their specific gravity; or, when confined, raises or depresses their elasticity to the same amount.

From this cause, the barometer alone, will no longer be the exact measure of the progressive density; but for this purpose, its indications must

* See Ency. Brit., Supplement, Article CLIMATE, p. 188.

be associated with those of the thermometer. The mercurial column will be shortened $\frac{1}{480}$th. of its length at the several stages, for each degree of depression due to the elevation; and its fall for equal altitudes will differ by that quantity from the geometrical progression. The following table gives the height of the barometer, the specific gravity, and the temperature for the scale of heights before proposed.

TABLE III. *Shewing the Fall of the Barometer, at different Heights in an Atmosphere, decreasing in temperature in the preceding Progression.*

Height in feet.	Barometer.	Sp. Gravity.	Temp.
0	30.000	1.00000	32.
5000	23.949	.82656	14.8
10000	19.106	.68321	− 3.1
15000	15.229	.56472	−22.4
20000	12.044	.46677	− 43.6
25000	9.579	.38582	− 67.5
30000	7.566	.31890	− 95.1

Such then must be the constitution of an atmosphere of perfectly dry air, surrounding a sphere of the temperature of 32°—perfect equality of pressure producing perfect rest, the specific gravity, pressure and temperature decreasing upwards, according to the above scale, and each being definite for the elevation. These calculations have been made upon the supposition of equal gravity at all heights; a supposition, which is not exactly accordant with fact: the difference, however, is extremely small and wholly unimportant to the general argument.

If the temperature of the sphere be now conceived to rise generally and equally in all its parts, a new adjustment of the gaseous strata must ensue. An increase of elasticity will take place, and the total height will be increased. The expansion must necessarily proceed from below upwards; for the impulse, being equal and simultaneous in each of the columns, into which we may suppose the atmosphere divided, they mutually confine one another in every other direction.

As there is no increase or decrease of ponderable matter in any of the vertical sections, the total pressure will remain as before, and the barometer at their bases will not be affected: but as a different distribution of the weight in the different horizontal sections takes place, its height will be altered in every other situation.

Fig. II.

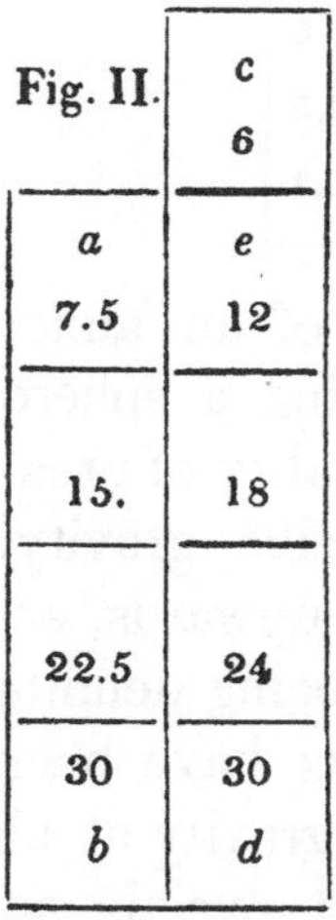

Let *a b* represent a column of fluid, whose total weight is 30, and whose four sections, taken at equal altitudes, are each 7.5: the scale of the progressive heights will be = 30. = 22.5 = 15. = 7.5. Let *c d* be the same column expanded by $\frac{1}{4}$ its length. Its total weight will be 30 as before; but the weights of its sections, taken at the same altitudes, will be reduced to 6, and the same progressive heights will be increased to 30. = 24 = 18 = and 12. The quantity of matter remains the same, but a greater proportion of it is distributed in the upper parts. But if, in proportion as the expansion takes place, the fluid should over flow at its upper surface, so that the length of the column may remain the same, then would its total weight from *e* to *d* be reduced to 24, and that of its several progressive heights to 18=12 and 6.

The first case presents us with an analogy adapted to our present purpose, the second we shall have occasion to apply as we proceed.

Upon the supposition which has been made, *viz.*, of a general rise in the temperature of the sphere, the temperature of the atmosphere will be raised throughout its mass by internal circulation; but the proportion due to the several degrees of elevation will be preserved. The following Table exhibits the arrangement which would take place from an increase of 16 degrees of heat in the sphere:

TABLE IV. *Shewing the effect upon the Barometer of a general Increase of Temperature in the Atmosphere.*

Height in feet.	Barometer.	Sp. Gravity.	Temp.
0	30.000	.96668	48
5000	24.072	.80402	31.4
10000	19.338	.66878	14.1
15000	15.525	.55629	− 4.3
20000	12.409	.46273	−24.5
25000	9.915	.38489	− 47
30000	7.852	.32016	−62.3

The height of the barometer at the base of the column, which denotes its total weight, remains the same as in Table III., but increases at the various stages of altitude: the force of expansion having effected a different distribution of the ponderable matter, and raised a greater proportion to the upper regions.

Through this succession of changes, the atmosphere again attains a state of equilibrium and repose; and the action being equal all over the sphere, the adjustment is soon effected.

Let us next suppose that the temperature of the sphere, round which the atmosphere is diffused, instead of being equal in all its parts, increases by equal degrees from the Poles to the Equator; and let us assume that the temperature of the former is 0°, and the temperature of the latter 80°. The height of the barometer is still to be taken as 30.000 inches upon all parts of the surface. The following Table will exhibit the pressure, density, and temperature, at the two extreme points of such an arrangement, together with their gradual diminution for equal ascents:

TABLE V. *Shewing the comparative Densities and Elasticities of two Columns of Air, of different Temperatures, at different Elevations.*

Height in feet.	Barometric Column.		Specific Gravity.		Temperature.	
	Poles.	Equator.	Poles.	Equator.	Poles.	Equator.
0	30.000	30.000	1.06666	.90000	0	80
5000	23.597	24.342	.86935	.75737	− 18.5	64.4
10000	18.587	19.779	.70856	.63735	− 37.8	48.4
15000	14.591	16.060	.57752	.53640	− 58.8	31.4
20000	11.411	13.043	.47071	.45150	− 82.1	12.8
25000	8.900	10.521	.38365	.37980	− 109.1	− 7.6
30000	6.906	8.483	.31270	.31980	− 140.3	− 30.7

In considering this arrangement, we may remark, first, that at the surface of the sphere, the elasticity of the air, as measured by the Barometer, remaining the same, its specific gravity is very much greater at the poles than at the equator; and hence it is clear, that the atmospheric column must be proportionately shorter at the former, than at the latter point.

The further conclusion follows, that this heavier fluid must, by the laws of Hydrostatics, press upon and displace the lighter; and a current will be established from the poles to the equator.

Our second remark is, that this difference of gravity becomes less as we ascend from the surface, and at a certain point is neutralized: while, on the other hand, the elasticity, which is equal at the surface, varies with the height; and the Barometer stands higher, at equal elevations, in the equatorial than in the polar column. This disproportion increases with the elevation; and at some definite height, must more than compensate the unequal density of the lower strata, and occasion a counter-flux from the equator to the poles.

It will be convenient to consider these differences of gravity, and elasticity, as distinct and antagonist powers; and to measure their forces, if possible, upon the same scale. This may readily be done, by considering, that the pressures of equal columns are as their specific gravities, that is to say, .90000 : 1.06666 : : 30.000 : 35.553, which gives an excess of 5.553 inches of mercury, as the measure of the excess of gravity in the case proposed.

This excess of gravity, we have seen, is unopposed at the surface of the sphere by any excess of elasticity; so that it is the exact measure of the force with which a polar atmosphere would press upon an equatorial, supposing the two in juxta-position. It is also the measure of the pressure which would be required at the equator to equalize its density with that of the poles. If we could imagine this by any means effected, the Barometer at the former station must rise to 35.55; and the current would be reversed, and flow with the same force from the equator to the poles; this current being now occasioned by excess of elasticity, as it was before caused by excess of gravity An increased pressure of 2.77 inches would produce a state of perfect repose on the surface; the resulting augmentation of elasticity and gravity being jointly equal to the former excess of gravity. These forces being reciprocal in their action, whatever mechanical cause acts upon the one, must equally affect the other. The following Table exhibits the excess of the two powers, together with their balance for the heights before assumed.

TABLE VI. *Shewing the Force of the Polar and Equitorial Currents, at different Elevations.*

Height in feet.	Elasticity.	Density.	Balance.	
0	−0.00	+5.55	+5.55	
5000	−0.74	+3.73	+2.99	Lower Current from the Poles to the Equator.
10000	−1.19	+2.38	+1.19	
15000	−1.47	+1.37	−0.10	
20000	−1.63	+0.64	−0.99	Upper Current from the Equator to the Poles.
25000	−1.62	+0.12	−1.50	
30000	−1.58	−0.20	−1.78	

The lower, or polar current, upon this supposition, extends to the height of about two miles and a half, gradually diminishing in force; and at that height, gives place to an upper or equatorial current, which increases in strength the higher we ascend.

The velocities of these currents may also easily be calculated: for as the velocity of air rushing into a vacuum*, is found by multiplying the square root of the height of the homogeneous atmosphere, expressed in feet, by 8, so will their rate be found, in the number of feet per second, by multipying, by 8, the square root of the height of a column of air, equivalent to their respective forces. Thus the force of the polar current being 5.55 inches, the height of an equipondrant column of air would be 4856 feet, and $\sqrt{4856} \times 8$ would give, in round numbers, 557 feet per second, or 379¾ miles per hour. The rate of the equatorial current, at the height of 30000 feet, by a similar calculation, would be about 350 feet per second.

But our hypothesis assumes, not that the equatorial and polar columns are in contact, but that the temperature graduates equally between the points; and, if we divide the hemisphere into bands of 10 degrees each, the pressure of the first upon the second will be equal to the second upon the third, and so on; that is to say, the greatest force of the lower current will be, by our calculation, 0.617 inches, and that of the upper 0.240 inches, for every 10 degrees of latitude. Their respective velocities, equalizing them for each degree of heat, may also

* Young's Natural Philosophy, Vol. I., page 279.

be approximated as follows. The extreme differences of 80 degrees, we have seen, are equivalent to 5.55 inches and 2.16 inches; and the intermediate differences of about 9 degrees to 0.617 inches, and 0.240 inches. The differences for each degree will, therefore, be .068 inches and .026 inches, which give a velocity of 61 feet per second for the lower current, and 38 feet for the higher; or about 41 and 25 miles per hour.

This interchange of the polar and equatorial atmospheres, must tend towards an equalization of temperature; and, in fact, would, in time, produce an equal diffusion of the heat of the sphere itself, were not a cause included in our supposition, to provide for the permanency of its existing state.

As we have calculated that these currents, with such respective degrees of force, are the consequence of the equal height of the Barometer all over the surface of our sphere, so we conclude, that this equal height is maintained by this constant and regular flow; and any irregularity communicated to the currents would immediately be shewn by a change in the mercurial column. Let us imagine, for an instant, that any cause (no matter at present whence originating) should retard the velocity of the polar current, without at first affecting the equatorial, it is obvious that the Barometer would fall at the equator, and rise at the poles; for the balance of forces would be disturbed by the want of compensation for the matter removed at one extremity and accumulated at the other.

The subject may derive some illustration from the following figure:—

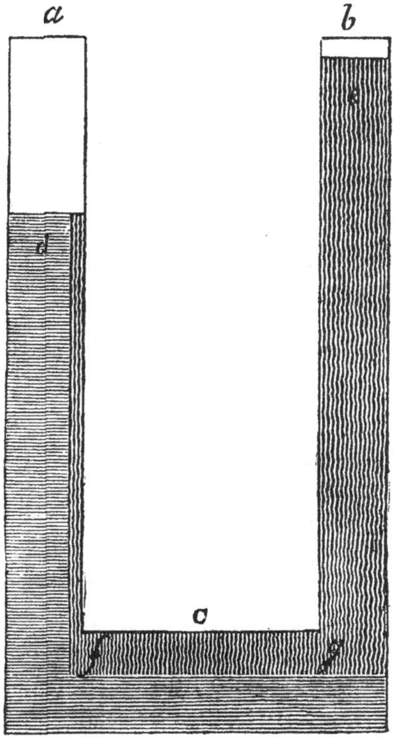

Let *a b c* represent a bent tube, in which a column of oil in one leg is balanced by a column of water in the other. The former will be longer than the latter, and their surfaces will stand respectively at *e* and *d*. Now a current of oil will pass along the upper part of the level portion of the tube *f g*, and will ascend through the water in the leg *a*, while a current of water will occupy the lower part of the level, and these will continue to flow till the lower part of the tube *f g* be filled with the heavier fluid. During no part of this interchange of places will any alteration arise in the equal pressure of the columns; for the particles, notwithstanding their horizontal motion, never for an instant cease to gravitate: the longer, indeed, will be shortened, and the shorter lengthened; but their total weights will at all times remain the same. If we could by any means devise a method of retarding the motion of the water, while that of the oil continued the same, the latter would soon flow from the leg *b*, and be accumulated upon the top of the water in the leg *a*; and the weight of the column in *a* would be as much increased as that of the column in *b* would be diminished.

Thus, then, our first *postulatum* constitutes an atmosphere which is necessarily at rest in all its parts: our second, occasions one as necessarily in continual and equal motion—let our third be a sphere increasing in heat, unequally, from the poles to the equator. The extremes are to be 0° and 80° as before, and at the middle point, or the latitude of 45, we will suppose the temperature to be the exact mean of 40°-: but from that centre, the increase towards the equator, is to be by a rapidly and equally decreasing rate, for equal distances; and the decrease towards the poles, by a similar progression. The temperatures, for every 10 degrees of latitude, from the poles to the equator, may then be as follows:

Pole	Lat. 80	Lat. 70	Lat. 60	Lat. 50	Lat. 45
0	3. 2	9. 6	19. 2	32.	40

Lat. 40	Lat. 30	Lat. 20	Lat. 10	Equator
48	60. 8	70. 4	76. 8	80.

The height of the Barometer is still to be 30.000 inches every where upon the surface.

TABLE VII. *Shewing the Elasticity, specific Gravity, and Temperature, for every Ten Degrees of Latitude, of an Atmosphere of dry Air surrounding a Sphere unequally heated from the Poles to the Equator, together with the Decrease of each, due to different Elevations.*

Height.	Poles			Latitude 80			Latitude 70			Latitude 60			Latitude 50		
Feet.	Elast.	S. Grav.	Temp.	Elast.	S. Grav.	Temp.	Elast.	S. Grav.	Temp.	Elast.	S. Grav.	Temp.	Elast.	S. Grav.	Temp.
0	30.000	1.06666	0	30.000	1.06038	3.2	30.000	1.04685	9.6	30.000	1.02707	19.2	30.000	1.00000	32
5000	23.597	.86935	−18.5	23.652	.86542	−15.2	23.707	.85684	− 8.5	23.793	.84427	1.5	23.949	.82656	14.8
10000	18.587	.70856	−37.8	18.630	.70637	−34.3	18.724	.70140	−27.3	18.893	.69405	−16.9	19.106	.68321	− 3.1
15000	14.591	.57752	−58.8	14.642	.57654	−55.1	14.775	.57407	−47.7	14.962	.57061	−36.8	15.229	.56472	−22.4
20000	11.411	.47071	−82.1	11.484	.47057	−78.2	11.617	.46991	−70.2	11.827	.46904	−58.8	12.044	.46677	−43.6
25000	8.900	.38365	−109.1	8.965	.38408	−104.7	9.102	.38463	−96 3	9.314	.38558	−83.8	9.579	.38582	−67.5
30000	6.906	.31270	−140.3	6.978	.31352	−135.7	7.100	.31483	−126.5	7.302	.31699	−112.7	7.566	.31890	−95.1

Height.	Latitude 40			Latitude 30			Latitude 20			Latitude 10			Equator		
Feet.	Elast.	S. Grav.	Temp.	Elast.	S. Grav.	Temp.	Elast.	S. Grav.	Temp.	Elast.	S. Grav.	Temp.	Elast.	S. Grav.	Temp.
0	30.000	.96668	48	30.000	.93960	60.8	30.000	.91978	70.4	30.000	.90625	76.8	30.000	.90000	80
5000	24.072	.80402	31.4	24.215	.78533	44.6	24.279	.77135	54.5	24.319	.76160	61.1	24.342	.75737	64.4
10000	19.338	.66878	14.1	19.531	.65639	27.9	19.675	.64693	38.1	19.738	.64017	44.9	19.779	.63735	48.4
15000	15.525	.55629	− 4.3	15.739	.54863	10.	15.898	.54258	20.7	16.012	.53806	27.7	16.060	.53640	31.4
20000	12.409	.46273	−24.5	12.673	.45856	− 9.4	12.811	.45507	1.7	12.974	.45220	9.3	13.043	.45150	12.8
25000	9.915	.38489	−47	10.162	.38327	−31.2	10.342	.38166	−19.4	10.467	.38010	−11.6	10.521	.37980	− 7.6
30000	7.852	.32016	−75.3	8.135	.32035	−55.9	8.313	.32010	−43.2	8.424	.31948	−35.	8.483	.31980	−30.7

TABLE VIII. *Shewing the Force of the Currents for different Heights at every Ten Degrees of Latitude.*

Height.	Latitudes 90 & 80			Latitudes 80 & 70			Latitudes 70 & 60			Latitudes 60 & 50			Latitudes 50 & 40		
Feet.	Elast.	S. Grav.	Bal.	Elast.	S. Grav.	Bal.	Elast.	S. Grav.	Bal.	Elast.	S. Grav.	Bal.	Elast.	S. Grav.	Bal.
0	~	+.178	+.178	~	+.387	+.387	~	+.575	+.575	~	+.810	+.810	~	+1.034	+1.034
5000	−.055	+.112	+.057	−.055	+.246	+.191	−.086	+.367	+.281	−.156	+.531	+.375	−.123	+.693	+.570
10000	−.043	+.062	+.019	−.094	+.142	+.048	−.169	+.214	+.045	−.213	+.325	+.112	−.232	+.449	+.217
15000	−.051	+.028	−.023	−.133	+.070	−.063	−.194	+.101	−.093	−.260	+.176	−.084	−.296	+.261	−.035
20000	−.073	+.004	−.069	−.133	+.021	−.112	−.210	+ 025	−.185	−.217	+.068	−.149	−.365	+.126	−.239
25000	−.065	−.013	−.078	−.137	−.015	−.152	−.212	−.028	−.240	−.265	−.007	−.272	−.336	+.029	−.307
30000	−.072	−.023	−.095	−.122	−.039	−.161	−.202	−.062	−.264	−.264	−.057	−.321	−.286	−.036	−.322

Height.	Latitudes 40 & 30			Latitudes 30 & 20			Latitudes 20 & 10			Latitudes 10 & 0			
Feet.	Elast.	S. Grav.	Bal.	Elast.	S. Grav.	Bal.	Elast.	S. Grav.	Bal.	Elast.	S. Grav.	Bal.	
0	~	+.854	+.854	~	+.648	+.648	~	+.447	+.447	~	+.208	+.208	Lower Polar Current.
5000	−.143	+.597	+.454	−.064	+.456	+.392	−.040	+.322	+.282	−.023	+.141	+.118	
10000	−.193	+.408	+.215	−.144	+.309	+.165	−.063	+.222	+.159	−.041	+.094	+.053	
15000	−.214	+.245	+.031	−.159	+.197	+.038	−.114	+.150	+.036	−.048	+.056	+.008	
20000	−.264	+.133	−.131	−.138	+.114	−.024	−.163	+.095	−.068	−.069	+.024	−.045	Upper Equatorial Current.
25000	−.247	+.051	−.196	−.180	+.052	−.128	−.125	+.051	−.074	−.064	+.010	−.054	
30000	−.283	−.008	−.291	−.178	+.008	−.170	−.111	+.021	−.090	−.059	−.011	−.070	

To face Page 25.

The material originally positioned here is too large for reproduction in this reissue. A PDF can be downloaded from the web address given on page iv of this book, by clicking on 'Resources Available'.

Table VII. furnishes us with the elasticity, specific gravity, and temperature of such an atmosphere, calculated upon these *data*, for every 10 degrees of latitude, from the surface, by equal altitudes, to the height of 30000 feet. Table VIII. exhibits the excess of the lateral pressure of each column upon that which adjoins, arising from the balance of forces.

It will be observed, that the currents still set as before, and at nearly the same altitudes, but with unequal velocities in different parts of their courses. The pressure, the density, the temperature, and the velocity, are all definite, for the latitude, and for the elevation; and it is by the exact balance alone of these circumstances, that the Barometer is maintained at an unvarying height, at the surface of the sphere.

We may also remark, that a change of temperature, which equally pervades a column of air throughout its length, may effect an adjustment of density without disturbing the equiponderant mercurial column situated at its base: the force, however, of the compensating currents will be altered; and, under some circumstances, their courses even may be changed. Let us imagine that the temperature of latitude 50, as it stands in the preceding Table, is altered by some cause not affecting the neighbouring columns; and that the temperature rises from 32° to 60°.8 at its base, and equally pervades its whole length: the force of the current will be increased from latitude 60 to 50, in its original direction, while that from 50 to 40 will be reversed; as will appear more clearly from the following Tables, in which the change is made according to this assumption.

TABLE IX. *Shewing the Alteration of Specific Gravity and Elasticity in an Atmospheric Column, from an Increase of Temperature at the Surface of the Sphere in a given Latitude.*

Height.	Latitude 60			Latitude 50			Latitude 40		
	Elasticity.	S. Gravity.	Temp.	Elasticity.	S. Gravity.	Temp.	Elasticity.	S. Gravity.	Temp.
0	30.000	1.02707	19.2	30.000	.93960	60.8	30.000	.96668	48.
5000	23.793	.84427	1.5	24.215	.78533	44.6	24.072	.80402	31.4
10000	18.893	.69405	− 16.9	19.531	.65639	27.9	19.338	.66878	14.1
15000	14.969	.57061	− 36.8	15.739	.54863	10.0	15.525	.55629	− 4.3
20000	11.827	.46904	− 58.8	12.673	.45856	− 9.4	12.409	.46273	−24.5
25000	9.314	.38558	− 83.8	10.162	.38327	−31.2	9.915	.38489	−47.
30000	7.302	.31699	−112.7	8.135	.32035	−55.9	7.852	.32016	−62.3

TABLE X. *Shewing the Alteration of Direction and Force in the Atmospheric Currents, from the same Cause.*

Height.	Latitudes 60 and 50			Latitudes 40 and 50		
	Elasticity.	Sp. Gravity.	Balance.	Elasticity.	Sp. Gravity.	Balance.
0	——	+2.560	+2.560	——	+0.840	+0.840
5000	−0.422	+1.722	+1.300	−0.143	+0.580	+0.437
10000	−0.638	+1.100	+0.462	−0.193	+0.380	+0.187
15000	−0.770	+0.710	−0.060	−0.214	+0.240	+0.026
20000	−0.846	+0.300	−0.546	−0.264	+0.130	−0.134
25000	−0.848	+0.070	−0.778	−0.247	+0.050	−0.197
30000	−0.833	−0.009	−0.842	−0.283	−0.006	−0.289

Here we perceive that the wind, which had blown on the surface from latitude 60 to 50 with a force of 0.810 inches, is now increased to 2.560 inches, and that which set from latitude 50 to 40, with a force of 1.034 inches, now blows from latitude 40 to 50 with a force of 0.840. A corresponding change of velocity and direction ensues in the upper currents, and the compensation of pressure thus takes place.

From the nature and essential properties of a permanently-elastic fluid, it follows that any cause tending to diminish gradually its specific gravity at the base of a column, or to augment it at its summit, must affect it throughout its length; so that, if its heat be slowly increased below, its temperature must rise from one extremity to the other.

But although such a change may take place, as has just been demonstrated, without affecting the length of the equiponderant column of mercury situated at the lower extremity, the barometer will rise at all higher stations. By comparing together Tables VII. and IX., this effect will be easily appreciated. The augmentation of temperature in latitude 50 from 32° to 60°.8 takes place on the surface, while the mercurial column remains at 30 inches; but at the height of 5000 feet it rises from 23.949 inches to 24.215 inches, making a difference of 0.266 inches. This difference increases to a certain extent with the elevation. Corresponding changes for less alterations of heat are readily perceived by comparing together the different latitudes of Table VII.

The cases which have hitherto been proposed have all been of the same nature: the alterations of temperature have been imagined to take place in the sphere itself, and from it to have been slowly communicated to the atmosphere, through which they have spread under the regular modifications due to the increasing capacities of its successive strata. Let us next consider the effect which would be produced by the heating of any of the upper layers, from some temporary cause not originating in, or extending to, the lower. For this purpose, in the column appropriate to latitude 30, in Table VII., at the fifth station, or the height of 20,000 feet, we will suppose an increase of heat to take place of 10 degrees. This increase will extend upwards, but the inferior portions remain of their original temperature. Now, the first effect of this change will be, an augmentation of elasticity in the upper beds of the atmosphere; which, exerting its force upon the high equatorial current, will accelerate its due velocity on one side, and retard it on the other. The expanding air, not being laterally confined by a proportionate expansion of the neighbouring sections, will not accumulate above; but will flow off, and cease its vertical pressure upon that column. The upper regions will therefore be rarefied, and become lighter, and pressing with less weight upon the lower, the barometer will fall at the surface of the sphere in proportion to the amount of the expansion. This effect has been already explained in the second application of Fig. 2.

Let us illustrate this action, by first representing the column, so partially changed in temperature, upon the supposition that no such compensation takes place.

TABLE XI.—*Shewing that a partial Alteration of Temperature and Specific Gravity in an Atmospheric Column must affect the Density generally by mechanical Adjustment.*

Height.	Elasticity.	Sp. Gravity.	Temperature.	
			Regular.	Irregular.
0	30.000	.93960	60.8	60.8
5000	24.215	.78533	44.6	44.6
10000	19.531	.65639	27.9	27.9
15000	15.739	*.53710	10.0	*20.
20000	12.673	*.44890	— 9.4	* 0.6
25000	10.162	*.37520	—31.2	—*21.2
30000	8·135	*.31360	—55.9	—*45.9

That such a succession of densities would result is certain; for they are due to the given pressures and temperatures: and it is also certain that such a succession could not exist in nature; for it is contrary to the fundamental law of geometrical progression. But if we suppose the Barometer to fall, as represented in the following Table, the regular series is maintained.

TABLE XII.—*Shewing the Fall of the Barometer, which would be occasioned by a partial Alteration of Temperature in the upper part of an Atmospheric Column.*

Latitude 10.				
Height.	Elasticity.	Sp. Gravity.	Temperature.	
			Irregular.	Regular.
0	*29.37	.91990	60.8	60.8
5000	*23.70	.76890	44.6	44.6
10000	*19.12	.64270	27.9	27.9
15000	15.73	.53710	* 20	10.0
20000	12.67	.44890	* 0.6	— 9.4
25000	10.16	.37520	—*21.2	—31.2
30000	8.13	.31360	—*45.9	- 55.9

The density of an elastic fluid is the result of its gravity acting upon its elasticity, and by the reaction of these two powers any change in the vertical column is instantaneously communicated throughout its entire length, and no inequality of density can for a moment exist.

Let us now imagine that the local accession of heat, instead of pervading at once the whole of either horizontal section, commences at some definite point, and gradually extends itself in depth. To render the march of this effect intelligible, we will consider its operation at several stages of its progress. We will first endeavonr to appreciate the influence of an increase of 5 degrees of heat, at the height of 5000 feet in the same column of latitude 30. The disturbing cause now affects the lower current, and the expanding air, not being checked by a simultaneous increase of elasticity in the adjoining columns, rushes forwards with accelerated velocity. The diminution of density occasioned by the excessive drain, is distributed throughout the column by mechanical adjustment. The results are as follow :

TABLE XIII.—*Shewing the Effect upon the Barometer of a partial Increase of Temperature at an elevation of* 5000 *Feet.*

Latitude 30.				
Height.	Elasticity.	Sp. Gravity.	Temperature. Irregular.	Temperature. Regular.
0	*29.68	.92973	60.8	60.8
5000	24.21	.77713	*49.6	44.6
10000	*19.32	.64957	27.9	27.9
15000	*15.57	.54829	10.0	10.0
20000	*12.53	.45373	− 9.4	− 9.4
25000	*10.05	.37921	−31.2	−31.2
30000	* 8.05	.31690	−55.2	−55.2

The gradual extension of the same increase of

heat to the height of 10000 feet, produces the following arrangement:

TABLE XIV.—*Shewing the Effect upon the Barometer of an Extension of the Increase of Temperature to* 10000 *Feet.*

Latitude 30				
Height.	Elasticity.	Sp. Gravity.	Temperature. Irregular.	Temperature. Regular.
0	*29.37	.91990	60.8	60.8
5000	*23.95	.76890	*49.6	44.6
10000	19.32	.64270	*32.9	27.9
15000	*15.41	.53710	10.	10.
20000	*12.40	.44890	— 9.4	— 9.4
25000	* 9.94	.37520	—31.2	31.2
30000	* 7.96	.31360	—55.9	—55.9

Thus it appears that the fall of the Barometer would be proportionate to the extent to which the rise of temperature would reach in this progressive manner. A small increase, thus operating, produces the same amount of depression, as if a greater expansion had been exerted in a more limited space. The following Table presents the effect upon the Barometer of a gradual rise of two degrees of temperature, from 5000 to 25,000 feet.

TABLE XV. *Shewing the Effect upon the Barometer of a small partial Increase of Temperature, gradually extending itself throughout the Column.*

Height.	Latitude 30, 1st Change.			Latitude 30, 2d Change.			Latitude 30, 3d Change.			Latitude 30, 4th Change.			Latitude 30, 5th Change.		
	Elasticity.	S. Grav.	Temp.	Elasticity	S. Grav.	Temp.	Elasticity	S. Grav.	Temp.	Elasticity	S. Grav.	Temp.	Elasticity.	S. Grav.	Temp.
0	*29.87	.9355	60.8	*29.74	.9314	60.8	*29.61	.9273	60.8	*29:49	.9232	60.8	*29.37	.9192	60.8
5000	24.21	.7819	* 46.6	*24.11	.7785	* 46.6	*24.01	.7751	* 46.6	*23.91	:7717	* 46.6	*23.81	.7683	* 46.6
10000	*19.45	.6546	27.9	19.45	.6518	* 29.9	*19.37	.6490	* 29.9	*19.29	.6462	* 29.9	*19.21	.6434	* 29.9
15000	*15.68	.5463	10.	*15.62	.5440	10.	15.62	.5417	* 12.	*15.56	.5394	* 12.	*15.50	.5371	* 12.
20000	*12.62	.4565	− 9.4	*12.57	.4545	− 9.4	*12.52	.4525	− 9.4	12.52	.4505	*− 7.4	*12.47	.4485	*− 7.4
25000	*10 12	.3816	−31.2	*10.08	.3800	−31.2	*10.05	3784	−31.2	*10.01	.3768	−31.2	10.01	.3752	*−29.2
30000	* 8.10	.3190	−55.9	* 8.07	.3177	−55.9	* 8.04	.3164	−55.9	8.01	.3151	−55.9	* 7.98	.3138	55.9

In the full effect of these three examples, represented in Tables XII., XIV., and the last column of Table XV., the progression of density is the same; and the barometer falls to the same amount at the base of the column.

The limit of the action of the irregular expansion of aërial column is fixed, not only by its amount and progress, but also by the compensating effects of the contiguous sections. Its influence upon the lateral currents is calculated below.

TABLE XVI. *Shewing the Effect of the preceding Changes upon the Force and Direction of the Currents.*

	First Modification.						Second Modification.						Third Modification.					
	Latitudes 40 & 30			Latitudes 20 & 30			Latitudes 40 & 30			Latitudes 20 & 30			Latitudes 40 & 30			Latitudes 20 & 30		
Height	Elast.	S. Grav.	Bal.	Elast.	S. Grav.	Bal.	Elast.	S. Grav.	Bal.	Elast.	S. Grav.	Bal.	Elas.	S. Grav.	Bal.	Elas.	S. Grav.	Bal.
0	+.63	+1.44	+2.07	+0.63	0.	+0.63	+.63	+1.44	+2.07	+.63	0	+0.63	+.63	+1.44	+2.07	+.63	0	+0.63
5000	+.37	+1.09	+1.46	+0.58	+0.08	+0.66	+.12	+1.09	+1.21	+.73	+0.08	+0.81	+.26	+1.09	+1.35	+.47	+0.08	+0.55
10000	+.21	+0.81	+1.02	+0.55	+0.140	+0.69	+.0	+0.81	+0.81	+.35	+0.14	+0.49	+.12	+0.81	+0.93	+.46	+0.14	+0.60
15000	−.21	+0.60	+0.39	+0.16	+0.18	+0.34	+.11	+0.60	+0.71	+.48	+0.18	+0.66	+.02	+0.60	+0.62	+.39	+0.18	+0.57
20000	−.26	+0.43	+0.17	+0.14	+0.20	+0.34	+.0	+0.43	+0.43	+.41	+0.20	+0.61	−.06	+0.43	+0.37	+.34	+0.20	+0.54
25000	−.25	+0.30	+0.05	+0.18	+0.21	+0.39	−.03	+0.30	+0.27	+.40	+0.21	+0.61	−.10	+0.30	+0.20	+.33	+0.21	+0.54
30000	−.28	+0.21	−0.07	+0.18	+0.22	+0.39	−.11	+0.21	+0.10	+.35	+0.22	+0.57	−.13	+0.21	+0.08	+ 33	+0.22	+0.55

From Latitude 40 to 30, it will be observed, that the force of the polar current is greatly increased; while from 30 to 20, it is reversed. The different modifications of the heating process produce different adaptations of the upper currents, which the comparison of the several tables will sufficiently explain.

It may readily be imagined, that irregularities thus introduced into these compensating movements, the consequence of diminished mechanical pressure, must of themselves be liable to produce changes of temperature in the columns, foreign to the natural gradation; and that, amongst others, the atmosphere, in its upper parts, may be liable to greater depression of heat than would be due to the elevation alone. A gradual process of cooling taking place in the higher portions of a body of air, would communicate itself to the whole mass, in an analogous manner to the equal diffusion which would ensue from the slow communication of heat to the lower parts; that is to say, without producing any effect upon the Barometer, at the surface of the sphere, or any irregularity in the gradation of temperature. But where the change is effected suddenly, by the admixture of a large body of cold air, a mechanical effect is produced by the increased pressure of the mass; and the equilibrium of density takes place before the proper adjustment of temperature. An atmosphere hence results, whose heat decreases in a proportion greater than is due to the decrease of density. The effect is analogous to that which arises from an irregular increase; and the Barometer must rise to equalize the specific gravity. The following Table has been constructed upon the supposition of a depression of temperature of ten degrees, taking place in the column of air at latitude 30, and extending at once from the height of 1000 feet to 15000.

TABLE XVII. *Shewing the Rise of the Barometer, occasioned by the sudden partial Reduction of Temperature in an Atmospheric Column.*

Height.	Latitude 30.			
	Elasticity.	Sp. Gravity.	Temperature.	
			Irregular.	Regular.
0	*30.626	.95917	60.8	60.8
5000	*24.715	.80166	44.6	44.6
10000	19.531	.67007	*17.9	27.9
15000	15.739	.56003	* 0.	10.
20000	*12.933	.46805	− 9.4	− 9.4
25000	*10.371	.39127	−31.2	−31.2
30000	* 8.305	.32697	−55.9	−55.9

The corresponding effects upon the upper and lower currents are deduced below.

TABLE XVIII. *Shewing the Effect upon the Currents of the preceding Change.*

Height.	Latitudes 30 and 40			Latitudes 30 and 20		
	Elasticity.	Sp. Gravity.	Balance.	Elasticity.	Sp. Gravity.	Balance.
0	+0.626	−0.234	+0.392	+0 626	+1.284	+1.910
5000	+0.643	−0.075	+0.568	+0.336	+0.989	+1.325
10000	+0.193	+0.041	+0.234	−0.144	+0.765	+0.621
15000	+0.214	+0.120	+0.334	−0.159	+0.568	+0.409
20000	+0.524	+0.172	+0.696	+0.122	+0.419	+0.541
25000	+0.456	+0.207	+0.449	+0.029	+0.313	+0.342
30000	+0.453	+0.223	+0.676	−0.008	+0.224	+0.216

It is not required here to point out all the means by which such changes of heat as we have represented may be effected; or to trace further the endless modifications of densities and currents which would result from their different applications: it is sufficient, at present, to have shewn that, supposing them to arise, certain general consequences must follow. Neither in the tables which I have constructed must absolute accuracy be expected; the decimal mode of calculation which I have adopted, did not admit of precision, without a degree of labour, which would have been disproportionate to the object which I have in view. The different series of densities are at all times in geometrical progression to the heights, and they should be precisely compounded of the pressures and temperatures with which they are joined, reckoning as 1 the specific gravity of air at 32°, and under a pressure of 30 inches of Mercury of the same temperature. These conditions the tables generally fulfil to one or two places of decimals; but as the calculations have been made one under the other, the remainders, which were of no consequence in the original sums, have become appreciable quantities by the various successive multiplications and divisions. The errors, however, are in no case so large as to interfere with the general principles which it is my aim to establish.

We have hitherto contemplated these changes with reference to the particular column of the atmosphere in which they had their origin; we must now endeavour to trace their effects upon those with which they are connected. We must recollect, that it has been established, as a principle, that the equal height of the Barometer, in every situation upon the surface of the sphere, was dependant upon the maintenance of the equatorial and polar currents, with a certain determinate velocity in the different parts of their courses, and that no disproportionate alteration or interruption in these could take place without a corresponding effect upon the mercurial column. Now, upon a reference to Tables VII. and VIII., it will be found, that, to keep the Barometer at 30.000 inches under Latitude 40, a current is required of the force of 0.854 inches towards Latitude 30, counterbalanced by one in the contrary direction, of the force of 0.291 inches at the elevation of 30000 feet: but by the unequal alteration of temperature shewn in Table XIV., the current at the surface is increased to 2.07 inches, and continues with diminishing force to the height of 30000 feet in the same direction. It is clear, therefore, that a much greater drain takes place upon this Latitude, without an adequate compensating supply,—the Barometer must, therefore, fall throughout the column. This fall, it will be observed, may take place without any disturbance of the temperature. The atmosphere incumbent upon Latitude 20, will be similarly affected by the same change of temperature at Latitude 30. In its original state, the lower polar current flows upon the surface with a force of 0.648 inches, and feeds this column with a supply of air. It is balanced at the height of 30000 feet by an equatorial current of 0.170 inches.

The course of the former is now reversed, and the drain is increased in the contrary direction. A rapid fall of the Barometer must, therefore, ensue.

On the other hand, an increased afflux of air, beyond the usual supply, to any portion of the atmosphere, occasioned by the expansion of any of the neighbouring parts, must cause an increase of density; and the equiponderant column will, of course, be lengthened. It is easy to perceive, that these secondary effects must widely extend the influence of the original disturbing cause; and it is obvious, that every depression of the Barometer must be accompanied by an equivalent rise in distant parts of the elastic medium, and *vice versâ*. The local impulse extends its influence in this, as in all other fluids, by the laws of undulation. The mean pressure, at any moment of time, of all the waves upon the surface of the sphere, will be the pressure of the atmosphere at rest, and the average of a large number of oscillations at any particular spot will approximate to the same quantity.

I have thus attempted to shew that the proximate cause of the fall of the Barometer, at the surface of the sphere, in an atmosphere constituted as our postulates have required, may be an increase of elasticity in its upper parts, beyond what is due to their respective elevations; and of its rise an analogous decrease in the same situations. These changes I have endeavoured to exhibit, as affecting directly the columns themselves in which the temperature varies, and remotely the adjoining columns, from their influence upon the lateral currents. This influence we have hitherto contemplated as extending only in the direction from the poles to the equator or from the equator to the poles ; as if the changes of temperature which we have supposed, had extended, under the same parallel of latitude, round the sphere. The course of our argument now requires that we should shortly consider the same changes, as bounded in their longitudinal as well as in their latitudinal extent. For this purpose, we must suppose our sphere to be divided into sections of 10 degrees at right angles to the former division.

Now, the arrangement, represented in Tables VII. and VIII., resulted from the temperature of the sphere itself, which we supposed to increase in heat in a regular progression from the poles to the equator; and the set of the aërial currents, under such circumstances, must necessarily be from north to south, and from south to north. But let us imagine that the local increase of heat, which is represented in Tables IX. X., is not only confined to 10 degrees of latitude, but also to 10 degrees of longitude: it will then be obvious that there will be currents established at right angles to the former winds; and they will all tend to compensate the irregularity which has been introduced. The following Tables present the results of the calculation of these eastern and western currents.

TABLE XIX. *Shewing the Effects upon the Atmospheric Columns of a general Alteration of Temperature in the direction of the Longitude.*

	Longitude 20 and 360			Longitude 10.		
Height.	Elasticity.	Sp. Gravity.	Temp.	Elasticity.	Sp. Gravity.	Temp.
0	30.000	1.00000	32.	30.000	.92960	60.8
5000	23.949	.82656	14.8	24.215	.78533	44.6
10000	19.106	.68321	−3.1	19.531	.65639	27.9
15000	15.229	.56472	−22.4	15.739	.54863	10.
20000	12.044	.46677	−43.6	12.673	.45856	−9.4
25000	9.579	.38582	−67.5	10.162	.38327	−31.2
30000	7.566	.31890	−95.1	8.135	.32035	−55.9

TABLE XX. *Shewing the Force of the Currents occasioned by the preceding Alterations.*

	Longitudes 360 or 20 and 10.		
Height.	Elasticity.	Sp. Gravity.	Balance.
0	0	+1.929	+1.929
5000	−.266	+1.301	+1.035
10000	−.425	+0.837	+0.412
15000	−.510	+0.496	−0.014
20000	−.629	+0.251	−0.378
25000	−.583	+0.076	−0.507
30000	−.569	−0.043	−0.612

But that portion of the atmosphere which thus rushes to supply the place of the air which has been rarefied within the prescribed limits, is already, as we have seen, in motion from the pole to the equator: its course will, therefore, be intermediate between the two forces which impel it, and it will reach its destination with a northern or southern deflection. It will not be necessary to enter into the calculation of other cases of disturbance with the same limitations of longitude, as it is obvious that analogous effects must follow; and easterly or westerly currents, differently modified in direction and force, must result, whenever partial alterations of density take place, either from the immediate effects of expansion, or from change of mechanical pressure. The examples which have been adduced will be sufficient to illustrate the nature and operation of certain general principles, whose application is almost infinite.

There is, however, one more view of the subject which may assist us in our after-application of these particulars to the intricacies of atmospheric changes. Referring back again to Tables VII. and VIII., let us now suppose an increase of ten degrees of temperature to take place along the whole extent of any given meridian, and a decrease of equal amount at the opposite point; and let all the meridians on each side be similarly affected in a ratio between the two. The distribution of heat upon these two lines, and the two intermediate, would be as follows, for every ten degrees of latitude.

TABLE XXI. *Shewing the Distribution of Heat all over the Sphere, upon the supposition of a gradual Increase of Heat between the opposite Meridians.*

	Longitude 270	Longitude 0	Longitude 90	Longitude 180
Lat. 90	0	10.	0	−10.
Lat. 80	3.2	13.2	3.2	− 6.8
Lat. 70	9.6	19.6	9.6	− 0 4
Lat. 60	19.2	29.2	19 2	+ 9.2
Lat. 50	32.	42.	32.	22.
Lat. 40	48.	58.	48.	38.
Lat. 30	60.8	70.8	60 8	50.8
Lat. 20	70.4	80.4	70.4	60.4
Lat. 10	76.8	86.8	76.8	66.8
Lat. 0	80.	90.	80.	70.

This increase of heat, we are further to imagine to take place, throughout the respective columns, in so gradual a manner as not to affect the barometer at their bases.

Then will there be two currents established upon the surface of the sphere, in opposite directions on either side of the cold meridian towards the hotter, with a force of 1.304 inches; or rather, the body of the air, which was before in motion from north to south, will now be deflected with this force to the east and west; and the whole lower atmosphere, excepting upon these lines where the effect would be null, will move from the poles to the equator, with a greater or less bend to the east and west. If the cause producing this variation of heat be supposed to move round the sphere from east to west, then will every meridian, in succession, be subjected to alternations of eastern and western currents.

All our reasoning has hitherto been applied to a sphere at perfect rest in itself: we will now give it motion, and suppose it to turn upon its poles with a certain regular velocity from west to east. We will, however, continue the supposition of equal gravity, and put out of consideration, for the present, the effect of centrifugal force. Since this rotatory motion must be greatest at the equator, and is directed eastward, the air, in its passage from the poles, not having attained the maximum velocity, will have a relative motion westwards; and hence, the combined motion of the wind will be directed in the northern hemisphere from north-east to south-west; and in the southern, from south-east to north-west. Whenever this apparent tendency coincides with an actual impulse in the same direction, derived from other sources, it will augment its force; and when opposed to one in a contrary direction, it will tend to neutralize it. Thus, in the supposition which has been made above, of an accession of temperature upon the whole of any one meridian, the current, which we found would thence arise from the east towards that meridian, would be increased by this further mechanical impulse; while the western current on the opposite side would be decreased, if not annihilated.

But as the lower polar current would thus have a

relative western direction, with regard to the motion of the sphere itself, so the upper equatorial current would have an absolute movement in the contrary direction. The particles of air, which are transported from the polar regions to the equator, have not time to assume the velocity of the different parallels of latitude as they reach them; and are, therefore, necessarily behind them, as they revolve. To other bodies, therefore, possessing that velocity, they oppose a resistance which appears to come from the eastern quarter. Those, however, which are transported above from the equator to the poles, have an excess of absolute motion from west to east above those parts of the globe towards which they are carried. Now, as the heating power, which is the main-spring of all the motions of the atmosphere, is supposed to be in the sphere itself, it follows that the upper parts, which are most remote from it, will become cooled; while those which are nearer to, or in contact with it, maintain their proper temperature. As they cool, they of course become specifically heavier and descend; their place being supplied by the subjacent warmer strata. Another kind of circulation is thus established in a direction perpendicular to the horizontal currents which we have been considering, and in no wise interfering with them. If we consider, therefore, the motion of any one particle of air, we shall find that its course has an angular direction compounded of these two motions.

Fig. IV.

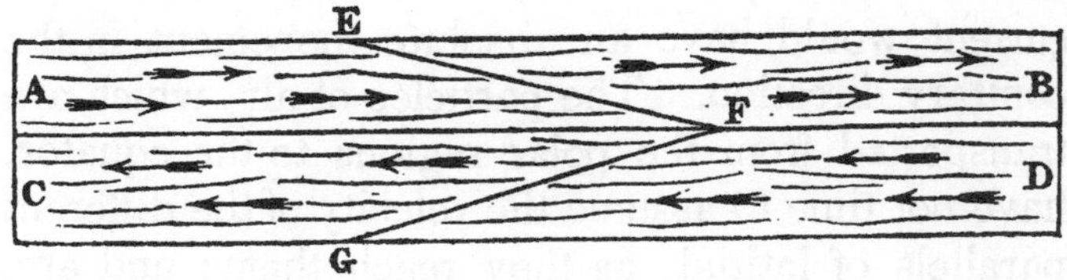

Let A B, Fig. IV., represent an upper, and C D an under, current, moving with equal velocities in opposite directions. Let us take any particle in the higher E, and suppose it to assume a density greater than is due to its situation, then will its course be from E to F, in an inclined direction, resulting from the two forces which solicit it. If we follow it further, we shall find that one of these forces ceases its action, and gives place to another in a contrary direction; its path will then be from F to G, resulting from the other force, which continues its action, and the new impulse which it now receives. In this manner may an interchange of particles be kept up between the two principal currents, without at all interfering with their courses. We have, however, just observed, that the upper equatorial current is endued with a movement of rotation from west to east greater than that of the polar latitudes towards which it is carried: this impulse, being unopposed, must be borne by the particles in their descent from the higher to the lower stream; and the consequence must be, that the latter will be deflected from its course; and the northern current will receive a westerly direction at the point where this influence reaches its stream with sufficient power.

We have thus, by gradual stages, obtained some insight into the properties of an atmosphere of permanently-elastic fluid, surrounding and gravitating towards a sphere of unequal temperature, increasing from the poles to the equator, and revolving upon its axis with equal and definite velocity. Its state of equilibrium, which it must always be striving to attain, by whatever obstacles opposed, is maintained by two grand systems of currents, equally balanced, varying in force and direction, and originating partly from differences of density, and partly from relations to the rotatory movement. The principal circulations are in a horizontal direction from the poles to the equator, in the lower system; and from the equator to the poles, in the upper: and in a vertical direction, a constant interchange of particles between the superior and inferior strata. These motions are effected by means of differences of temperature and consequent differences of density. Subordinate to them, are two partial and local currents, which, as they arise from the rotation of the axis of the sphere, are in a direction at right angles to the former, and opposed to each other. The conjunction of these forces produce certain deflections from the primary directions; but, as the upper and lower systems are oppositely affected throughout, they compensate each other's motions, and their combined pressure is the same in every part. In this nicely-balanced order of things, we have seen how slight irregularities of temperature might produce great disturbances, and we have traced various expansions and contractions, which acting unequally upon the antagonist currents, would destroy the adjustment of their several velocities. Accumulations in some parts, and corresponding deficiencies in others, would hence arise, the amount of which would be weighed by the barometer. These, in seeking to regain their proper level, and struggling to restore the equilibrium, would give rise to temporary and variable winds, which would modify the regular currents, and often reverse their courses.

Having established these several particulars, as I think, upon fixed and acknowledged principles, I shall now proceed to investigate the second division of my proposed inquiry.

PART II.

On the Habitudes of an Atmosphere of Pure Aqueous Vapour.

In the first part of this treatise I have considered the habitudes of an atmosphere of perfectly dry and permanently elastic fluid: the second branch of the investigation leads me to the consideration of an atmosphere of pure unmixed aqueous vapour. I shall first contemplate it, in this as in the former case, as surrounding a sphere of uniform temperature throughout, which we must now suppose to be covered with water. This temperature we will fix at 32°, as in our first hypothesis of the permanently-elastic fluid. It will not be necessary here to make any distinction between water in its fluid, and in its solid, state: ice, as well as water, it has been well ascertained, yields vapour of elasticity proportionate to its temperature; the general term may, therefore, be employed without impropriety.

Now the elastic force of steam, for the different degrees of heat within the range of atmospheric temperature, has been determined with very great precision. The following Table, extracted from the works of Mr. Dalton, includes the results as far as they are necessary to our present purpose:—

TABLE XXII. *Shewing the elastic Force of Vapour for every Degree from 0° to 90°.*

Temp.	Pressure.	Temp.	Pressure.	Temp.	Pressure.	Temp.	Pressure.
0	0.064	22	0.139	45	0.316	68	0.676
1	.066	23	.144	46	.328	69	.698
2	.068	24	.150	47	.339	70	.721
3	.071	25	.156	48	.351	71	.745
4	074	26	.162	49	.363	72	.770
5	.076	27	.168	50	.375	73	.796
6	.079	28	.174	51	.388	74	.823
7	.082	29	.180	52	.401	75	.851
8	.085	30	.186	53	.415	76	.880
9	.087	31	.193	54	.429	77	.910
10	.090	32	.200	55	.443	78	.940
11	.093	33	.207	56	.458	79	.971
12	.096	34	.214	57	.474	80	1.000
13	.100	35	.221	58	.490	81	1.040
14	.104	36	.229	59	.507	82	1.070
15	.108	37	.237	60	.524	83	1.100
16	.112	38	.245	61	.542	84	1.140
17	.116	39	.254	62	.560	85	1.170
18	.120	40	.263	63	.578	86	1.210
19	.124	41	.273	64	.597	87	1.240
20	.129	42	.283	65	.616	88	1.280
21	.134	43	.294	66	.635	89	1.329
		44	.305	67	.655	90	1.360

According to this Table, with the temperature at 32°, the equiponderant column of mercury would be 200 inch; and it would be the same at every part of the surface. The density of the vapour, like that of the gaseous atmosphere, and for the same reasons, must decrease in a geometrical progression for equal perpendicular distances, and the temperature will decline with it. The ratio, however, of its diminution will be very different.

To exemplify the calculation, we will take the height of 10000 feet. We must first find the height of an homogeneous atmosphere of such vapour equivalent to .200 inch of Mercury. Its specific gravity, compared to dry air, is as 2.317 to 557.800, or as the weights of a cubic foot of each respectively; therefore, 2.317 : 557.800 :: 10500 : 2527794 then 2527794 × 0.2 inch = 505558 inches = 42129 feet,

	Height of Homogeneous Vap.		Given height.		Modulus of Logarithm.		Diff. of Log. of Densities.
and	42129	:	10000	::	.4342945	:	.1030868

$$\text{Log. of } \underset{\text{Density, at } 32^{\circ}}{.200} = .3010300$$
$$- .1030868$$
$$= .1979432 = \text{Log. of } \underset{\text{Density, at } 25^{\circ}}{.157}$$

At the height, therefore, of 10000 feet, the mercurial column, which the atmosphere would support, would only be .157 inch, and the constituent temperature of vapour of this degree of elasticity is 25°. In this manner the following Table was constructed of the elasticity, density, and temperature of such an atmosphere as we are now contemplating, at different heights:—

TABLE XXIII. *Shewing the Decrease of Density and Temperature in an Atmosphere of Aqueous Vapour, of the force of .200 inch at different Elevations.*

Height in Feet.	Elasticity.	Density.	Temp.
0	0.200	1.000	32
5000	.177	.890	28.5
10000	.157	.790	25
15000	.140	.708	22
20000	.124	.636	19
25000	.110	.577	16
30000	.100	.518	13

With such an arrangement, there would be perfect equilibrium, and consequently perfect rest, all over the sphere. No precipitation or evaporation would take place, and the atmosphere would remain transparent and undisturbed. Such also must be the state to which an atmosphere of vapour would strive to attain, notwithstanding any obstacles which might be opposed to it. Hence we may also infer that, if condensation were to take place in any part of such an atmosphere, evaporation must follow in other parts to maintain the balance of forces ; and conversely, that evaporation must be accompanied by precipitation.

Should the temperature of the sphere rise gradually and equally over all its surface, the elasticity of the steam would increase with it, without disturbance ; and, following its own law of decrease for its different elevations, would remain perfectly transparent.

In considering the second modification of circumstances, that, namely, of the temperature of the sphere increasing from the poles to the equator, we must first observe, that a pure unmixed atmosphere of vapour could not follow such a gradation. The elasticity of the whole would be determined by that of the lowest point; and the water would distil from the hottest point to the coldest, with such rapidity, as to occasion strong ebullition at the former. The condensation of vapour may be effected, not only by decrease of temperature, but by increase of pressure: it is not necessary, therefore, that it should pass from the hottest to the coldest point to be precipitated, which would be a gradual process, but the elastic force, arising from an increase of density at one extremity, would instantly be felt at the other; the impression being conveyed as through a spring. The best illustration of this effect may be derived from the Cryophorous of Dr. Wollaston, in which the force of the vapour is so much reduced by the cold applied to one extremity of the instrument, as speedily to produce congelation at the other by the rapidity of the consequent evaporation. For our present purpose, however, we must put out of our consideration the rapidity of this action, and imagine the passage of the vapour from one point to another to be so mechanically retarded, as to enable it to assume the gradations due to the heat of the sphere. We may then estimate the relative force and pressure of two of the perpendicular columns at different stations. The following table represents the state of the vapour at the equator supposing the temperature 80° as before.

TABLE XXIV. *Shewing the Decrease of Density and Temperature in an Atmosphere of Aqueous Vapour of the force of 1.000 Inch at different Elevations.*

Height.	Elasticity.	Density.	Temp.
0	1.000	4.571	80.
5000	.897	4.115	76.5
10000	.804	3.682	73.
15000	.722	3.366	70.
20000	.648	3.026	67.
25000	.581	2.723	63.
30000	.521	2.487	60.

By comparing these results with the last Table, it will be observed, that, unlike the case of the permanently-elastic fluid, both the density and elasticity increase greatly with the temperature; and the consequence must be that the equatorial columns must press upon the polar throughout their length. A circulation will hence arise very different from that of the aërial currents. The vapour will flow in a mass from the equator to the poles, and, being necessarily condensed in its course, will return from the poles to the equator in the form of water. Great evaporation will constantly be going on at the latter station, and condensation at every other: so that the atmosphere, excepting at the equator, would be rendered turbid by perpetual clouds and rain. As in the case of the permanently-elastic fluid, the temperature of the sphere would, by this process, soon become equalized. did not our hypothesis provide for its permanency: the equatorial parts would be quickly cooled by the evaporation, and the polar warmed by the heat evolved during the condensation.

It is further worthy of attention, that, the elasticity of vapour increasing nearly in a geometrical proportion for equal increments of heat, the decrease of temperature in ascending in this atmosphere will be in arithmetical proportion only. The diminution is very nearly three degrees for every 5000 feet.

Upon the hypothesis of the gradation of temperature before assumed, in the case of the gaseous atmosphere, the following Table will represent the corresponding elasticity and density of the vapour, at the surface of the sphere, for every ten degrees of latitude.

TABLE XXV. *Shewing the Force and Density of an Atmosphere of Aqueous Vapour, for every Ten Degrees of Latitude, surrounding a Sphere unequally heated.*

Poles.			Latitude 80			Latitude 70			Latitude 60			Latitude 50		
Elas.	Density	Temp	Elas.	Density	Temp	Elas.	Density	Temp	Elas.	Density	Temp	Elas.	Density	Temp
064	0.340	0	.072	0.380	3.2	.089	0.466	9.6	.125	0.641	19.2	.200	1.000	32

Latitude 40			Latitude 30			Latitude 20			Latitude 10			Equator.		
Elas.	Density	Temp	Elas.	Density	Temp	Elas.	Density	Temp	Elas.	Density	Temp	Elas.	Density	Temp
.351	1.700	48	.539	2.547	60.8	.731	3.403	70.4	.900	4.143	76.8	1.000	4.571	80

Under these circumstances, the equatorial regions will remain perfectly transparent, while rain will continue to fall in every other situation, in proportion to the densities of the respective places and the decrease of temperature ; and the supply of vapour will be entirely kept up by the evaporation at the equator. This circulation may be regarded as a species of regulated distillation. The height of the barometer decreases rapidly towards the poles from 1 inch to 064 inch, and the quantity of condensation is definite for each latitude ; for the resistance to the passage of the vapour is supposed to be constant and equal.

Let us now imagine that the temperature of any particular latitude is raised to the level of that which adjoins : then will condensation cease at that particular place, evaporation will commence, and the atmosphere will become transparent. The quantity of water precipitated will be proportionally increased on the other side. If, on the contrary, the temperature be lowered to the standard of the latitude next above, the precipitation will be increased, and the higher latitude will be cleared. The following Table represents Latitude 30 under these two conditions.

TABLE XXVI. *Shewing the State of the Atmospheres arising from Alterations of Temperature in any intermediate Columns.*

Latitude 40			Latitude 30			Latitude 20		
Cloudy			Clear			Cloudy		
.351	1.700	48	.731	3.403	70.4	.731	3.403	70.4
Clear			Cloudy			Cloudy		
.351	1.700		.351	1.700		.731	3.403	70.4

Again—if the mechanical retardation of the flowing vapour, which we have imagined, were subject to variation, the quantity of evaporation and precipitation would be proportionate to the velocity of its passage: thus, supposing the evaporation from a given surface at a given temperature, and under a certain resistance, to be three grains per minute, it would be increased to six grains with half the resistance. It would be easy to apply these consequences to analogous cases, but it will not be required to trace them more particularly. The changes at the surface affect the whole of the superincumbent column equally, and the temperature of the vapour follows its own law of decrease. But what will be the consequence, if the vapour should be forced to adapt itself to a progression of temperature different from that of its own; and if from some cause or other (no matter at present whence originating) the heat of the upper regions should diminish at a greater rate than is due to the natural gradation?

Let us, for instance, suppose that the heat of the water upon the sphere is 80°, but that, at the height of 5000 feet above the surface, a temperature exists of 64° 4, which from that point follows the former decreasing scale. The water will have a tendency to throw off vapour of the same constituent heat as its own temperature; but the pressure above, being rendered too little by the influence of the forced degree of cold, to preserve the necessary elasticity below, the atmosphere will only possess the tension due to the lower degree; that is to say, the constituent temperature of the vapour will be only

67°.9. Evaporation must therefore ensue below, and its concomitant precipitation will take place above. The calculation of these effects has furnished the following tabular representation of their connexion:

TABLE XXVII. *Shewing the Effect upon the Atmosphere of Vapour of a forced gradation of Temperature.*

Height.	Elasticity.	Constituent Temp. of Vapour.	Sensible Temp.	State of Atmosphere.
0	.673	67.9	80	Clear
5000	.606	64.4	64.4	Cloudy
10000	.542	61	61	Clear
15000	.490	58	58	"
20000	.443	55	55	"
25000	.401	52	52	"
30000	.363	49	49	"

The consequence of this situation of things will be, that a cloud will be formed at the height which has been named: for the atmosphere will be forced upwards by the nascent vapour below, and will be condensed at this point. The cloud, however, supposing the process to be sufficiently gradual, would not extend very far downwards, for the water, during its precipitation, would be redissolved by the excess of heat in the lower regions, so that they might remain transparent and undisturbed. The ultimate effect would be that the temperature would be slowly equalized, and the balance of force restored. The water, in its circulation backwards and forwards, would act as a carrier of the heat, which it would abstract from the lower parts by its evaporation, and give out to the upper by its condensation. The atmosphere would thus gradually recover its state of equilibrium and repose. The upper regions, upon this supposition, remain clear, for there the regular gradation is undisturbed.

This part of the subject may, perhaps, derive some illustration from the following analogy and figure.

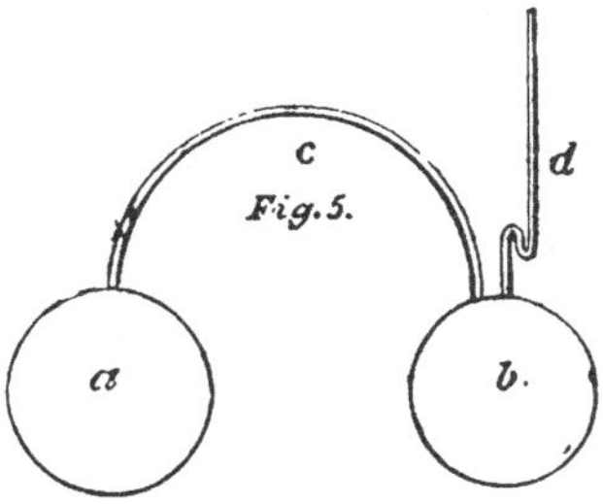

Let *a* and *b*, Fig. 5, represent two glass globes connected together by the tube *c* :—*d* is a mercurial gauge for the purpose of measuring the elasticity of any included vapour. Let us suppose the apparatus to be free from air, and water to be included in *a*; and let the temperature of both the globes be 80° The column of mercury supported in *d* will then be equal to 1 inch, and no evaporation or condensation can take place. Let us now imagine the globe *b* to be suddenly cooled down to 32°. The mercury in the barometer *d* will instantly fall to .200 inch; the water will rapidly evaporate, and be condensed in *b*, and if both the temperatures be maintained, will entirely pass from *a* to *b*. The difference, between the force of vapour at 80° and 32°, will thus become the measure of the force of evaporation.

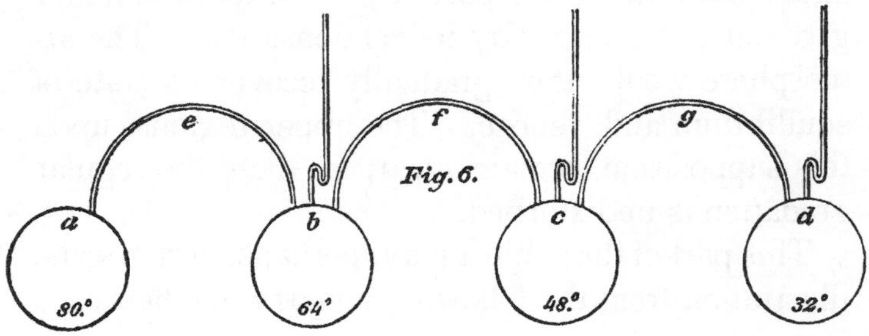

Fig. 6.

Now let Fig 6 represent four globes, connected together in the same way as the two in Fig. 5. The whole apparatus is supposed to be free from air, and the ball *a* to contain water. The temperature of each is to be maintained respectively at 80°, 64°, 48°, and 32°. The gauges would then all denote a tension of .200 inch, and the water would distil over rapidly into *d*. The vapour, in its passage, meets with no obstruction; and the effect is necessarily the same as in the last case, where no intermediate receivers were placed. But, should the vapour be retarded in its course by any obstacle of a mechanical nature, as if the connecting tubes *e f* and *g* were packed with sand or cotton, the result would be very different. At the commencement of the process the rising steam would assume the elasticity necessary to enable it to pass the intervening obstructions into the globe *d*, and, whatever this elasticity might be, when in *d*, it would be reduced to the force of .200 inch. With this force it would, therefore, press upon the surface of the water in *a*. The nascent steam has now, not only the mechanical impediment to overcome, but this additional pressure; so that its elasticity, and consequently its temperature, must rise in proportion. In *c*, however, it necessarily assumes the temperature of that vessel. A greater degree of elasticity, from *c* to *a*, now presses upon the fluid, and the force of the generated steam must again rise. It is again partly condensed in *b*, whose temperature will not support the higher degree, and the increased tension is exerted from *b* to *a*. The different gauges, *b* *c* and *d*, will thus denote the respective elasticities .597 inch, .351 inch, and .200 inch, appropriate to their several temperatures. The differences of force will denote the rate of evaporation in each.

Such mechanical obstruction, as we have supposed, would oppose itself to the free passage of vapour in motion, but would exert no pressure or influence upon it in a state of rest; and if we were to imagine that all the water had distilled over from *a*, and were included in *d*, the gauges would all stand permanently at .200 inch.

Let us now apply this to our atmosphere.—We have already proposed the simple case of a sudden decrease of heat at one stage of its height, by which condensation was produced; the elasticity was thereby reduced to the degree appropriate to that temperature at that elevation, and evaporation commenced from the surface. This evaporation was proportionate to the difference between the elasticity of vapour of the temperature of the sphere, and the elasticity of the superincumbent mass. We will now suppose that the rapid decrease of temperature continues throughout the column, and that at the following stages of its height it is forced to adapt itself to the annexed progression.

TABLE XXVIII.

Height.	Temperature.
0	80.
5000	64.4
10000	48.4
15000	31.4
20000	12.8
25000	− 7.6
30000	−30.7

The elasticity could not then exceed 043 inch upon the surface: the evaporation would consequently be excessive, and its force would almost amount to explosive violence; while the condensation above would be proportionate, and the precipitation would resemble a water-spout in its effects. We must now provide (and by what means we will not now stop to inquire) some obstacle by which the course of the vapour may be retarded in its ascent, in a similar way to that which we have imagined in the glass globes,—then may the condensation take place gradually and at different heights. The relative distances of these points of precipitation will depend upon the force of the vapour, and the greater or less facility with which it overcomes the mechanical obstruction. For the scale of temperature laid down, the following table would represent an adequate balance of evaporation and condensation, with the appropriate degrees of elasticity between the points.

TABLE XXIX. *Shewing the Effects of a further forced Progression of Temperature upon the Atmosphere of Vapour.*

Height.	Sensible Temperature.	Constituent Temp. of Vap.	Elasticity.	State of Atmosphere.	Force of Evaporation.
0	80	67.9	.673	Clear	327
5000	64.4	64.4	× .606	Cloudy	467
10000	48.4	19.	124	Clear	
15000	31.4	16.	.112	Clear	
20000	12.8	12.8	× .100	Cloudy	80
25000	− 7.6	−20.7	.027	Clear	
30000	−30.7	−30.7	× .020	Hazy	

The last division of this Table gives the relative force of evaporation at the different points, supposing the total effect, if unopposed and sudden, to be 1000, and the same numbers will represent the comparative amount of precipitation, at the several intervals of condensation, or the relative densities of the three clouds. In this manner the struggle between the elasticity of the stream, and the condensing power of the cold, is divided and moderated; and the whole process becomes so gentle as quietly to restore the balance of force and temperature, provided the counteracting cause be not of a permanent nature. The moisture, falling gradually back into the excess of heat below, is converted into vapour of higher force, which pressing more upon the inferior strata, proportionably raises their densities.

From these considerations, it would appear, that, in any single column considered by itself, clouds of greater or less densities, and evaporation of greater or less force, must be the consequence of a temperature decreasing in a more rapid progression than is due to the law of aqueous vapour.

While the atmosphere is in the state represented in Table XXIX., let us now contemplate the effects of a general reduction of sensible temperature upon the constituent temperature, and the different points of precipitation. We will suppose the fall to take place gradually, and to amount to 10 degrees. In the first place, the elastic force upon the surface will not be diminished, but will approach the point of precipitation within three degrees. A plane of condensation will be established between the surface and the height of 5000 feet. So, likewise, the vapour from 9000 to 14,000 feet will not be disturbed, but the second plane of condensation will descend from 18,500 feet, to an intermediate position between that elevation and 14,000 feet. The shifting of these planes would not be sensible at the surface ; for the light precipitations, which would accompany their slow subsidence, would be expended in equalizing the temperature.

We must next contemplate these various phenomena, hitherto considered as confined to a single column, in connexion with adjacent sections. Let us take, as an illustration, the equatorial column of 80° temperature, in the state in which we have just considered it, and the adjoining one of 76°. 8. The flow of the lateral currents may be determined by the following Table:—

TABLE XXX. *Shewing the State of the Atmosphere occasioned by the Intermixture of Lateral Currents.*

Height.	Latitude 10				Latitude 0			
	Sensible Temperature.	Constituent Temperature.	Elasticity.	State of Atmosphere.	Sensible Temperature.	Constituent Temperature.	Elasticity.	State of Atmosphere.
0	76.8	51	.388	Clear	80	67.9	.673	Clear
5000	61.1	48	.351	Clear	64.4	64.4	.606	Clear
10000	44.9	45	.316	Cloudy	48.4	19	.124	Clear
15000	27.7	12	.096	Clear	31.4	16	.112	Clear
20000	9.3	9.3	.087	Cloudy	12.8	12.8	100	Clear
25000	−11.6	−32	.019	Hazy	− 7.6	−27	.027	Clear
30000	−35.	−35	.016	Hazy	−30.7	−30.7	.020	Clear

In this Table, the first point of condensation above, in the equatorial division, is supposed to take place at the height of 5000 feet; while, at latitude 10, it is fixed at 10,000 feet; and it will be seen, that up to the former elevation, the vapour of the first column is of much greater elasticity and density than that of the latter: it will, consequently, flow towards it with considerable force. No cloud will be formed, as before, at the point of condensation, for the supply arising from the evaporation at the surface, will be carried off in a lateral direction; or, if previously formed, would soon be dissipated by the same action. Nor would the transparency of latitude 10 be affected up to this height; for the current which it would receive, would, in constituent temperature, still be below what its sensible heat would maintain. But above this line a dense cloud would be precipitated. A counter-flow of small extent towards the equator will be established at 10,000 feet; and above this, again, the pressure will return to the first direction. The constituent temperature of the returning current, being below the temperature of the elevation, the transparency of the equatorial column will be preserved throughout. These lateral currents are supposed to take place under the same mechanical retardation as the ascending vápour.

It would be easy to multiply and vary these illustrations; but enough has been done, to shew generally that the necessary condition of transparency, in any vertical section of an atmosphere of pure vapour, in which, from some extraneous cause, the temperature diminishes faster than the natural progression, is, that the quantity generated from the evaporation, necessarily accompanying such circumstances, should be carried off to adjoining regions.

Our hypotheses have hitherto been framed upon the assumption, that the sphere round which the aqueous atmosphere has been diffused, was covered with water, whence a continual supply of vapour would flow equivalent to every increase of temperature. Let us now suppose that water is only partially diffused, and that the uncovered portions are absolutely dry. Vapour, out of the contact of water, is affected in the same way as the permanently elastic fluids, by variations of temperature; that is to say, it contracts or expands $\frac{1}{480}$th part of its volume for each degree of change, above its point of precipitation, by Fahrenheit's scale. If a current, therefore, were to pass over a dry space, heated to a degree higher than itself, the same changes in its constitution would take place, in miniature, as we have already traced in the dry atmosphere. Its density would diminish, while its elasticity would remain the same upon the surface, and be increased at higher stations. The following Table presents us with the different degrees of elasticity in a column of the constituent temperature of 80° heated to 90°

TABLE XXXI. *Shewing the Effects of Expansion upon Vapour heated beyond its Dew-point.*

Height.	Elasticity.	Constituent Temperature.	Sensible Temperature.
0	1.000	80	90
5000	.899	76.5	86.5
10000	.808	73	83
15000	.727	70	80
20000	.653	67	77
25000	.587	63	73
30000	.528	60	70

Such modifications would necessarily ensue in the cases which have already been considered, of the constituent temperature falling below the sensible heat; but, by comparing this Table with Table XXIV., it will be seen, that the total effect at the greatest extreme does not exceed .007 inches; and it will be unnecessary, in the general view which we are now taking of the subject, to introduce a correction to such a small amount.

In the case of vapour becoming heated in this manner, out of the contact of water, it may reach its point of deposition at a high elevation without producing any sensible cloud; for, although it would be slowly precipitated, it would be instantly restored to the elastic form by the excess of heat in the inferior strata; and no accumulation could be formed for want of supply from the dry surface below. A slight haziness might possibly be the result.

Let us now imagine, a stream of vapour, of known

density, filtering its way laterally through the resisting obstacle which we have supposed, from one part of the sphere, which is covered with water of a certain temperature, to another which is perfectly dry, and of equal or superior temperature. As it arrives at the latter point, it will diffuse itself rapidly over the dry space; and its elasticity, being no longer confined by an incumbent atmosphere of like density, will be reduced, and it will assume that force which its own diffusion will enable it to maintain. Or a stream of vapour, of high elasticity, flowing into a space where there already exists an atmosphere of inferior force, will be reduced in density to that of the general mean. In Table XXXII. are represented two contiguous columns of vapour; the first incumbent upon water of the temperature of 70°, and the second upon dry land of the temperature of 80° It is obvious that the former will flow into the latter, but will no longer be distinguishable by its constituent temperature of 60°, but will be reduced to the standard of the second column 32°. The elasticity, however, of this column will rise with the diffusion of the first.

TABLE XXXII.—*Shewing the Effects of the Diffusion of Vapour of High Elasticity from Water, over a dry Surface.*

Water.			Dry Earth.		
General Tem.	Const. Temp.	Elasticity.	General Temp.	Const. Temp.	Elasticity.
70	60	524	80	32	.200

But the surface upon which an atmosphere of any particular density rests, may be neither water, nor yet perfectly free from it, but we will imagine it to be earth differently embued with moisture and variously heated—then a partial supply, varying in quantity in different places but of the same degree of density, would take place, and clouds of more or less opacity would be formed, at corresponding situations, in the planes of deposition above. The following Table will tend to illustrate these positions.

TABLE XXXIII. *Shewing that the Elasticity of Vapour, yielded by different Surfaces variously heated, is governed by the incumbent Atmosphere.*

		Water, Temperature 60.8			Moist Earth, Temperature 70			Dry Earth, Temperature 80		
Height.	General Temp.	Constant Temp.	State of Atmosphere.	Force of Evap.	Constant Temp.	State of Atmosphere.	Force of Evap.	Constant Temp.	State of Atmosphere.	Force of Evap.
0	60.8	34	Clear	.368	34	Clear	507	34	Clear	786
5000	44.6	31	Clear		31	Clear		31	Clear	
10000	27.9	28	Cloudy	Density .368	28	Cloudy	Density .092	28	Hazy	
15000	10	− 6.4	Clear		− 6.4	Clear		− 6.4	Clear	
20000	− 5.4	− 9.4	Cloudy	.126	− 9.4	Hazy	.021	− 9.4	Clear	
25000	−31.2	−31.2	Ha zy	.020	−31.2	Clear		−31.2	Clear	
30000	−55.9	−55.9	Clear		−55.9	Clear		−55.9	Clear	

The same general temperature is here supposed to prevail momentarily in every part of an atmosphere of equal force; resting upon a surface covered with water in one part, with moist earth in another, and dry in a third, and varying moreover in heat in the three situations. The first point of precipitation is placed at 10000 feet. The water upon which the first part of the column rests, is of the same degree of heat, as the general temperature at the surface; the force of evaporation is 368, and as the supply is equal to the force, the density of the cloud is 368. The moist earth, upon which the second portion rests, is of the temperature of 70°, which makes the force of evaporation 507: but less steam being given off from the earth, than from the water, the quantity of the precipitation is proportionally diminished. It is calculated in the Table at one fourth. The dry surface, which supports the third portion, is heated to 80°, and yields no vapour: the evaporating force, which is equal to 786, is wholly unapplied, and no cloud can therefore be maintained The higher points are subject to the same modifications. The temperature of the evaporating surface regulates the quantity of water raised in vapour, the tension of the pre-existing atmosphere determines its elasticity.

I shall here conclude the separate consideration of the properties of aqueous vapour, applicable to my design. It is impossible not to perceive that I have already been obliged to anticipate some particulars, which would have fallen more consistently with my intended division under the third section: but I have restricted the inquiry as much as possible to the single object in view. It will, however, have been anticipated, that the mechanical retardation of the motions of the vaporous atmosphere, together with the forced progression of temperature, which have been so often referred to, belong to the state of mixture with the gaseous atmosphere; and I may now further connect the preceding remarks by observing, that, in the operations of the aqueous steam, a power is developed fully adequate to produce the disturbances of temperature hypothetically proposed in the examination of the permanently-elastic fluid.

PART III.

On the Habitudes of an Atmosphere of permanently-elastic Fluid, mixed with aqueous Vapour.

Having separately considered some of the properties of simple atmospheres of permanently-elastic fluid and of aqueous vapour, most essential to the object of our inquiry, we may now proceed to investigate the compound qualities of a mixture of the two, and their mutual relations so combined.

The properties, which each possessed in its separate state, will be retained in this connexion unchanged, and the two fluids will exercise no further action upon each other, than a mechanical opposition when in motion. The particles of steam, in penetrating the interstices of the permanently-elastic fluid, experience the same species of retardation as exists in their flowing through the pores of sand or cotton. When a state of equilibrium is attained, this mutual action ceases, and the particles of each press only upon those of their own kind. There are, therefore, two principal points of view, under which such a mixture may be regarded,—one, in which the particles are in a state of equipoise amongst themselves; and the other, where they are seeking an equilibrium by means of intestine motion. With respect to the first, there is no distinction between such a complete mixture and that of two or more permanently-elastic fluids; and it may be regarded like a mixture of gases, as an homogeneous fluid.

We will now inquire what would be the natural state of such an atmosphere, surrounding a sphere of uniform temperature throughout?

Before we can answer this question satisfactorily, we must consider what are the effects of mixing known measures of the gases, with vapour of different degrees of force.—The first effect is increase of bulk, under equal pressure in the permanently-elastic fluid; not in proportion to the measure of vapour added to it, but in proportion to its elasticity. Thus, if we mix a cubic foot of dry air, of the temperature of 212° and of the elasticity of 30 inches, with as much steam as would rise in the space of a cubic foot at the same temperature, and consequently of the force of 30 inches, the mixture would occupy the space of two cubic feet for 30 inches : 60 inches :: 1 : 2.

So, if we mix a like measure of air, of the temperature of 32° and of the elasticity of 30 inches, with as much vapour as would form in the same space at the temperature of 32°, and, consequently, of the force of only 0.200 inch, the bulk of the gas will only be increased .00666 of a cubic foot.

For 30.000 : 30.200 :: 1. : 1.00666

The case is exactly analogous to what would happen from the mixture of gases, of different degrees of elasticity. A cubic foot of air, of the elasticity of 30 inches, added to another cubic foot of the same elasticity, will have its volume doubled; but a cubic foot of air, of the elasticity of only .200 inch, being added to a cubic foot, of the elasticity of 30 inches, will only increase its volume .00666 of a cubic foot.

The second result is, that the specific gravity of the gas is decreased; but not exactly in proportion to its expansion: for while the vapour dilates its parts, it adds its own weight to the mixture. But this weight, though increasing with the elasticity, being, in all cases, less than that of an equal bulk of common air, decrease of density must follow. The diminution becomes greater with every increment of temperature.

Let us imagine an homogeneous atmosphere of air, of the temperature of 77°, and 30 inches pressure: its specific gravity, compared to air at 32°, would be .90626 : 1.00000. Let us suppose this to be mixed with an atmosphere of vapour of the same temperature, and .910 inch force: the specific gravity of the mixture under equal pressure would be .89312: its total pressure would amount to 30.910 inches; and its height would be increased from 28775 feet to 30260 feet. But, as we have seen before, the mean temperature of an homogeneous atmosphere must fall, in assuming that gradation of density which is essential to its natural state; the quantity of vapour must, therefore, suffer a proportionate reduction. If we were to assume the scale of decrease, represented in Table VII., for a temperature at the surface of 77°, the corresponding diminution of its force would be as follows:—

TABLE XXXIV. *Shewing the small Quantity of Vapour, which could exist in an Atmosphere of Air, supposing it saturated throughout.*

Height.	Temp.	Elasticity.
0	77	910
5000	61	.542
10000	45	.316
15000	27.5	171
20000	9.3	.088
25000	−11	.042
30000	−35	.016

The average quantity, therefore, of the vapour to this height, could not exceed 0.297 inch. We must further consider that the altitude to which we have hitherto followed our speculations, has comprised but two-thirds of the total height of our atmosphere; the remaining third may, without any risk of error, be considered, from the lowness of its temperature, to be totally free from vapour. The mean pressure, therefore, of steam would thus be reduced to .198 inch; and supposing such a state of circumstances to be possible, the barometer would only rise from 30 inches to 30.198 inches in the atmosphere surrounding a sphere of the temperature of 77°, by a change from absolute dryness to perfect moisture.

Nor would such a state of saturation constitute, by any means, a natural condition of such an atmosphere as we are contemplating. Even if a general mixture could be effected in the proportions which we have above imagined, the elasticity of the steam at the bottom of the column would be greater than the weight which the upper strata would confine; so that being urged upwards it would be condensed by the temperature of the air, which decreases faster than the progression due to the vapour and the barometer would not rise to the height just stated.

A state of complete mixture, in which all the particles would be at rest amongst themselves, cannot, therefore, exist in the natural atmosphere; and it can only, consequently, be regarded, in the second point of view distinguished above, namely, as in a state of intestine motion.

To place these particulars in a clearer light, let us trace the progress of vapour just beginning to form in a perfectly dry atmosphere. For this purpose we will imagine the temperature of the sphere to be 77° The first arrangement will be such as is represented under latitude 10, in Table VII. Let us now imagine water suddenly to overflow the surface, and evaporation will instantly commence. No atmosphere of vapour exists to impede its progress, the nascent steam will, therefore, merely assume the degree of tension necessary to overcome the *vis inertiæ* of the air which obstructs its motion. What this force may be, we have not, perhaps, sufficient *data* to determine. We must, for the present, fix it arbitrarily, and assume that, at the temperature of 77°, and pressure of 30 inches, it amounts to .200 inch. The constituent heat of vapour of this elasticity is 32°, so that at the height of about 13,500 feet it would meet with its point of condensation. An aqueous atmosphere of such degree of force being now established, fresh resistance to this amount is made to the progress of evaporation; and the elasticity of the rising steam must be doubled. Its constituent temperature is thus raised to 52°, and it cannot, therefore, pass the height of 7500 feet, without decomposition. The resistance upon the surface now amounts to .601 inch, to overcome which, vapour at 65° must be emitted. The first point of precipitation, in ascending from the surface, would thus be fixed at about 3600 feet. We may now further remark, that the diffusion of vapour does not cease at the height of 13500 feet, to which point we had first traced it; but the mechanical obstruction is proportionably reduced, and it is carried by successive stages to more lofty regions, where its tenuity is so much increased that it speedily eludes all observation.

With regard to the various points of condensation, it is probable, as was before remarked in the atmosphere of pure steam, that no cloud would be formed at any of them. The process of evaporation would be so gentle under these circumstances, that little above six grains of water would be raised per minute from the surface of a square foot; so that, as the gradual precipitation of this quantity took place between the different stages, it would instantly be re-dissolved by the excess of heat into which it would naturally incline to fall. The circulation thus becomes a process of equalization, by which the temperature of the upper regions is raised: the heat which is abstracted below by evaporation is evolved by condensation, the pressure of the vapour is increased, and all the changes tend to that distribution of heat which we formerly contemplated as the natural state of an unmixed atmosphere of steam.

The average quantity of vapour, which would exist upon the hypothesis which we have just assumed, while the atmosphere maintained its proper progression of temperature, may be roughly approximated as follows:—A stratum, of the force of .616 inch, extends to the height of 3600 feet; another, of the force of .401 inch, reaches 3900 feet further; a third, of only .200 inch, stretches almost as far as both the former together; making a total of 13500 feet. The mean, therefore, to this point is nearly

$$\frac{.616}{4} + \frac{.401}{4} + \frac{.200}{2} = .354 \text{ inch.}$$

For the further distance of 17500 feet, we cannot greatly err in taking .064 inch, as the mean pressure, making the average to the height of 31000 feet, .209 inch. One-third of the atmosphere beyond this being considered free from vapour, reduces the mean to 139 inch.

The following diagram may possibly tend to elucidate the effects of an unequal addition of matter, or of unequal expansion in various parts of the same column of fluid.

Fig. 7.

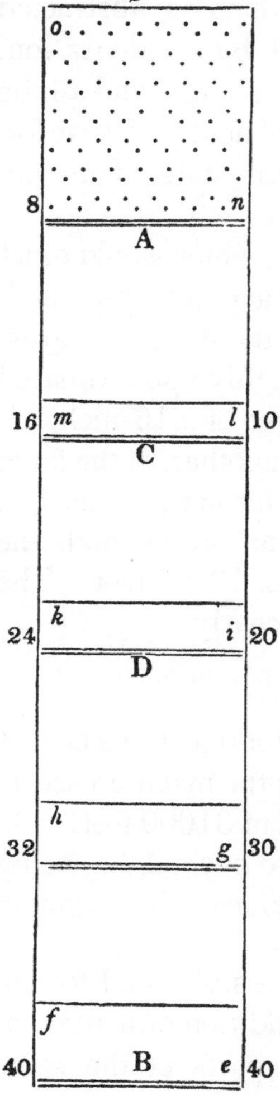

Let A B represent the column whose height and weight we will call 40. Its four sections A B C D are each equal to 10. Let us suppose an addition of matter to take place in B, *e f* equal to 4; in D, *g h* equal to 3; in C, *i k* equal to 2; and in A, *l m* equal to 1. The total increase of weight and pressure at the bottom will be 10, or one-fourth of the original amount: the same as if the total amount had been equally distributed throughout the mass, or added at once to the top of the column, as *n o*.

Again—if in the same column an unequal expansion were to take place in the four different sections, $ef = 4$, $gh = 3$, $ik = 2$, and $lm = 1$, the total increase of bulk *n o* would be the same as if the expansion had every where been equal, and the weight at the base remaining the same, that of the sections would be altered from 10 . 20 . 30 . 40, to 8 . 16 . 24 . 32 and 40.

In the case of the atmosphere, which we are considering, both these changes are combined: the barometer rises at the surface of the sphere, and the weight of the several strata is still further changed. The following Table exhibits the state of the barometer at equal altitudes, before and after the admission of vapour, in an atmosphere surrounding a sphere of the uniform heat of 77° ; together with the temperature appropriate to the elevation and the dew-point.

TABLE XXXV.—*Shewing the State of the Barometer, at equal Altitudes, in an Atmosphere of Air, before and after the Admission of Vapour.*

Height.	Temperature.	Barometer. Atmos. without Vapour.	Barometer. Atmos. with Vapour.	Dew-Point.
0	76.8	30.000	30.139	65
5000	61.1	25.214	25.348	52
10000	44.9	21.193	21.318	32
15000	27.7	17.812	17.928	9
20000	9.3	14.970	15.079	0
25000	−11.6	12.583	12.682	−35
30000	−35.	10.578	10.667	−35

Such, then, would be the new state of things, from the admission of water to the surface of the sphere: a state, however, which, notwithstanding the equality of the superficial heat, could not be one of permanent rest. A perpetual struggle would ensue between the temperature due to the density of the air, and the constituent temperature of the vapour, accompanied by perpetual evaporation below, and simultaneous condensation above. No winds or lateral currents would be established, but an increasing circulation in a vertical direction.

It will be observed, that the total alteration of pressure is but small, even in the almost extreme case which has been selected, and the changes of density still less; it is in the unequal distribution, and its consequent effects, that we must seek for the principal influence of vapour in the mixed atmosphere.

Changing now our hypothesis of the sphere of equal temperature, for that of the sphere of unequal temperature, increasing from the poles to the equator, we will again assume, that the barometer stands every where at the same height upon the surface; which height we will suppose, as before, to be 30 inches. The state of the vapour in the different columns of the mixed atmosphere, is to be imagined to be in the same proportion as in the atmosphere which we have just been considering—that is to say, its constituent temperature, at the surface of the sphere, is to be within 11 degrees of the temperature of the several zones. The whole arrangement is thus represented in the following Table:

TABLE XXXV. *Shewing the Specific Gravity, Elasticity, Temperature, and Dew-point, of a mixed Atmosphere of Air and Aqueous Vapour, at different Altitudes and different Latitudes from the Pole to the Equator.*

Height.	Poles.					Latitude 80					Latitude 70					Latitude 60					Latitude 50				
	Total Pressure.	Pressure of Vap.	Specific Gravity.	Temp.	Dew Point.	Total Pressure.	Pressure of Vap.	Specific Gravity.	Temp.	Dew Point.	Total Pressure.	Pressure of Vap.	Specific Gravity.	Temp.	Dew Point.	Total Pressure.	Pressure of Vap.	Specific Gravity.	Temp.	Dew Point.	Total Pressure.	Pressure of Vap.	Specific Gravity.	Temp.	Dew Point.
0	30.000	.044	1.06666	0	− 11	30.000	.047	1.06038	3	− 8	30.000	.057	1.04685	9	− 2	30.000	.085	1.02607	19	8	30.000	.134	.99835	32	21
5000	23.597		.86935	− 18		23.652		.86542	− 15		23.707	.033	.85684	− 8	−17	23.793	.047	.84427	1	− 8	23.949	.079	.82563	14	6
10000	18.587		.70856	− 37		18.630		.70637	− 34		18.724		.70140	−27		18.893		.69405	−17		19.106	.045	.68321	− 3	− 9
15000	14.591		.57752	− 58		14.642		.57654	− 55		14.775		.57407	−47		14.969		.57061	−36		15.229		.56472	−22	
20000	11.411		.47071	− 82		11.484		.47057	− 78		11.617		.46991	−70		11.827		.46904	−58		12.044		.46677	−43	
25000	8.900		.38365	−109		8.965		.38408	−104		9.102		.38463	−96		9.314		.38558	−83		9.579		.38582	−67	
30000	6.906		.31270	−140		6.978		.31352	−135		7.100		.31483	−126		7.302		.31699	−112		7.566		.31890	−95	

Height.	Latitude 40					Latitude 30					Latitude 20					Latitude 10					Equator				
	Total Pressure.	Pressure of Vap.	Specific Gravity.	Temp.	Dew Point.	Total Pressure.	Pressure of Vap.	Specific Gravity.	Temp.	Dew Point.	Total Pressure.	Pressure of Vap.	Specific Gravity.	Temp.	Dew Point.	Total Pressure.	Pressure of Vap.	Specific Gravity.	Temp.	Dew Point.	Total Pressure.	Pressure of Vap.	Specific Gravity.	Temp.	Dew Point.
0	30.000	.237	.96358	48	37	30.000	.375	.93463	61	50	30.000	.507	.91278	70	59	30.000	.616	.89764	77	65	30.000	.698	.89018	80	69
5000	24.072	.139	.80230	31	22	24.215	.221	.78245	44	35	24.279	.328	.76703	54	46	24.319	.401	.75621	61	52	24.342	.443	.75133	64	55
10000	19.338	.068	.66800	14	2	19.531	.112	.65503	28	16	19.675	.162	.64489	38	26	19.738	.200	.63762	45	32	19.779	.229	.63436	48	36
15000	15.525	.037	.55629	− 4	− 14	15.739	.046	.54855	10	− 8	15.898	.068	.54180	20	2	16.012	.087	.53703	27	9	16.060	.100	.53520	31	13
20000	12.409		.46273	− 24		12.673	.034	.45856	− 9	−17	12.811	.046	.45500	1	− 8	12.974	.064	.45147	9	0	13.043	.071	.45068	12	3
25000	9.915		.38489	− 47		10 162		.38327	−31		10.342		.38166	−19		10.467	.015	.38010	−11	−35	10.521	.032	.37980	− 7	−16
30000	7.852		.32016	− 62		8.135		.32035	−56		8.313		.32010	−43		8.424		.31948	−35		8.483		31980	−30	

To face Page 82.

The material originally positioned here is too large for reproduction in this reissue. A PDF can be downloaded from the web address given on page iv of this book, by clicking on 'Resources Available'.

By comparing this synoptic view with that presented by Table VII., it will be seen, that the specific gravity and elasticity of the air is but very slightly affected by this intermixture of aqueous vapour; so slightly, indeed, that the course and velocity of the currents, as represented in Table VIII. may be considered, without any chance of disturbing our main argument, as unaltered; and their balance to be that by which the barometer is maintained at an unvarying height. It will also be remarked, that, while the great aërial ocean is divided into two distinct strata, flowing in opposite directions from north to south and from south to north, the aqueous part, which is nearly confined to the lower current, presses in a contrary direction. The adjustment of these particulars remaining as now supposed, the compensating winds flow on, in the courses which have been described, and the balance remains undisturbed.

The admixture of vapour, which we have hitherto considered, has not yet affected the gradation of temperature, resulting from the decreasing density of the atmosphere in its upper parts; the process of evaporation, however, which has been described, must, in time, necessarily induce such an alteration. The stream, as it reaches its point of condensation, must give out its latent heat, and during its precipitation, combining with a fresh proportion, it again ascends and again evolves it in the middle regions. It may thus be considered as carrying caloric from the surface of the sphere to higher strata; and it is obvious how a considerable section of any one column may thus have its temperature equalized and fully saturated with aqueous particles. The currents thus become affected both by the expansive powers of the vapour and of the extricated heat—causes, the influence of which, so applied, must be partial, and cannot reach the higher regions. The unequal action must produce a fall in the barometer, as has been before explained.

As, on one hand, this effect upon the barometer is produced by the augmentation of the aqueous vapour; so, on the other, a rapid increase of the latter may be produced by a fall in the former. The mechanical resistance of the air must, of course, be increased by its motion in opposition. When this is stopped, as it soon is, by a trifling fall of the mercurial column, the vapour will rush forward with its whole force, retarded only by its filtration through the quiescent air; and the temperature of the higher latitude being unable to support its elasticity, precipitation must follow. From the operation of these causes, the temperature of the latitude is partially affected, the density of the air is still further reduced, and the aërial current is reversed. The course of the vapour is thus greatly accelerated, and abundant precipitation will follow.

The progress of the precipitated moisture, from the time when its first streaks would visibly shoot across the air, to the time when it would descend in rain upon the globe, is not without its interest. In proportion to the density of the vapour, no doubt, must be the magnitude of the condensed particles. When first formed in the higher elevations, the cloud would probably assume a light *cirriform* appearance; in lower regions the precipitation would be more dense, and the attraction of aggregation stronger; the mass would subside gently to a lower station, where the density of the air would oppose a greater resistance to its descent. Here, in a higher temperature, the cloud would again begin to be dissolved, and would assume a rounded and compact form; and thus the equalization of the temperature, and the diffusion of the vapour, would be carried on from several points at once. The different beds obey the impulse of the winds, and, as they sail along, enlarge the circumference of their action: till, at length, the natural equilibrium of the atmosphere can be no further curbed. The precipitations increase, the strata of the clouds inosculate, and the air no longer buoys up their load.

It will be convenient here to subjoin a synoptic view of the force of the aërial current, and the counter-pressure of the vapour in a mixed atmosphere, surrounding a sphere unequally heated in the manner already set forth.

TABLE XXXVII. *Shewing the Force of the different Currents in a mixed Atmosphere of Air and Vapour, between the Poles and the Equator.*

Height	Latitudes 90 & 80		Latitudes 80 & 70		Latitudes 70 & 60		Latitudes 60 & 50		Latitudes 50 & 40	
	Wind.	Vapour.	Wind.	Vapour.	Wind.	Vapour.	Wind.	Vapour.	Wind.	Vapour.
0	+.178	−.003	+.387	−.010	+.575	−.028	+.810	−.049	+1.034	−.103
5000	+.057		+.191		+.281	−.014	+.375	−.032	+ .570	−.060
10000	+.019		+.048		+ 045		+.112		+ .217	−.023
15000	−.023		−.063		−.093		−.084		− .035	
20000	−.069		−.112		−.185		−.149		− .239	
25000	−.078		−.152		−.240		−.272		− .307	
30000	−.095		−.161		−.264		−.321		− .322	

Height.	Latitudes 40 & 30		Latitudes 30 & 20		Latitudes 20 & 10		Latitudes 10 & 0	
	Wind.	Vapour.	Wind.	Vapour.	Wind.	Vapour.	Wind.	Vapour.
0	+.854	−.138	+.648	−.132	+.447	−.109	+.208	−.082
5000	+ 454	−.082	+.392	−.107	+.282	−.073	+.118	−.042
10000	+.215	−.044	+.165	−.050	+.159	−.038	+.053	−.029
15000	+.031	−.009	+.038	−.022	+ 036	−.019	+.008	−.013
20000	−.131		−.024	−.012	−.068	−.018	−.045	−.007
25000	−.196		−.128		−.074		−.054	
30000	−.291		−.170		−.090		−.070	

I have expressed my doubts above, whether we have sufficient *data* to determine the amount of the resistance which the pores of the gaseous constituents of the atmosphere offer to the passage of vapour in motion. Experiments certainly are wanting to elucidate this relation; the observations, however, of De Saussure and Dalton throw some light upon the subject. The resistance alluded to, may be regarded as two-fold; first, in connexion with the permanently-elastic fluid at rest, and secondly, with it in motion.

With regard to the state of rest, the opposition with which vapour passes through air, is, in proportion to its density. De Saussure concluded, from his experiments, that a diminution in the density of one-third doubled the rate of evaporation.

With regard to the state of motion, a breeze in opposition to the stream of vapour, must retard its progress as much as one in the same direction favours it. Much obscurity envelopes this inquiry from the vagueness of the terms employed in denoting the velocity of the air. Mr. Dalton has determined that the rate of evaporation, in a perfect calm, being denoted by 120, that of a brisk wind is 154, and of a high wind 189. The retardation of opposing currents, of the same respective forces, may therefore be reckoned in proportion.

Some important conclusions follow from these propositions, however wanting in precision. It is impossible, in the present state of our knowledge, to determine the absolute velocity with which vapour travels under any of the circumstances mentioned; but the relative rate of different parts of the same column may be approximated. Thus, taking latitude 30, laid down in the last table, the current which blows in the direction of latitude 40, may be deemed high, and retards the motion of the vapour towards latitude 20 accordingly. At the height of 10,000 feet, the density of the air is reducėd one-third, and the velocity is consequently doubled: to which, we must also add, that the opposing current, at the same elevation, declines in strength, whereby the force is again increased in the proportion of 189 to 154. More vapour, therefore, probably would pass at this elevation than at the surface; although its excess of elasticity is only 044 inches at the former station, and .138 inches at the latter. Whenever a deep stratum of air has had its temperature and vapour equalized, in the manner before described, it is easy to conceive that the aqueous atmosphere may travel in its upper parts with considerable velocity, in a course directly opposed to the wind at its lower. The approximation may be carried a little further, perhaps, as follows. The effect of a brisk wind, in accelerating evaporation, is equal to an increase of about three-tenths of the elasticity; that of a high wind to six-tenths. The retarding influence of the polar current, in its regular state, may therefore be apportioned to the different latitudes in Tables XXXV. and XXXVI., as follows:—From the poles to latitude 80 $=\frac{1}{10}$ of the elasticity, to lat. 70 $=\frac{2}{10}$, lat. 60 $=\frac{4}{10}$, lat. 50 $=\frac{5}{10}$, lat. 40 $=\frac{6}{10}$, lat. 30 $=\frac{5}{10}$, lat. 20 $=\frac{3}{10}$, lat. 10 $=\frac{2}{10}$, and from lat. 10 to the equator $\frac{1}{10}$. The following Table then, represents the efficient force of the vapour in a lateral direction, calculated for the surface of the sphere, and for the altitude of one-third the density.

TABLE XXXVIII. *Shewing the efficient lateral Force of Vapour between the Poles and Equator at the Surface of the Sphere, and at the Altitude of one-third the Density.*

Height.	Latitudes 90 & 80		Latitudes 80 & 70		Latitudes 70 & 60		Latitudes 60 & 50		Latitudes 50 & 40	
	Balance of Force.	Effects of Wind and Density.	Balance of Force.	Effects of Wind and Density.	Balance of Force.	Effects of Wind and Density.	Balance of Force.	Effects of Wind and Density.	Balance of Force.	Effects of Wind and Density.
0	−.003	−.002	−.010	−.008	−.028	−.017	−.049	−.025	−.103	−.042
10000	−.001	−.002	−.004	−.008	−.008	−.017	−.012	−.026	−.023	−.045

Height.	Latitudes 40 & 30		Latitudes 30 & 20		Latitudes 20 & 10		Latitudes 10 & 0	
	Balance of Force.	Effects of Wind and Density.	Balance of Force.	Effects of Wind and Density.	Balance of Force.	Effects of Wind and Density.	Balance of Force.	Effects of Wind and Density.
0	−.138	−.069	−.132	−.093	−.109	−.088	−.082	−.074
10000	−.044	−.072	−.050	−.090	−.038	−.070	−.029	−.058

I must here repeat, that these tables are not meant to impose an air of precision upon the subject, which the present state of our knowledge does not warrant; but to assist our conceptions of the general effects of so many conflicting causes. The last table will give some idea of the retardation of force, in the vapour, occasioned by the wind, at the surface of the sphere, and also of the increase of velocity occasioned by diminished pressure in the upper regions. It is easy to understand that, whenever the aërial current coincides with the direction of the vapour, the progress of the latter is accelerated in the same proportion.

The permanency of the Barometric pressure, on the surface of the sphere, is dependant, as we have seen, upon the equal balance of the aërial currents, its fluctuations have been traced to the destruction of this equipoise, by unequal and local expansions and condensations. One of the chief causes of these latter, there can be no doubt, is the increase and the decrease of the aqueous vapour, counteracting the natural progression of temperature, by the caloric evolved in its condensation: but there is another, to which no allusion has yet been made, which must necessarily be powerful in this operation. It has hitherto been supposed, for the sake of simplifying the subject, that the source of heat has been in the sphere itself; and that all the regular changes of temperature have emanated from its surface. This so far agrees with the condition of the atmosphere of the earth, with which it is our final object to identify our various hypotheses; for, while in a transparent state, the sun's rays pass through the air without materially affecting it, and expend their energy upon the surface of the globe. But, if the atmosphere become cloudy and opaque, the rays of heat, emanating from an external source, are in great part absorbed before they reach the surface, and an increase of temperature and elastic vapour, must take place, in the middle regions. Another source is hence derived of partial and powerful expansion.

To this we may also add, the property which the clouds possess, of preventing the radiation of heat from the surface beneath them, and the greater conducting power of damp, than of dry, air.

Amongst the literally numberless modifications of circumstances, to which an atmosphere, of the nature we have been considering, is liable, there are yet two or three, to which it will be necessary shortly to refer. The surface of the sphere, has hitherto been chiefly considered as perfectly plain, and either thoroughly dry, or every where covered with water,—we will now contemplate it as covered with water, to the extent of three fourths of its superficies, and the remaining fourth of dry earth, uneven and intersected by eminences. This intermixture of land and water, at once introduces inequalities of temperature, of a different character from those that have been hitherto considered. They arise chiefly, from the greater rapidity both of heating and cooling, in the dry surface, dependant upon the peculiar constitution of the watery element. It will not be essential to our purpose to trace them into details. As the processes by which their impressions are communicated to the incumbent air, are slow and gradual, they mostly affect the different columns in an equable manner; so that their influence upon the currents resolves itself into the cases which have been already proposed, of total and regular expansion. With respect to the vapour, however, the case is different. It is evident, from principles before established, that the parts of the atmosphere which are immediately over the dry spaces, will not remain free from its admixture; for the elasticity of the surrounding medium will soon supply the vacuum. The rapidity of this equalization will depend upon the mechanical obstruction of the air, being increased or diminished by an adverse or favourable wind. When once diffused over the land, it would be more subject to condensation; and the amount of precipitation must be restored from the expanse of waters.

Unevenness of surface would also tend to modify the atmosphere, in an inferior manner. Any elevation would obviously partake of the temperature due to the stratum of air, into which it rose; but the action must be reciprocal, and as the heating surface is raised to higher regions, those regions must be proportionally and unequally affected.

But these, and similar particulars, belong more especially to the history of the terrestrial atmosphere; and we are now arrived at that point where we may discontinue the synthetical process, and proceed to prove the accuracy of our inductions, by the method of analysis.

PART IV.

Examination of the Particular Phenomena of the Atmosphere of the Earth.

The test of our theories must be their application to meteorological observations. Of these, I shall now proceed to select a few of the most general, and the best authenticated; and, after a concise statement of each fact, will endeavour to shew how far it is reconcileable with our preceding conclusions. If the principal phenomena of this intricate branch of natural philosophy should be found to derive any elucidation from the manner of viewing them here adopted, it may be worth while hereafter to review the registered changes of different climates in a more particular manner than is consistent with the nature of this Essay.

The first fact to which I shall address myself is, that

I. *The mean height of the Barometer at the level of the sea, is the same in every part of the globe.*

Equality of pressure is one of the fundamental laws of Hydrostatics, and, consequently, we have seen that it is one of the first conditions of an atmosphere at rest. We have also seen, that, when acted upon by disturbing causes, the restoration of this equilibrium is the object of all the motions excited. Where the cause is permanent and equal, the effect produced is exactly adequate to maintain an unfluctuating balance; and equality of pressure is attained by means of regular currents. Where the cause is temporary and partial, it has not time to effect a general adjustment, and partial disturbances are the consequence; but the local effects, which are in reality mere undulations of the medium, must always be, as much in excess on one side as they are in defect on the other, and oscillate round the same point of equilibrium. The balance of fluctuations will, therefore, still exhibit equality of pressure.

II. *The Barometer constantly descends in a geometrical progression for equal ascents in the atmosphere, subject to a correction for the decreasing temperature of the elevation.*

This observation applies, whatever be the height of the mercurial column at the surface of the sea, and however remote from its mean state.

The density of the air is the result of the action and re-action of its gravity and elasticity; and between the two forces, there must be an exact balance, that is to say, the weight of the air which tends to compress it, and the elasticity by which it endeavours to expand, must be equal. Now the force of gravity being exerted in a perpendicular direction, any increase or decrease in the two antagonist powers must instantly pervade the whole of the perpendicular column in which it takes place; so that, under every circumstance of disturbance, the geometrical progression of the density will be maintained. A local change in any section of the atmosphere is thus instantly distributed throughout its total height; while the general compensation of the system is more gradually produced by lateral movements.

The method of correcting the height of the mercurial column, for the temperature, usually adopted in Barometrical mensurations, is by no means correct. It consists in estimating the temperature of the air, by taking an arithmetical mean between the heights of the thermometer at the upper and lower stations, upon the supposition of the uniform diffusion of heat in the column intercepted between them. Although its adoption, in cases of moderate elevation, is attended by no very sensible error, its insufficiency is very manifest at great altitudes. M. De Luc, when he proposed this correction, was sensible of its imperfection; and General Roy observes, that "one of the chief causes of error in barometrical computations proceeds from the mode of estimating the temperature of the column of air from that of its extremities; which must be faulty, in proportion as the height and difference of temperature are great." The preceding speculations naturally suggest some ideas upon this subject, which it would occupy too much time to attempt to develop here. It is sufficient for our present purpose, to establish that some correction for temperature is necessary.

III. *The mean temperature of the earth's surface increases gradually from the poles to the equator.*

Upon this fact it is not necessary to enlarge. It

is well known to be the result of the unequal impression of the sun's rays. At the middle point the temperature has been found to be the exact mean; and from that centre the heat diminishes rapidly northwards, and increases with equal rapidity towards the south. The incumbent atmosphere of course follows the same gradation.

IV. *The mean temperature of the atmosphere decreases from below upwards, in a regular gradation.*

The fact is sufficiently established, by numerous observations.—Mr. Dalton was the first to demonstrate, that the natural equilibrium of heat in an atmosphere is, when each atom of air in the same perpendicular column, is possessed of the same quantity of heat; and, consequently, that such an equilibrium results, when the temperature gradually diminishes in ascending. This is the natural consequence of the increased capacity for heat, derived from rarefaction. When the quantity of heat is limited, the temperature must be regulated by the density. Professor Leslie* also, by some delicate experiments, determined the expression of this law of progression; and has shewn that, reckoning the density of the air, at the surface of the sphere, as unit, the difference between the density at any given altitude, and its reciprocal, being multiplied by 45, will express the mean diminution of temperature in degrees of Fahrenheit's scale. This rule accords exactly with numberless experiments made in different parts of the globe. Observations, likewise, prove that the fact is only elicited from the mean results; the actual daily temperature

* Ency. Brit. Sup., Article CLIMATE, p. 188.

being found to oscillate at certain distances on each side of this centre.

V. *The Barometer, at the level of the sea, is but very slightly affected by the annual or diurnal fluctuations of temperature.*

This will be apparent from every register, whose mean observations are consulted with this view. The vicissitudes of day and night, and the changes of the seasons, are produced by very gradual processes of heating, and cooling; the alterations of temperature have time to pervade the whole column, and the balance is preserved by the equalization of the effect. The heating power, in general, being the surface of the earth, the arrangement is in the most perfect form for effecting this end. The air above becomes cooled from its position, and of greater density than is due to its elevation; while that below is expanded by its contact with the heated body, and a rapid interchange of situation must necessarily ensue.

This cooling of the atmosphere by radiation, must be distinguished from that decrease of temperature, which arises from increased capacity. The amount of the latter is definite for the height, and it is a constituent element of the density due to the elevation; but by the former power, the air is cooled below this standard, and becomes in its successive strata specifically heavier. That this emission of heat into space is constantly going on, is evident from the permanency of the mean temperature, which, notwithstanding the constant accession which it receives from the sun, ever remains the same.

But although any undue acceleration of either of the great principal currents is prevented by this mixture to any extent, yet, in most parts of the globe, a tendency has been observed in the Barometer to fall during the day, and to rise at night * ; which would seem to indicate that the lower stratum becomes rather more affected by the diurnal temperature than the upper. This exception, however, is perfectly reconcileable to theory and to observation ; for it is well known, that the surface of the earth becomes more rapidly cooled by radiation, during the absence of the sun, than the air ; a small stratum, therefore, which is in immediate contact with it, becomes unduly affected ; and, as the influence, from its nature, cannot be extended upwards, an increase in the weight of the whole column takes place. The returning heat dissipates this accession by restoring the natural progression of temperature.

VI. *The Barometer, in the higher regions of the atmosphere, is greatly affected by the annual and diurnal fluctuations of temperature.*

This observation is easily confirmed in various ways ; but for the present, I shall refer for its correctness to those valuable registers, which are simultaneously kept at Geneva, and the summit of Mont St. Bernard. Upon the average of four

* See the Essay upon the Horary Oscillations of the Barometer.

years, at those stations, I find, that from sun-rise to 2 P.M., the upper Barometer gains upon the lower .037 inch, and from winter to summer .260 inch. This will be deemed direct and satisfactory proof of the proposition, although the places where the observations were made, are not exactly suited to exhibit the fact in its most striking form. The city of Geneva is situated at a great height above the level of the sea, and must, therefore, itself partake of the barometric change referred to. To this we may add, that experiments, made upon elevations of the surface of the earth, can by no means be looked upon exactly in the same light, as if they had been performed at equal elevations, in the atmosphere above the level of the sea. In the former case, the heating surface is raised into the higher regions, and cannot fail to produce a modification of circumstances, different from those which the simple problem supposes. As the existence, however, of an effect, and not its amount, is here required to be proved, the example will suffice.

It is a consequence which naturally, and infallibly, flows from that general alteration of temperature, which does not affect the Barometer at the level of the sea: for as the expansion and contraction do not alter the total weight of the aërial column, it is clear that they must change the relative weights of its different sections.

VII. *The heating and cooling of the atmosphere, by the changes of day and night, take place equally throughout its mass.*

This is fully established by the same series of

observations. Some apparent exceptions will be found, upon close examination, to confirm our conclusions. The mean ranges of the thermometers, for the twenty-four hours, exactly correspond, during the winter months, in the two situations: but in the summer, the lower exceeds the higher, 5 or 6 degrees. The reason of this discrepancy will be obvious, when we recollect that the convent of St. Bernard is situated within the influence of perpetual snows: much of the rising heat is, therefore, expended in their liquefaction, in the months when the daily temperature rises above the term of congelation. This is a striking example of the difference, which we have before referred to, between observations made upon lofty stations of the earth's surface, and at equal altitudes in the atmosphere above the sea. The impression of the surface, whether it be a heating or cooling power, must be communicated to the surrounding air.

This influence may also be traced in another way, and presents an illustration of a different point. The cooling power of the snow must produce an unequal effect upon the currents; and accordingly we find that a proportionate effect is indicated by the Barometer, at the base of the column, which stands a little higher in the summer months than in the winter.

VIII. *The average quantity of vapour in the atmosphere decreases from below upwards, and from the equator to the poles.*

This consequence is obviously derivable from the

preceding laws of temperature, and is, moreover, amply confirmed by experiment.

IX. *The condensation of elastic vapour into cloud raises the temperature of the air.*

In confirmation of this theoretical and practical conclusion, the following observation of M. de Luc may be adduced.

"Pendant que je réfléchissois sur l'apparition subite des nuages, je découvris un petit amas de vapeurs, du côté du nord, à 3 ou 400 pieds au-dessous de moi: Je le considérois avec attention, et je remarquois d'abord que son volume augmentoit sensiblement, sans qu'il me fût possible d'appercevoir d'où lui venoient ses accroissements. Je vis ensuite qu'au lieu de s'abaisser à mesure qu'il grossissoit, et qu'il paroissoit même devenir plus dense, il s'élevoit au contraire. Le vent le poussoit vers moi. Il m'atteignit enfin, et m'environna tellement que je ne vis plus ni le ciel ni la plaine. Je pensai au même instant, à observer mon thermomètre, qui étoit suspendu en plein air, éxposé au soleil et que j'avois vu auparavant à $+ 4\frac{2}{3}$ (42° Fah.). Je présumois que l'action du soleil étant interceptée par ce nuage mon thermomètre devoit baisser et je fus très surpris de le voir au contraire à $+ 5\frac{1}{2}$ (45° Fah.). Le nuage, qui continuoit à monter obliquement vers le sud, abandonna bientôt le lieu où j'étois, le soleil reparut mais, malgré son action, le thermomètre rédéscendit." De Luc, *tom.* iii., *p.* 251.

X. *Another remarkable phenomenon is, that there exists a general tendency in the wind to blow from north-east and south-east towards the equator, in latitudes below* 30°.

The main-spring of all the grand movements in the atmosphere is, no doubt, the regular gradation of temperature, which exists along the different latitudes of the earth's surface. It was Hadley who first attributed to its right cause, *viz.*, the excess of the rotatory velocity of the equator, the flexure towards the west, which the great polar currents receive, and which is known by the name of the trade-winds *. These winds, as they approach the equator, gradually lose their northerly and southerly directions, and blow directly from the east. This may be explained by the meeting of the currents from the two poles, which have thus their opposite impulses balanced; when nothing remains, but that excess of *inertia* which leaves them behind the revolution of the globe. For the same reason, they lose much of their energy in this situation, and the neutral line is subject to frequent calms.

XI. *While the trade wind blows upon the surface of the earth, a current flows in the contrary direction, at a great elevation in the atmosphere.*

This necessary consequence of the theory of the trade winds, rested for a long time upon theoretical conclusions only: the eruption, however, of the volcano, in the Island of St. Vincent, in the year

* Phil. Trans. 1735.

1812, placed the fact beyond dispute. The Island of Barbadoes is situated considerably to the east of St. Vincent, and, between the two, the trade wind continually blows, and with such force, that it is with considerable difficulty, and only by making a very long circuit, that a ship can sail from the latter to the former. Notwithstanding this, during the eruption at St. Vincent, dense clouds were formed at a great height in the atmosphere above Barbadoes, and a vast profusion of ashes fell upon the island. This apparent transportation of matter against the wind, caused the utmost astonishment amongst the inhabitants, and the certainty of the fact cannot but be considered as of the utmost interest to the science of meteorology.

XII. *The mean height of the Barometer is not affected by the trade winds.*

This is a proof that the quantity of air, which passes below from the poles to the equator, must be exactly balanced by an equal quantity flowing above in the opposite direction.

XIII. *Between the latitudes* 30° *and* 40°, *both in the northern and southern hemispheres, westerly winds prevail.*

The atmosphere, from the processes of heating and cooling, which it is incessantly undergoing, and from the position of the heating surface, must, as we have seen, be subject to a constant circulation from below upwards; but the direction of the falling particles is diverted from the exact perpendicular direction by the influence of the great lateral currents: they, therefore, reach the earth in an angular course. Thus, the molecules, situated over the equator, which become cooled by their remoteness from the earth, descend from their increase of specific gravity; but in their descent are carried by the set of the superior stream towards either tropic. Bearing with them, as they do, the excess of the equatorial velocity, from west to east, they no sooner enter the lower current than they impart this movement to it, and the wind is modified accordingly. The spaces included between latitudes 30° and 40°, appear, from observation, to be the regions where this influence first takes effect, and primarily prevails. If it be asked, why this effect, depending apparently upon as constant a cause as the trade winds themselves, is less certain and steady in its occurrence; the answer will lead us to other causes, which affect both it, and sometimes the eastern winds, and which will presently come under our notice. We may add to our remarks, at present, that the westerly winds, within the above-mentioned limits, are much more regular and constant in the southern hemisphere than in the northern. It may further be observed, that the restriction itself of the east winds, within the 30th degree of latitude, is owing to this counteracting influence, and that the strictness of their limits can be explained upon no other hypothesis.

XIV. *The western coasts of the extra-tropical continents have a much higher mean temperature than the eastern coasts.*

This difference is extremely striking between

the western coast of North America and the opposite eastern coast of Asia. It is explained by the heat evolved in the condensation of vapour, swept from the surface of the ocean by the western winds. This general current, in its passage over the land, deposits more and more of its aqueous particles, and by the time that it arrives upon the eastern coasts, is extremely dry: as it moves onwards, it bears before it the humid atmosphere of the intermediate seas, and arrives upon the opposite shores in a state of saturation. Great part of the vapour is there at once precipitated, and the temperature of the climate raised by the evolution of its latent heat.

XV. *A wind generally sets from the sea to the land during the day, and from the land to the sea during the night, especially in hot climates.*

The land and sea-breezes are amongst the most constant of the phenomena of the inconstant subject with which we are occupied. The land becomes much more heated by the action of the sun's rays than the adjacent water; and the incumbent atmosphere is proportionably rarefied: during the day, therefore, the denser air of the ocean rushes to displace that of the land. At night, on the contrary, the deep water cools much more slowly than the land, and the reverse action takes place. As these changes proceed gradually, the height of the barometer is not affected by them.

XVI. *The trade winds, in the neighbourhood of the western coasts of the large continents, in their course, have their direction changed.*

This is an effect of the same nature as that of the

land and sea-breezes. Those parts of Africa and America which lie between the tropics, become intensely heated by the action of a vertical sun: the columns of the atmosphere, which rest upon them, must therefore be highly rarefied, and the more temperate air of the surrounding seas will press upon them. This influence is so decided as to overcome the tendency of the east wind; and on the western coasts of both continents a wind from the west prevails. This is, again, an instance of a complete perpendicular change from a permanent cause, and the total pressure is unaffected.

Of the same nature are the Monsoons of the Indian Ocean, and other periodical winds. They are occasioned by a particular distribution of land and water, acted upon by the periodical changes of the sun's declination. While the sun is vertical to the places where they occur, the land becomes heated, and the air expanded, and the wind flows toward the coasts. As the sun retires towards the opposite point of its course, the land cools faster than the surrounding seas, and the course of the winds is westward. The simplest way of regarding the sun's motion in declination, as affecting the temperature of the various latitudes, is, to suppose a motion of the whole system; by which the line of greatest heat, and the two points of greatest cold, maintaining their relative distances, vibrate on either side of the earth's equator and poles.

None of these changes affect the barometer.

It may now be understood, that it is the intermixture of land and water, joined with other disturbing causes, which prevents the extra-tropical western winds from being as regular in their course as the tropical trade winds, and a slight inspection of the chart will demonstrate, that the northern hemisphere, including the great continents of Europe, Asia, and North America, must be more under this influence than the southern, which is comparatively free from such effects.

XVII. *Rain seldom occurs in the constant trade winds, but abundantly and constantly in the adjoining latitudes.*

Between the tropics, the elasticity of the aqueous vapour reaches its maximum amount, and within these limits only, rises to any extent into the upper current of the atmosphere. Its own force, therefore, which is laterally exerted, is assisted by the equatorial wind, and it flows to the north and south as fast as it rises within the zone. No accumulation can, therefore, be formed; and the temperature being remarkably steady, seldom varying more than two or three degrees, precipitation can but seldom occur.

The continental parts, however, of the same regions, being liable to greater vicissitudes of heat, are subject to rainy seasons, which are periodical, like the Monsoons of the same climates, and are governed, as they are, by the progress of the sun in declination. The condensation, while it lasts, is in proportion to the density of the vapour, and is violent beyond any thing that is known in temperate climates. The alternate seasons of fine weather are distinguished by cloudless skies and perfect serenity.

The extra-tropical latitudes, on the contrary, beyond the bounds of the trade winds, are, at all times, exposed to great precipitations. The vapour in its course is subjected to a rapidly decreasing temperature, and the condensation is fed by a constant supply. We are thus led to the consideration of a temperate zone, and a variable climate.

XVIII. *Between the tropics the fluctuations of the Barometer do not much exceed ¼ of an inch, while beyond this space they reach to 3 inches.*

The great characteristics of the tropical regions are constancy of temperature, and freedom from aqueous precipitation. I speak now more particularly of the included oceans, which are so extensive, compared with the land, as necessarily to stamp the character of the climate. The phenomena, on the latter, are so limited as scarcely to affect the total result, and are to be regarded more in the light of exceptions, in whatever points they differ from the general rule.

Now variations of temperature alone have been satisfactorily proved not to affect the mercurial column: and it is in the aqueous condensation, that we shall probably find the cause of barometrical changes. The vapour, as we have seen, passes north and south from the equatorial parts, and reaching the extra-tropical regions, is precipitated. The effect of this precipitation must be to destroy the progression of temperature in the vertical columns, by equalizing the heat of the strata exposed to its influence. But, as this process is carried on chiefly in the lower current, and cannot, from its very nature, equally affect the column, the total weight will be reduced by the consequent irregular expansion.

XIX. *In the temperate climates the rains and the winds are variable.*

One of the great causes, of the variableness of the wind, has already been pointed out in the greater abundance of land in the extra-tropical parts of the northern hemisphere. Another cause obviously must originate in the variations of barometrical pressure. The rain must depend very much upon the changes of the wind, and the retardation or acceleration which they offer to the progress of the vapour. But another cause arises from the unequal supply, which the process of evaporation receives from the irregular surface of the globe. This cannot be placed in a stronger light than by the following considerations. The Caspian Sea, which is placed in the centre of the largest continent of the world, receives the precipitations of an immense tract of the atmosphere by means of the rivers which flow into it, and drain the neighbouring countries. The whole of this supply is again returned by evaporation, and its waters have no other means of escape. The lakes of North America, situated in nearly the same parallel of latitude, and at the same altitude, receive the drains of a much less space; but annually roll an immense volume of water to the ocean. We are thus furnished with an hygrometer upon a large scale, by which we may judge of the state of saturation of the two atmospheres. The difference can arise from no other cause than the proximity of the surrounding seas in the latter situation, which furnish an inexhaustible source of vapour, which is deficient in the other. It is for the same reason that less vapour is contained in the atmosphere above a continent, than above the ocean, although more rain falls in the former situation than in the latter under the same latitudes, owing to the greater vicissitudes of temperature. Much of the aqueous atmosphere which is formed from the great deeps, is thus drawn off towards the continents, where a scarcity of water occasions an inadequate pressure of the vapour.

XX. *As we advance towards the Polar Regions we find the irregularities of the wind increased; and storms and calms repeatedly alternate, without warning or progression.*

The very instructive "Account of the Arctic Regions," for which we are indebted to Captain Scoresby, and the interesting Journal of Captain Parry, have made us well acquainted with the interesting regions of perpetual ice and snow. In our hypothetical statement of the progression of the earth's temperature, we have supposed no greater cold to prevail than that of 0° at the poles, but the experience of our intrepid Navigators has proved that a cold of 50° greater intensity sometimes prevails in latitudes still far removed from the 90th degree. The density of the air is, of course, proportionately increased, and its sources of inequality multiplied. When the sun is above the horizon, it produces comparatively little effect upon the icy mountains, while the neighbouring seas are warmed by its unceasing influence. The extremes of heat and cold will sometimes prevail within a very limited compass ; and forcible winds will blow in one place, when, at a distance of a few leagues, gentle breezes prevail. " Ships, within the circle of the horizon, may be seen enduring every variety of wind and weather, at the same moment; some under close-reefed top-sails, labouring under the force of a storm; some becalmed and tossing about by the violence of the waves; and others plying under gentle breezes from quarters as diverse as the cardinal points." The fluctuations of the barometer are also great and sudden, proving what theory would have induced us to conclude, that the irregularities of these regions extend to the higher strata of the atmosphere.

XXI. *In the extra-tropical climates, a fall in the Barometer almost always precedes a period of rain, and indicates an acceleration or change of the aërial currents.*

As the proximate cause of the fall of the barometer is an accumulation of aqueous vapour, and a consequent unequal expansion of the atmospheric columns, it is obvious that this alone would increase the probability of a proportionate precipitation ; but it is not the only reason of the effect. The fall of the barometer indicates a decrease of density in the aërial currents, and, consequently, a decrease of the resistance to the passage of the vapour. A constant stream will thus rush in with increasing force, augmenting by its condensation the cause of its velocity, till a current sets in from some other quarter, and restores the equilibrium.

XXII. *Barometers, situated at great distances from each other, often rise and fall together with great regularity.*

This proves that the cause of the variations must be very extended in its influence; and from all our preceding inquiries the conclusion *à priori* would be the same. The proximate cause is one of unlimited extent, and in so fluid a medium the remoter influence must be widely felt. It has been observed that this unison of action extends further in the direction of the latitude than in that of the longitude, and the remark materially confirms our theory; for, as the grand currents of the atmosphere flow in the direction of the meridians, any irregularity in their courses would most readily be propagated in the same line.

XXIII. *More than two currents may often be traced in the atmosphere at one time by the motions of the clouds, &c.*

The great fluctuations of the atmosphere have been referred to modifications and disturbances in the courses of two principal currents, but from the very nature of the disturbing causes which we have been considering, it is obvious that these must often include subordinate currents and minor systems of compensation within themselves. When we recollect, not only that every mountain and sea, but that every hill, and lake, and river, (not to descend to the more minute and numberless influences of artificial arrangements,) produce very appreciable impressions, it is obvious that we must consider the great streams as made up of inferior circulations; which, like the eddies of a river, or the waves of the sea, do not affect the main flux of the currents.

XXIV. *The force of the winds does not always decrease as the elevation increases; but, on the contrary, is often found to augment rapidly.*

A slight inspection of Table VIII. will shew that this remark is in exact correspondence with the theory. It will there be seen that, in the regular course of the currents, the lower wind dies away gradually as we ascend, and at a certain height gives place to a gentle breeze in a contrary direction: this increases in force with the height, till, notwithstanding the rarity of the air, its impulse is so great as to produce a very strong wind. Some of the hypothetical cases, of the disturbance of this regular order, present still more marked instances of winds increasing upwards.

XXV. *The variations of the Barometer are less, in high situations, than in those at the level of the sea.*

The range of the barometer for any given latitude, should be in inverse proportion to its elevation in the atmosphere: for, as we have seen, under all circumstances, the law of the progression of density must be maintained. Any cause, therefore, producing a given change at an altitude whose density is one half, must produce double the effect at the level of the full pressure; and, on the other hand, a rise of the barometer, of two-tenths of an inch at the base of a column, will only be felt as one-tenth at the height due to half the density.

The British Islands are situated in such a manner as to be subject to all the circumstances which can possibly be supposed to render a climate irregular and variable. Placed nearly in the centre of the temperate zone, where the range of temperature is very great, their atmosphere is subject, on one side, to the impressions of the largest continent of the world, and, on the other, to those of the vast Atlantic Ocean. Upon their coasts the great stream of aqueous vapour, perpetually rising from the western waters, first receives the influence of the land, whence emanate those condensations and expansions which deflect and reverse the grand system of equipoised currents. They are also within the reach of the frigorific effects of the immense barriers and fields of ice, which, when the shifting position of the sun advances the tropical climate towards the northern pole, counteract its energy, and present a condensing surface of immense extent to the increasing elasticity of the aqueous atmosphere. Amidst all the uncertainty and seeming confusion arising from this complication, our general principles may still be recognised ; and I would fain hope that the more they are studied, the more obvious they will appear, as in this case, above all others, "*exceptions prove the rule.*'

XXVI. *In Great Britain, upon an average of ten years, westerly winds exceed the easterly in the proportion of* 225 *to* 140.

From the geographical situation of the island, we

should be led to suppose that the atmosphere must come within the influence of the descending equatorial particles, and that winds from the west would prevail; and this conclusion is confirmed by the observation. Of those from the east, the northerly exceed the southerly in the proportion of about 74 to 54; leaving but a very small proportion indeed, which blow from the most irregular point, *viz.*, the south-east. The north-east may be regarded as those which, coming from the north, have not attained the velocity of rotation due to the latitude.

XXVII. *Upon the same average, the northerly winds are to the southerly as* 192 *to* 173.

By this classification we most readily detect the influence of local disturbing causes. Were it not for these, the northerly current would prevail throughout the year: but the condensation of vapour to the north and north-east of this situation is so great as to cause perpetual diminutions of the aërial columns, and consequent deflections of the currents. In the central parts of Europe, the northern winds are much more regular; and there, especially in summer, the Etesian breeze constantly prevails. Of the winds from the south, it may also be remarked, that the westerly exceed the easterly in the proportion of 104 to 54.

XXVIII. *Northerly winds almost invariably raise the Barometer; while southerly winds as constantly depress it.*

The northerly current is the natural course of the

air, and, where the regular order has been disturbed, in consequence of diminished local pressure, a return to it restores the equilibrium. Coming from the frozen arctic regions, it speedily reduces the accumulation of vapour, stops the supply, and dissipates the concomitant heat, from which originated the depression; and, if it flow with a velocity beyond its regular rate, causes a reduction of temperature below the due progression, and augments the total weight. The southerly wind, on the contrary, facilitates the passage of the vapour, which by its unremitting condensation unceasingly increases the cause of depression.

XXIX. *The most permanent rains of this climate come from the southern regions.*

The supply of vapour, which occasions rain, may be traced to two sources:—One is the evaporation of the latitude itself, where it is precipitated; and the other, the stream which is perpetually struggling to advance from the equatorial zone. These causes sometimes act conjointly, and sometimes separately. Rain from the first, is derived from sudden falls of temperature, produced by cold currents, or the changes of the seasons, and assumes the form chiefly of showers of greater or less continuance. The expenditure of vapour is but slowly supplied, and the precipitation occurs at intervals. Rain, from this source, is always accompanied by a declining temperature. When, on the contrary, in consequence of diminished pressure, the tropical current reaches in succession the colder parallels, the supply continues in a perpetual flow; the temperature is raised, the depression of the Barometer increases, and rain descends with little intermission.

XXX. *The mean height of the Barometer varies but little with the changes of the seasons.*

Many persons have supposed that the fluctuations of the Barometer are owing to the greater or less weight of the aqueous particles, contained in the atmosphere, at one time than at another. If, however, our theory be correct, the difference of pressure between a perfectly dry atmosphere, and one saturated with moisture, cannot much exceed .150 inch: the difference of the seasons must, therefore, be even less than this amount. But, small as it is, it may nevertheless be detected by the system of averages. Mr. Howard, in his invaluable work, upon the climate of London, has brought out the difference of pressure, for the several seasons, upon a calculation of ten years' observations, as follows:—

Brumal period above the autumnal .021 inch, Vernal period, above the Brumal, .030 inch, Estival, above the Vernal, .045 inch, and Autumnal below the Estival, .096 inch. These results are in exact accordance with our theoretical conclusions.

The subject of atmospheric vapour has hitherto been less studied than its importance would seem to require. Observations upon its variations are very deficient, owing to the uncertainty and imperfection of the means generally adopted for measuring its effects. A few general conclusions which have suggested themselves to me, from the constant use for three years of a more perfect instrument for that purpose, and some valuable results derived from the experiments of my friend, Captain Sabine, during a twelve-months' residence between the Tropics, will add no small weight to our synthetic deductions.

XXXI. *The elasticity of the aqueous vapour does not decrease gradually as we ascend in the atmosphere, in proportion to the gradual decrease of the temperature and density of the air; but the dew-point remains stationary to great heights, and then suddenly falls to a large amount.*

To this conclusion I had been led, by my theoretical speculations, long before I had any hopes of being able to confirm it, by direct experiment. I have, however, at last succeeded in obtaining complete evidence of its correctness. As the fact is altogether new in Natural History, and as it is essential to the theory which I have been endeavouring to establish, forming indeed the test by which it may be most correctly judged, I shall, I trust, be excused for enlarging somewhat upon it.

The first experiments to which I shall refer, as bearing upon this point, are those of Mr. Green, the Aëronaut. An account of this gentleman's ascent in a Balloon, from Portsea, on the 6th of September, 1821, is given in the 12th volume of the Journal of the Royal Institution (p. 114). Amongst other instruments of research, he took up with him one of my Hygrometers. He unfortunately omitted to take the point of deposition, before he commenced his ascent; but the omission is of less consequence as I happened to make an observation at the time, at no very great distance from the spot. At an elevation of about 9890 feet, he found the dew-point at 64°, exactly the same as I ascertained it to be at the surface of the earth. At 11060 feet it had fallen to 32°, making a difference of 32 degrees in little more than 1100 feet. Here, then, we have presumptive evidence of an immense bed of vapour rising in its circumambient medium, unaffected by decrease of density or temperature till checked by its point of precipitation; and of an incumbent bed of not much more than one-third the density, regulated, no doubt, as the last, by its own point of deposition in loftier regions.

Captain Sabine, by his experiments upon mountains in tropical climates, has established the same fact in the most unexceptionable manner. At Sierra Leone, he ascertained that the dew-point of the vapour, at the level of the sea, was 70°; and that it was the same at the same hour upon the summit of the Sugar-loaf Mountain, 2520 feet above. At the Island of Ascension, the barometer, 17 feet above the level of the sea, stood at 30.165 inches—temperature of air 83°, and the dew-point 68°. On the summit of the mountain the barometer fell to 27.950 inches, and the temperature of the air to 70°, while the dew-point only declined to 66°.5; so that in a height of 2220 feet, the temperature of the air fell 13°, and the constituent temperature of the vapour 1°.5.

At Trinidad, the temperature of the air at the level of the sea was 82°, and the dew-point 77°; 1060 feet above, they were both 76.°5, and precipitation was going on.

At Jamaica, by the sea-side, the temperature of the air was 80°, and the point of deposition 73°; while, on the mountains, at a height of 4080 feet, they were both 68°.5. At a station, not five hundred feet higher, by experiment twice repeated, the point of deposition was found to be 49°, and the temperature of the air 65°.

These results are utterly irreconcileable with the idea of the aqueous particles in the atmosphere being suspended by any law analogous to that of chemical solution; and I am much mistaken if they may not be received as experimental confirmation of the theory of mechanical mixture.

Captain Sabine's experiments furnish also some evidence of that slight diminution of density in the upper parts of the beds of vapour which would arise from the decrease of their own pressure, and which I have anticipated in the second part of this essay. In the experiments at Jamaica, the dew-point fell about 4°.5 in 4080 feet. In Table XXIII., it will be seen, that I have calculated this diminution to be 3°.5 in 5000 feet, for an atmosphere of much less density.

In noticing here the demonstrative nature of the evidence in favour of the mechanical theory of the mixture of the gases with vapour, I cannot refrain from saying a few words with regard to the same theory as applied to the mixture of gases with one another. With a view of simplifying as much as possible a subject which, above all others, perhaps, is the most complicated in nature, I have spoken of the permanently-elastic fluid of the atmosphere as a simple gas, whereas, it is well known, to be compounded of two or three. The constancy of the proportions, in which these are found to be combined in every situation, is the never-failing theme of wonder and admiration, and it is perpetually referred to as evidence of chemical combination. I must own, that it strikes my mind in a very different manner, and I conceive that the fact is much more reconcileable to the mechanical than to the chemical theory.

If we suppose a consumption of the oxygen to take place, by the decomposition of the atmosphere, at any given spot, in what way is chemical affinity to act to restore the uniformity of the compound? No evolution of oxygen takes place, and it cannot be supplied by the surrounding portions; for the affinity of nitrogen for oxygen can never be supplied by the decomposition of nitrogen and oxygen, which are held together by the same affinity. On the other hand, if the oxygen and the nitrogen be two distinct elastic atmospheres, mutually permeating and pervading each other's interstices, the particles of each pressing only upon like particles, and only slightly opposing one another in their separate motions, then a local consumption of oxygen would be instantly supplied by a rush of the elastic fluid towards the spot where the equality of pressure had deen disturbed. In the same way, any partial supply of either gas would instantly be equalized; and that equal diffusion which is avowedly inexplicable upon chemical principles, is perfectly reconcileable to the principles of hydrostatics. I shall not now stop to insist upon the analogy between vapours and gasses, though the late beautiful experiments of Mr. Faraday have almost annihilated the distinction even in name; but shall revert to the subject, from which it may be thought that I have been wandering.

XXXII. *The tension of vapour given off in the process of evaporation is determined, not by the temperature of the evaporating surface, but by the elasticity of the aqueous atmosphere already existing.*

I have often endeavoured, by means of the hygrometer, to detect, within a limited circle, a difference in the elastic state of the vapour incumbent upon different surfaces of various temperatures, but without success: the rising vapour was always of the same quality, whether from water, vegetation, or ploughed land; in sun-shine, or in shade. For the same reason, the dew-point is but little affected by the increase of daily temperature from morning to afternoon, or by its subsequent declension at night. But one of the most remarkable confirmations of the fact was ascertained by Captain Sabine upon the coast of Africa. While the sea-breeze was blowing upon that station, the hygrometer denoted the dew-point to be about 60°, but when the wind blew strong from the land, it approached, in its characters, to a Harmattan; and the point of precipitation was not higher than 37°.5, the temperature of the air being 66°. Notwithstanding the heat of the evaporating surfaces in the interior of that continent, the burning sands of its deserts yield so little vapour, that it becomes attenuated by its diffusion, and there can be little doubt that the aqueous atmosphere incumbent upon it, and which, when wafted to the coast by the rapid motion of the air, constitutes the true Harmattan, is not of greater force than that which rests upon the polar seas; and that while the heat of the air sometimes approaches to 90°, the constituent temperature of the vapour is below 32°.

XXXIII. *The apparent permanency and stationary aspect of a cloud is often an optical deception, arising from the solution of moisture on one side of a given point, as it is precipitated on the other.*

No phenomenon is more common amongst mountains, or upon hills by the sea-side, than clouds upon the summits which appear to be perfectly immoveable, although a strong wind is blowing upon them at the time. That this should be the real state of the case, is clearly impossible, as so attenuated a body as constitutes the substances of the clouds must obey the impulse of the air. The real fact is, that the vapour, which is wafted by the wind, is precipitated by the cold contact of the mountain; and is urged forward in its course till, borne beyond the influence which caused its condensation, it is again exhaled and disappears. A slight inspection and consideration of the phenomena will be sufficient to convince any one of the correctness of this explanation. Reasoning from analogy, we may conclude, that the process which thus proceeds, under our eyes, upon the summits of the hills, likewise takes place on either side of the planes of precipitation in the heights of the atmosphere: the vapour is continually condensed, as continually re-dissolved in the act of precipitation, and the cloud appears to be unchanged and stationary.

XXXIV. *The quantity of vapour in the atmosphere in the different seasons of the year (measured on the surface of the earth, and near the level of the sea,) follows the progress of the mean temperature.* This result of observation might readily have been anticipated; for the rate of evaporation, and the quantity which the air can support, are both obviously dependent upon the same progression. But this connexion is not discoverable in short periods; and the changes of diurnal temperature do not materially affect the quantity of elastic vapour. The air, at night, generally reaches the point of deposition, even at the surface of the sea, but in a very gradual manner: and at the same time the supply from evaporation ceases. The progress of the vapour in fine weather may often be very satisfactorily traced by means of the clouds. During the heat of the day it rises from the surface of the land and waters, and reaches its point of condensation in greater or less quantities at different altitudes. Partial clouds are formed in different parallel planes, which always maintain their relative distances. The denser forms of the lower strata, as they float along with the wind, shew the greater abundance of the precipitation at the first point of deposition, while the feathery shapes and lighter texture of the upper, attest a rarer atmosphere. These clouds do not increase beyond a certain point, but often remain stationary in quantity and figure for many hours: but as the heat declines they gradually melt away; till at length, when the sun has sunk below the horizon, the ether is unspotted and transparent. The stars shine through the night with undimmed lustre, and the sun rises in the morning in its brightest splendour. The clouds again begin to form, increase to a certain limit, and vanish with the evening shades. This gradation of changes, which we so often see repeated in our finest seasons, might, at first, appear to be contrary to our principles; and that precipitations should occur with the increase of temperature, and disappear with its decline, would seem, at first sight, to be diametrically opposite to all our conclusions. But a little consideration will shew that these facts confirm our theory. The vapour rises and is condensed; but in its precipitation falls into a warmer air, where it again assumes the elastic form; and as the quantity of evaporation below is exactly equal to supply this process above, the cloud neither augments nor decreases. When the sun declines, the surface of the earth cools more rapidly than the air; evaporation decreases, but the dissolution of the cloud continues. The supply at length totally ceases, and the concrete vapour melts completely away. The morning sun revives the exhalations of the earth, and the process of nimbification again commences, and again undergoes the same series of changes. The fall of the temperature shifts a little the planes of deposition, but scarcely affects the total pressure of the vapour. The deposition of dew (the formation and concomitant circumstances of which have been so successfully analyzed in the elegant essay of the late Dr. Wells,) slightly diminishes the quantity; but the first touch of the sun's rays restores it to the "blue expanse."

When, however, the natural equilibrium has been disturbed, when the temperature of the air has become equalized through various successive strata by the beds of vapour with which they are embued, the decline of the day will often determine precipitation, and will increase its amount if already established. The result of experiment has also shewn that a greater amount of rain falls while the sun is below, than while it is above, the horizon.

XXXV. *The pressure of the aqueous atmosphere, separated from that of the aërial, generally exhibits directly opposite changes to the latter.*

As the quantity of vapour increases, it will mostly be found that the barometer falls; and it rises with its decrease. This observation, which is amply confirmed by tracing the lines of each upon a graduated paper, does not apply to the averages of the different seasons, but to the daily fluctuations. This fact, so utterly irreconcileable to the hypothesis which ascribes the rise and fall of the mercurial column to the weight of the aqueous particles, materially confirms that which attributes them to the unequal expansion of balancing currents. The prime source of this expansion we have supposed to be the elastic vapour, and this experience confirms the theory.

XXXVI. *Great falls of the Barometer are generally accompanied by a temperature above the mean for the season; and great rises by one below the same.* This is a confirmation of the same nature as the last, and inseparably connected with it. It is by the evolution of heat that the vapour principally acts. The mean temperature which balances all irregularities, must be the regular temperature of the climate, and *cæteris paribus*, that at which the currents must be most disposed to regularity: variations, on either side of this point, must produce corresponding retardations and accelerations; and these, if not general through the mass, annihilate the equipoise.

I have thus filled up the outline which I laid down for my inquiry, and I trust that it will be found that I have not wholly failed to elucidate some hitherto-obscure points of the history of the atmosphere. It tends to give me some confidence in the justness of my views, that when I first conceived the idea of conducting my researches synthetically, I anticipated but few of the conclusions to which the experiment has led me.

The principles which I have employed are the fruits of the researches of the most eminent philosophers; to have owned my obligations to whom would have loaded this essay with references. Their labours are become the foundation-stones of science, and the common property of those who may follow them in endeavours to perfect the edifice. There is one name, however, to which this branch of science is so vastly indebted, that to allude to it more particularly I feel to be an incumbent duty. It is that of the author of the "Essays upon the Constitution of mixed Gases." To Mr. Dalton we owe nearly all the light which we possess upon this interesting and difficult subject; and from his deep researches, particularly, I have drawn most freely. If, indeed, these my endeavours shall be found to be deserving of any consideration, it will be as illustrations of the Daltonian theory.

I have scrupulously adhered to the natural consequences of the premises which I have adopted, without previously inquiring how far they were consonant with the phenomena to be explained: in their after application to these latter, I hope that it will be found that I have not been unsuccessful. The fluctuations of the barometer, and most of the phenomena of wind and rain, appear to me to adapt themselves most happily to the theoretical conclusions.

Both the *synthetical* and *analytical* processes agree in the same grand conclusions, which may thus briefly be recapitulated:—

There are two distinct atmospheres, mechanically mixed, surrounding the earth; whose relations to heat are different, and whose states of equilibrium, considering them as enveloping a sphere of unequal temperature, are incompatible with each other. The first is a permanently-elastic fluid, expansible in an arithmetical progression by equal increments of heat, decreasing in density and temperature according to fixed ratios, as it recedes from the surface, and whose equipoise under such circumstances, would be maintained by a regular system of antagonist currents. The second is an elastic fluid, condensible by cold with evolution of caloric; increasing in force in geometrical progression with equal augmentations of temperature; permeating the former and moving in its interstices, as a spring of water flows through a sand-rock. When in a state of motion this intestine filtration is retarded by the *inertia* of the gaseous medium, but in a state of rest the particles press only upon those of their own kind. The density and temperature of this fluid have a tendency likewise to decrease, as its distance from the surface augments; but by a less rapid rate than that of the former. Its equipoise would be maintained by the adaptation of the upper parts of the medium, in which it moves, to the progression of its temperature, and by a current flowing from the hotter parts of the globe to the colder. Constant evaporation on the line of greatest heat and unceasing precipitation, at every other situation would be the necessary accompaniments of this balance. Now the conditions of these two states of equilibrium, to which, by the laws of Hydrostatics, each fluid must be perpetually pressing, are essentially opposed to each other. The vapour or condensible elastic fluid is forced to ascend in a medium, whose heat decreases much more rapidly than its own natural rate; and it is therefore condensed and precipitated in the upper regions. Its latent caloric is evolved by the condensation, and communicated to the air; and it thus tends to equalize the temperature of the medium in which it moves, and to constrain it to its own law. This process must evidently disturb the equilibrium of the permanently-elastic fluid, by interfering with that definite state of temperature and densitywhich is essential to its maintenance. The system of currents is unequally affected by the unequal expansion; and the irregularity is extended, by their influence, much beyond the sphere of the primary disturbance. The decrease of this elasticity above, is accompanied by an extremely important re-action upon the body of vapour itself: being forced to accommodate itself to the circumstances of the medium in which it moves, its own law of density can only be maintained by a corresponding decrease of force below the point of condensation; so that the temperature of the air, at the surface of the globe, is far from the term of saturation; and the current of vapour, which moves from the hottest to the coldest points, penetrates from the equator to the poles, without producing that condensation in mass, which would otherwise cloud the whole depth of the atmosphere with precipitating moisture. The clouds are thereby confined to parallel horizontal planes, with intermediate clear spaces, and thus arranged are offered to the influence of the sun, which dissipates their accumulations and greatly extends the expansive power of the elastic vapour. The power of each fluid being in proportion to its elasticity, that of the vapour compared with the air, can never, at most, exceed 1 : 30: so that the general character of the mixed atmosphere is derived from the latter; which, in its irresistible motions must hurry the former along with it. The influence, however, of the vapour upon the air, though slower in its action, is sure in its effects, and the gradual and silent processes of evaporation and precipitation govern the boisterous power of the winds. By the irresistible force of expansion unequally applied, they give rise to undulations in the elastic fluid ; the returning waves dissipate the local influence, and the accumulated effect is annihilated, again to be reproduced.

In tracing the harmonious results of such discordant operations, it is impossible not to pause, to offer up a humble tribute of admiration of the designs of a beneficent Providence, thus imperfectly developed in a department of creation where they have been supposed to be the most obscure. By an invisible, but ever-active, agency the waters of the deep are raised into the air, whence their distribution follows, as it were by measure and weight, in proportion to the beneficial effects which they are calculated to produce. By gradual, but almost insensible, expansions the equipoised currents of the atmosphere are disturbed, the stormy winds arise, and the waves of the sea are lifted up; and that stagnation of air and water is prevented, which would be fatal to animal existence. But the force which operates, is calculated and proportioned: the very agent which causes the disturbance bears with it its own check ; and the storm, as it vents its force, is itself setting the bounds of its own fury.

The complicated and beautiful contrivances, by which the waters are collected "above the firmament," and are at the same time "divided from the waters which are below the firmament," are inferior to none of those adaptations of INFINITE

WISDOM, which are perpetually striking the inquiring mind, in the animal and vegetable kingdoms. Had it not been for this nice adjustment of conflicting elements, the clouds and concrete vapours of the sky would have reached from the surface of the earth to the remotest heavens; and the vivifying rays of the sun would never have been able to penetrate through the dense mists of perpetual precipitation.

Nor can I here refrain from pointing out a confirmation, which incidentally arises, of the Mosiac account of the creation of that atmosphere whose wonders we have been endeavouring to unravel. The question has been asked, How is it that light is said to have been created on the first day, and day and night to have succeeded each other, when the sun has been described as not having been produced till the fourth day? The Sceptic presumptuously replies, this is a palpable contradiction, and the history which propounds it must be false. But, Moses records that God created on the first day, the earth covered with water, and did not till its second revolution upon its axis, call the firmament into existence. Now one result of the previous inquiry has been, that a sphere unequally heated and covered with water, must be enveloped in an atmosphere of steam, which would necessarily be turbid in its whole depth with precipitating moisture. The exposure of such a sphere to the orb of day would produce illumination upon it; that dispersed and equal light, which now penetrates in a cloudy day, and which indeed is "good:" but the glorious source of light could not have been visible from its surface. On the second day, the permanently-elastic firmament was produced, and we have seen that the natural consequences of this mixture of gaseous matter, with vapour, must have been, that the waters would begin to collect above the firmament, and divide themselves from the waters which were below the firmament. The clouds would thus be confined to definite plains of precipitation, and exposed to the influence of the winds, and still invisible sun. The gathering together of the waters on the third day, and the appearance of dry land, would present a greater heating surface, and a less surface of evaporation, and the atmosphere during this revolution would let fall its excess of condensed moisture; and upon the fourth day it would appear probable, even to our short-sighted philosophy, that the sun would be enabled to dissipate the still-remaining mists, and burst forth with splendour upon the vegetating surface*. So far, therefore, is it from being impossible that light should have appeared upon the earth before the appearance of the sun, that the present imperfect state of our knowledge, will enable us to affirm, that, if the recorded order of creation be correct, the events must have exhibited themselves in the succession

* I am indebted to Mr. Granville Penn's admirable "Estimate of the Mineral and Mosaical Geologies," for my first hint upon this subject. The greater part of this Essay had been written before I perused his work; and I was pleased to find that I had unconsciously proved the necessity of that turbid state of the aqueous atmosphere, previous to the creation of the firmament, upon the probability of which he has argued in such a masterly and convincing manner.

which is described. The argument therefore recoils with double force in favour of the inspiration of an account of natural phenomena which, in all probability, no human mind, in the state of knowledge at the time it was delivered, could have suggested; but which is found to be consistent with facts that a more advanced state of science and experience have brought to light. If, however, it were reasonable to expect that the ways of God should in all cases be justified to the knowledge, or rather the ignorance, of man, the boldest philosopher might well pause before he applied the imperfect test of a progressive philosophy to the determination of the momentous questions involved in these considerations.

I am aware that two grand principles have been passed over in the previous investigation, which cannot but have great power in modifying atmospheric changes: I allude to the agency of the electric fluid, and the influence of the moon. But their modes of operation are, at present, too obscure, and stand too much in need of experimental elucidation to allow of their being applied with the requisite precision; and to have referred to them before, would only have been unnecessarily to complicate the subject. Since the time when the mind and the nerve of a Franklin first conceived and executed the bold design of analyzing the lightnings of the heavens, but few have been found to follow in the glorious, but dangerous, path, which he has opened. The only regular series of observations I am acquainted with, are those by Mr. Read, published in the *Philosophical Transactions* for 1794. They exhibit much perseverance and ingenuity, and are of a highly interesting nature, but required to have been longer continued to have enabled him to draw general conclusions. It is to be regretted that none of the great scientific establishments, either at home or abroad, have taken up this important branch of meteorology. The aids of modern science would be well applied to elucidate the still obscure relations between electricity and meteorological phenomena. From some experiments of my own upon the subject, I am inclined to believe that the elasticity of vapour is increased when electrically charged; but I have nothing decisive to offer upon the point. Should this opinion be confirmed, it is obvious that the influence must be more rapid, partial and powerful, than that of any distributing cause which we have hitherto contemplated.

That the different phases of the moon have some connexion with changes in the atmosphere, is an opinion so universal and popular, as to be, on that account alone, entitled to attention. No observation is more general; and on no occasion, perhaps, is the almanack so frequently consulted, as in forming conjectures upon the state of the weather. The common remark, however, goes no further than that changes from wet to dry, and from dry to wet, generally happen at the changes of the moon. When to this result of universal experience we add the philosophical reasons for the existence of tides in the aërial ocean, we cannot doubt that such a connexion exists. The subject, however, is involved in much obscurity. Mr. Howard is the only one who has treated it with the consideration which it deserves.

In his book may be found much information upon it, the result of laborious investigation. It would be foreign to my purpose to enter at large upon this interesting ground, but the previous investigation suggests one particular view of it, which it may be useful shortly to state.

The action of the moon upon the aërial columns over which it passes, may be regarded as diminishing the force of gravity. This action must be greater in proportion as the moon approaches the earth; in proportion as it coincides with the analogous action of the sun; and in proportion as its passage over the meridian comes near to the perpendicular direction. The result of this diminution of gravity must be a general decrease of density; and its effect upon the lateral currents, an acceleration of the incoming, and a decrease of the outgoing streams. The loss of weight will thus be compensated, and the excess of elasticity hence derived, will lengthen the column. The final adjustment will, therefore, be assimilated to that which arises from an equal expansion by heat. Now the effect of the atmospheric tide has hitherto been sought for, and measured upon the surface of the earth, at the base of the column; and much conjecture and disappointment have ensued from not finding the effect as great, or as regular, as had been anticipated. But, if this view of the subject be correct, the total weight of the perpendicular column would not be affected so much as that of its horizontal sections; and the amount of the lunar influence should be sought in the variations of the differences of density between some high elevation and the level of the sea. The mean of a series of experiments carefully conducted with this view, when the moon is upon the meridian and at the horizon, would possibly exhibit the amount of the daily tides; their weekly increase and diminution; the influence of the moon's apogee and perigee; and that of its north and south declination. It has, however, I think, been proved that the influence is still felt at the surface of the earth; and the barometer, upon an average, stands lower at new and full moon, than at the quarters. This also would naturally be expected when it is considered that the attraction of the moon is an action upon the power of gravity, and acts instantaneously in the perpendicular direction; while the compensating effects upon the lateral currents is gradual.

Is it not possible that some of the remaining discrepancies of barometrical mensurations may be traced to this influence?

Should the speculations in which I have indulged in this Essay be found worthy of consideration, they will probably suggest some new modes of conducting meteorological experiments, and some alterations in the method of arranging the results. Any useful, practical result hence arising will amply repay the labour of the undertaking.

AN ESSAY

UPON THE CONSTRUCTION AND USES OF A NEW HYGROMETER.

" Nec non et in conviviis mensisque nostris vasa quibus esculentum additur sudorem repositoriis linquentia diras tempestates prænutinant."—*C. Plin. Nat. Hist.*, Lib. xviii.

In the year 1812, my attention was attracted by the passage above extracted from Pliny, which appeared to me, by the interpretation which I affixed to it, to point to a natural phenomenon which might be rendered subservient not only to prognostications of the weather, according to the suggestion of that accurate observer, but to some of the more refined purposes of modern science. I was, however, for some time doubtful how far the interpretation which had occurred to me could be borne out by the translation of the expression *esculentum;* as it was a necessary condition to this interpretation, that whatever was served up in the *vasa* should have been cold.

The passage is thus rendered into English in a very old translation which I consulted: " And to " conclude, and make an end of this discourse; " whensoever you see, at any feast, the dishes and " platters whereon your meat is served up to the "board, sweat or stand of a dew, and leaving that "sweat which is resolved from them either upon "dresser, cupboard, or table, be assured that it is "a token of terrible tempests approaching."

Translation of C. Pliny, by Philemon Holland. 1601.

Upon referring to several competent judges, they confirmed my conjecture, and agreed with me in thinking, that the dew or sweat, so accurately described as forming, in particular kinds of weather, upon vessels in which food was served up, could only have arisen from depression of temperature.

This, perhaps, will therefore be considered as one of the most curious cases upon record, in which the sagacity of the ancients anticipated an observation which has been held to be peculiarly demonstrative of the superior refinements of the present state of experimental philosophy, and may settle a disputed claim to the honour of priority of discovery amongst the existing race of natural philosophers.

However this may be, my mind was thus directed to the deposition of moisture which takes place upon certain bodies when brought into an atmosphere which is warmer than themselves; and following up the suggestion of Pliny, I readily conceived that the fact was connected with meteorological phenomena; and that experiments, founded upon it, might be devised to elucidate the relation of air to vapour. I shortly after applied myself seriously to the inquiry, and was soon satisfied of the accuracy of the conjecture.

The manner in which I proceeded at that time, was as follows: I made a mixture of two salts calculated to produce cold by their solution; I then arranged half a dozen drinking-glasses upon a board, each furnished with a thermometer, and poured water into one of them. I added a tea-spoonful of the freezing mixture, which invariably produced a copious dew upon the exterior of the glass. I emptied the contents of the first glass into the second, and so into the third, *&c.*, till the liquor, gradually acquiring heat by the process, arrived at such a temperature as no longer to produce any condensation upon the vessel. This point, as marked by the thermometer, was noted, and found to vary, very considerably, in relation to the temperature of the air, according to different states of the atmosphere.

I kept a journal of the weather for several months; registering the variations of the barometer, thermometer, De Luc's hygrometer, and the temperature at which moisture was condensed, and obtained some very interesting results.

I afterwards varied my apparatus in the following manner: I procured five small hollow cylinders of brass, three inches in diameter, and four inches in height, fitted with a small cock in the bottom of each. These were very highly polished, and placed in a frame, one immediately over another, so that by turning the cork, the contents of the upper would flow into that immediately beneath it. I put the cold liquid into the top cylinder; and when steam was produced upon its surface, suffered the solution to run into the next, and so into the third, *&c.*, till all condensation ceased; when the temperature was marked as before. I found this apparatus very sensible, the bright surface of the metal being visibly obscured by the slightest film of moisture. These experiments were, however, troublesome, and required much time to ensure accuracy. The results I forbear from particularly detailing, as they are superseded by the more exact observations which I have been enabled to make with the instrument which I am about to describe.

It was not till many months after I had commenced this course of inquiry, that I discovered that the mode of investigation which had been suggested to me by the observation of the Roman naturalist, was not as new as I had conceived it to be. The same principle had been applied by the Academicians del Cimento (the restorers of experimental philosophy, as they have been very properly called), to the purposes of hygrometry.

They took a glass vessel of a conical form, and kept it full of snow or pounded ice. This vessel was suspended in the open air with its point downwards, and the moisture which was condensed upon it, trickled down its sides, and dropped from the point of the cone. The frequency of the drops, was applied by them, as a measure of the humidity of the atmosphere. M. le Roi also, adopting the same idea, simplified its application by putting water into a glass vessel, and gradually lowering its temperature, by means of ice, till the appearance of a slight dew upon the surface denoted the point of saturation. The temperatutre of this point he measured, by means of the thermometer. He judged of the humidity of the air by the greater or less degree of depression necessary to produce precipitation. Lastly, Mr. Dalton, in his "Essay upon the force of steam or vapour from water and other liquids at different temperatures," (one of an interesting series, read before the Literary and Philosophical Society of Manchester, and published in the fifth volume of their Memoirs, which it would be difficult to match for originality and sound philosophical induction), thus describes his method of finding the force of the aqueous vapour:—

"I usually take a tall cylindrical glass jar, dry on the outside, and fill it with cold spring-water, fresh from the well; if dew be immediately formed on the outside, I pour the water out; let it stand awhile to increase in heat, dry the outside of the glass well with a linen cloth, and then pour the water in again: this operation is to be continued till the dew ceases to be formed, and then the temperature of the water must be observed. Spring water is generally about 50°, and will mostly answer the purpose the three hottest months of the year: in other seasons an artificial cold-mixture is required."

The discovery of want of originality damped for a time the ardour of a laborious pursuit; but I have ever since been impressed with the great utility of any contrivance which might enable an observer to mark with precision, neatness, and expedition, the constituent temperature of atmospheric vapour. Upon reading the account of the ingenious contrivance of Dr. Wollaston, which he has termed the Cryophorus, the subject again occurred to me; and I received from that instrument the hint, which, after many trials, led to the completion of my hygrometer.

Fig. 1, represents the instrument in its full dimensions; *a* and *b* are two thin glass balls of 1¼ inch diameter, connected together by a tube, having a bore about ⅛th inch. The tube is bent at right angles, over the two balls, and the arm *b c* contains a small thermometer *d e*, whose bulb, which should be of a lengthened form, descends into the ball *b*. This ball having been about two-thirds filled with ether, is heated over a lamp till the fluid boils, and the vapour issues from the capillary tube *f*, which terminates the ball *a*. The vapour having expelled the air from both balls, the capillary tube *f* is hermetrically closed by the flame of a lamp. This process is familiar to those who are accustomed to blow glass, and may be known to have succeeded after the tube has become cool, by reversing the instrument and taking one of the balls in the hand, the heat of which will drive all the ether into the other ball, and cause it to boil rapidly. The other ball *a* is now to be covered with a piece of muslin. The stand *g h* is of brass, and the transverse socket *i* is made to hold the glass tube in the manner of a spring, allowing it to turn and be taken out with little difficulty. A small thermometer *k l* is inserted into the pillar of the stand. The manner of using the instrument is this:—After having driven all the ether into the ball *b* by the heat of the hand, it is to be placed at an open window, or out of doors, with the ball *b* so situated as that the surface of the liquid may be upon a level with the eye of the observer. A little ether is then to be dropped upon the covered ball. Evaporation immediately takes place, which, producing cold upon the ball *a*, causes a rapid and continuous condensation of the ethereal vapour in the interior of the instrument. The consequent evaporation from the included ether, produces a depression of temperature in the ball *b*, the degree of which is measured by the thermometer *d e*. This action is almost instantaneous, and the thermometer begins to fall in two seconds after the ether has been dropped. A depression of 30 or 40 degrees is easily produced, and I have seen the ether boil, and the thermometer driven down below 0° of Fahrenheit's scale. The artificial cold, thus produced, causes a condensation of the atmospheric vapour upon the ball *b*, which first makes its appearance in a thin ring of dew, coincident with the surface of the ether. The degree at which this takes place is to be carefully noted. A little practice may be necessary to seize the exact moment of the first deposition; but certainty is very soon acquired. It is advisable, when the instrument has been constructed with a transparent ball, to have some dark object behind it, such as a house, or a tree; as the cloud is not so readily perceived against the open horizon. The depression of temperature is first produced at the surface of the liquid, where evaporation takes place; and the currents, which immediately ensue to effect an equilibrium, are very perceptible. The bulb of the thermometer *d e*, is not quite immersed in the ether, that the line of greatest cold may pass through it. In very damp or windy weather the ether should be very slowly dropped upon the ball, otherwise the descent of the thermometer will be so rapid as to render it extremely difficult to be certain of the degree. In dry weather, on the contrary, the ball requires to be well wetted more than once, to produce the requisite degree of cold. If at any time there should be reason to suspect the accuracy of an observation, it may easily be corrected by observing the temperature at which the dew upon the glass again disappears: the mean of the two observations (whose errors, if any, will lie in contrary directions,) will give the true result. It is obvious that care should be taken not to permit the breath to affect the glass. With these precautions the observation is simple, expeditious, easy, and certain.

Being desirous of ascertaining whether the superior power of metals in conducting heat, together with the high polish of which they are susceptible, might not be rendered conducive to the perfection of the hygrometer, I endeavoured to modify its form in such a way as to allow of their being employed in its construction. After some unsuccessful trials I completed one, of which the subjoined is an outline.

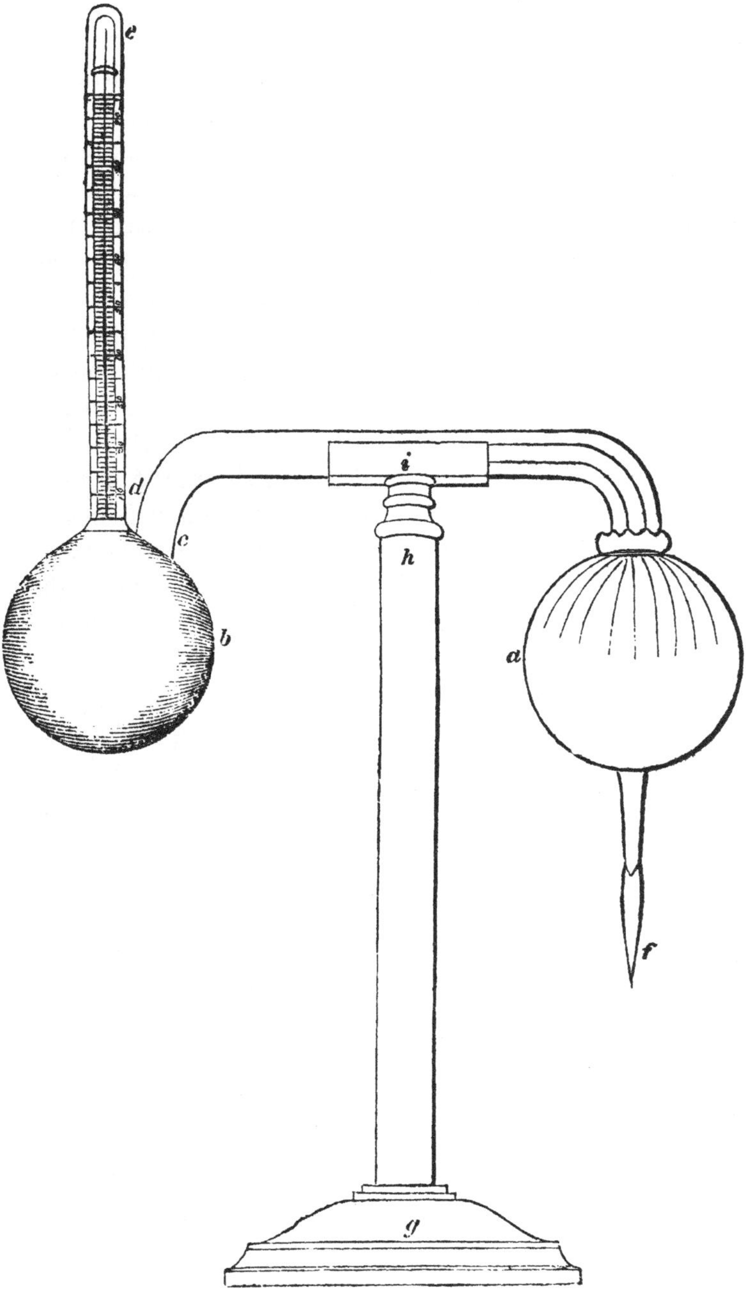
e
d
c
b
i
h
a
f
g

The balls *a* and *b*, together with their connecting tube, are made of very thin brass. To the orifice *f* is soldered a small piece of platinum tube, which, from its property of welding with glass, allows of the junction of a piece of glass tube, and, after the instrument has been boiled as before directed, may be hermetically closed in the usual way. The thermometer *d e* is so constructed that its bulb, which is enclosed in the ball *b*, is rather less than the diameter of its stem, which is made proportionably thick. It is ground air-tight into a collar of brass, made for its reception on the top of the ball. The ball *a* is covered with muslin, and the ball *b* is very highly polished. The advantages which I looked for in this construction of the instrument were two: First, I expected that an unpractised observer might more readily be able to mark with precision the instant of the first precepitation of the dew. The white mist is directly seen, whereas a little experience is required to obtain an equal degree of certainty with the transparent glass. Secondly, I imagined that its sensibility might be increased by extending, at pleasure, the scale of the thermometer *d e*. The divisions of the thermometer included in the glass instrument are necessarily small; but those of the external thermometer may be made of any required magnitude, without rendering the bulk of the whole inconveniently great.

It was also an important object to ascertain whether any hygrometric property of the glass, or difference between it and the metal in attraction of moisture, would have any appreciable effect upon the condensing power.

Long experience has, however, convinced me that the metallic hygrometer possesses no real superiority over the glass. The visibility of the deposition in the latter is rendered perfect by making the condensing ball of black glass, and viewing it by reflected light in the manner of a mirror; and I never could perceive any difference in the sensibility of the two instruments.

Thus much on the construction of the hygrometer: It is simple and easy. Its graduation depends upon no arbitrary or disputed determinations of wet and dry: it is liable to no deterioration from use, age, or accidental circumstances; and above all things, whenever, or by whomsoever made, it is incapable, in proper hands, of affording erroneous results. It may be more or less boiled; the *vacuum* may be more or less perfect; and it may, consequently, require the affusion of a larger or smaller quantity of ether to make it act: but (provided the thermometer be correct) the observation, when obtained, cannot deceive. Its determinations are, therefore, as strictly comparable one with another, under all circumstances, as those of the barometer or the thermometer.

In describing the various uses and applications of the hygrometer, I shall commence with the most popular; its use, namely, as a weather-glass.

When consulted with a view of predicting the greater or less probability of rain, or other atmospheric changes, two things are to be principally attended to—the difference between the constituent temperature of the vapour, and the temperature of the air; and the variation of the dew-point. In general, the chance of rain, or other precipitation of moisture from the atmosphere, may be regarded as in inverse proportion to the difference between the two thermometers: but in making this estimate, regard must be had to the time of day at which the observation is made. In settled weather the dryness of the air increases with the diurnal heat, and diminishes with its decline; for the constituent temperature of the vapour remains nearly stationary. Consequently, a less difference at morning or evening is equivalent to a greater in the middle of the day.

But to render the observation most completely prospective, regard must be had at the same time to the movement of the dew-point. As the elasticity of the vapour increases or declines, so does the probability of the formation and continuation of rain. An increasing difference, therefore, between the temperature of the air, and the temperature of the point of condensation, accompanied by a fall of the latter, is a sure prognostication of fine weather; while diminished heat, and a rising dew-point, infallibly portend a rainy season. When observations shall have been made and registered for a sufficient length of time, the mean results for the different periods of the year will afford accurate standards of comparison whereby to judge of the state of the vapour; and the three years' Journal appended to this Essay, will not be without its use in this respect. In winter, when the range of the thermometer during the day is small, the indication of the weather must be taken more from the actual rise and fall of the point of condensation, than from the difference between it and the temperature of the air. It must be remembered that a state of saturation may exist, and precipitation even take place in the finest weather, and under a cloudless sky; but this is when the diurnal decline of the temperature of the air, near the surface of the earth, falls below an unfluctuating term of precipitation; and it is probable, that at some period or other of the twenty-four hours, this term is always passed. The radiation of the earth, in the absence of the sun, cools the stratum of air in contact with it; and a light precipitation takes place, of so little density as totally to escape the observation of the eye. At other times it becomes visible, and assumes the appearance of mist or fog. Under such circumstances, the hygrometer will sometimes exhibit a different kind of action. If it be brought from an atmosphere of a higher temperature into one of a lower degree, in which condensed aqueous particles are floating, the mist will begin to form at a temperature several degrees higher than that of the air. The heat emanating from the ball of the instrument, dissolves the particles of water, and forms an atmosphere around it of greater elasticity than the surrounding medium; so that, when it is put in action, the point of deposition is proportionably raised. This action does not at all interfere with the determination of the real force and quantity of vapour; for, in all such cases, the full saturation of the atmospheric temperature must have place, and, consequently, the temperature of the vapour must be coincident with that of the air.

This kind of precipitation, which may often be detected by the hygrometer, when it would otherwise escape notice, far from being indicative of rain, generally occurs in the most settled weather. It is analogous to the formation of dew, and is dependant upon the same cause, the radiation of the earth, which can only take place under an unclouded sky.

A sudden change in the dew-point, is generally accompanied by a change of wind: but the former sometimes precedes the latter by a short interval; and the course of the aërial currents may be anticipated before it affects the direction of the weathercock, or even the passage of smoke.

My own experience, and the testimony of others, assure me that the hygrometer, thus applied, is more to be depended upon than any instrument that has yet been proposed. Even when its indications are contrary to those of the barometer, reliance may be placed upon them; but simultaneous observations of the two most usefully correct each other. The rise and fall of the mercurial column is, most probably, primarily dependant upon the state of the upper regions of the atmosphere, with regard to heat and moisture. Local *chemical* alterations of its density, thus partially brought about, are *mechanically* adjusted, and the barometer gives us notice of what is going on in inaccessible regions. A rise in the dew-point, accompanied by a fall of the barometer, is an infallible indication that the whole mass of the atmosphere is becoming embued with moisture, and a copious precipitation may be looked for. If the fall of the barometer take place at the same time that the point of precipitation is depressed, we may conclude that the expansion which occasions the former has arisen at some distant point, and wind, not rain, will be the result. But when the air attains the point of precipitation, with a high barometer, we may infer that it is a transitory and superficial effect, produced by local depression of temperature. Particular illustrations of these modified effects might easily be adduced in this place, but they will be more conveniently studied in the abundant observations of the annexed journal.

Thus does the Hygrometer mark with infallible precision the comparative degrees of moisture and dryness in the atmosphere, and by exhibiting them in degrees of the thermometer, refer them to a known standard of comparison, and speak in a language which every body understands. But its observations may be made applicable to a much wider field of research, and adapted to still more important objects. By means of tables, we can find with the utmost accuracy and ease the positive weight of aqueous vapour diffused through any given portion of space, and its force or elasticity as measured by the column of mercury which it is capable of supporting: we discover at once the proportion of moisture in any space to the quantity which would be required to saturate it, or what has been termed the true natural scale of the hygrometer: we can calculate with perfect ease the specific gravity of any mixture of air and aqueous vapour: and we can measure the force and quantity of evaporation. Upon the *data* employed in the construction of the tables, it will be necessary that I should premise a few reflections.

Mr. Dalton, in his valuable Essay before referred to, has detailed the results of a laborious series of experiments, by which he has ascertained with great precision the force of vapour from water at every degree between its freezing and its boiling points, and derived from them a *formula*, by which he extended the results from the freezing of mercury to the 325th degree of Fahrenheit's scale. Dr. Ure has since* entered upon the same investigation with a different modification of apparatus, calculated to avoid some irregularities to which Mr. Dalton's was exposed. He carried his actual experiments as high as 312°; and thus ascertained that Mr. Dalton's *ratio* of progression for the force of vapour, though apparently accommodated to the intervals between 32° and 212°, could not serve for the higher ranges. In the prosecution of the inquiry, he was led to the discovery of a very simple *ratio*, which admirably connected together the whole series of experiments. In a former essay upon this subject, I adopted Mr. Dalton's numbers; which, for the range of atmospheric temperature, exhibited not only a perfect adaptation to his own experiments, but also a surprising accordance with those of Dr. Ure: but reflecting that from these and other considerations, the rule from which they were derived could not be the law of nature, I have recalculated the tables from the *data* of Dr. Ure. It is gratifying to find that, for the purposes of the hygrometer, the difference after all is very inconsiderable.

The second column of Table I. exhibits the force of aqueous vapour, hence derived, in inches of mer-

* Phil. Trans. 1818., p. 338.

cury, at the temperature marked in the corresponding line of the first column.

Upon these two *data*, namely, the force and temperature of the vapour, are chiefly founded the calculations which have furnished me with the series of the third column, which contains the weight in grains of a cubic foot of the vapour at the corresponding temperature and pressure. The method of computing it is as follows:—Steam at 212°, and under a pressure of 30 inches of mercury, is, as nearly as possible, 1700 times lighter than an equal bulk of water at its maximum of density; and a cubic foot of water, at the temperature of 40°, weighs, according to the accurate investigations of Dr. Rice, 437272 grains: the weight, therefore, of a cubic foot of steam, at the above temperature and pressure is $\frac{437272}{1700}$, or 257.218 grains. Hence we may find the weight of an equal bulk of vapour of the same temperature under any other given pressure, suppose 0.560 in.: for the volume being in inverse proportion to the pressure,

Ins.		Ins.		Grs.		Grs.
30	:	0.560	::	257.218	:	4.801

the weight required.

Having now obtained the weight of a cubic foot of vapour, at a pressure of 0.560 in., and at a temperature of 212°, we may proceed to find its weight under the same pressure at any other temperature, suppose 60°. It has been fully established, that all aëriform bodies (vapours out of the contact of their respective fluids, as well as gases,) expand $\frac{1}{480}$th part of their volume for every accession of temperature, equivalent to one degree of Fahrenheit's scale; therefore, reckoning a volume of gas at 32° as unity, its volume at 60° is to its volume at 212°, as $1 + \frac{28}{480}$ is to $1 + \frac{180}{480}$, or :: 1.0583 : 1.3749. and the density and weight being in inverse proportion to the volume,

Vol. at 60.		Vol. at 212.		Grs.		Grs.
1.0583	:	1.3749	: :	4.801	:	6.222

the weight of the cubic foot of vapour at the temperature of 60° and under a pressure of .560 in.

It has also been proved by Mr. Dalton, that as much vapour of determined temperature is formed in a given bulk of air as in a vacuum of equal space; therefore, the above result gives the weight of vapour which can exist in a cubit foot of air at the temperature of 60° The fourth column of the table contains the proportionate expansion for the corresponding degrees.

TABLE I. *Shewing the Force, Weight, and Expansion of Aqueous Vapour, at different Temperatures, from* 0° *to* 95°.

Temp.	Force.	Weight of a Cubic Foot.	Expansion.	Temp.	Force.	Weight of a Cubic Foot.	Expansion.
0	0.068	0.856	.9334	26	0.176	2.096	.9875
1	0.071	0.892	.9355	27	0.182	2.163	.9896
2	0.074	0.928	.9375	28	0.188	2.229	.9917
3	0.077	0.963	.9396	29	0.194	2.295	.9938
4	0.080	0.999	.9417	30	0.200	2.361	.9959
5	0.083	1.034	.9438	31	0.208	2.451	.9980
6	0.086	1.069	.9459	32	0.216	2.539	1.0000
7	0.089	1.104	.9480	33	0.224	2.630	1.0020
8	0.092	1.139	.9500	34	0.232	2.717	1.0041
9	0.095	1.173	.9521	35	0.240	2.805	1.0062
10	0.098	1.208	.9542	36	0.248	2.892	1.0083
11	0.103	1.254	.9563	37	0.256	2.979	1.0104
12	0.107	1.308	.9584	38	0.264	3.066	1.0125
13	0.111	1.359	.9605	39	0.272	3.153	1.0145
14	0.115	1.405	.9625	40	0.280	3.239	1.0166
15	0.119	1.451	.9646	41	0.292	3.371	1.0187
16	0.123	1.497	.9667	42	0.304	3.502	1.0208
17	0.127	1.541	.9688	43	0.316	3.633	1.0229
18	0.131	1.586	.9709	44	0.328	3.763	1.0250
19	0.135	1.631	.9730	45	0.340	3.893	1.0270
20	0.140	1.688	.9750	46	0.352	4.022	1.0291
21	0.146	1.757	.9771	47	0.364	4.151	1.0312
22	0.152	1.825	.9792	48	0.376	4.279	1.0333
23	0.158	1.893	.9813	49	0.388	4.407	1.0354
24	0.164	1.961	.9834	50	0.400	4.535	1.0375
25	0.170	2.028	.9855	51	0.414	4.684	1.0395

TABLE I. *Continued.*

Temp.	Force.	Weight of a Cubic Foot.	Expansion.	Temp.	Force	Weight of a Cubic Foot.	Expansion.
52	0.428	4.832	1.0416	76	0.936	10.107	1.0916
53	0.444	5.003	1.0437	77	0.966	10.387	1.0937
54	0.460	5.173	1.0458	78	0.997	10.699	1.0958
55	0.476	5.342	1.0479	79	1.028	11.016	1.0979
56	0.492	5.511	1.0500	80	1.060	11.333	1.1000
57	0.508	5.679	1.0520	81	1.093	11.665	1.1020
58	0.526	5.868	1.0541	82	1.127	12.005	1.1041
59	0.543	6.046	1.0562	83	1.162	12.354	1 1062
60	0.560	6.222	1.0583	84	1.198	12.713	1.1083
61	0.577	6.399	1.0604	85	1.235	13 081	1.1104
62	0.594	6.575	1.0625	86	1.273	13.458	1.1125
63	0.615	6.794	1.0645	87	1.312	13.877	1.1145
64	0.636	7.013	1.0666	88	1.351	14.230	1.1166
65	0.657	7 230	1.0687	89	1 390	14.613	1.1187
66	0.678	7.447	1.0708	90	1.430	15.005	1.1208
67	0.699	7.662	1.0729	91	1.470	15.432	1.1229
68	0.722	7.899	1.0750	92	1 510	15.786	1.1250
69	0.745	8.135	1.0770	93	1.551	16.186	1.1270
70	0.770	8.392	1.0791	94	1.593	16.593	1.1291
71	0.796	8.658	1.0812	95	1.636	17.009	1.1312
72	0.822	8.924	1.0833				
73	0.849	9.199	1.0854	212	30.000	257.218	1.3749
74	0.877	9.484	1.0875				
75	0.906	9.780	1.0895				

It is at all times desirable to bring the results of calculation, however exact the *data* upon which they are founded, to the test of actual experience; and we have the ready means of so doing with regard to the above table. The indefatigable De Saussure, in his " Essais sur L' Hygrométrie," gives the results of a series of experiments, to determine the quantity of moisture which air is capable of dissolving at certain temperatures. The means which he employed were simple. He thoroughly dried the air of a large glass balloon, of known capacity; and then suspended in it a small piece of linen, which had been moistened and accurately weighed. He ascertained the point of saturation by means of a manometer, which ceased to move when the term of extreme humidity had been obtained, and then withdrawing the linen, he instantly noted its loss of weight. He thus found that at the temperature of 15°.16 Reau. a French cubic foot of air took up 11.0690 grains of water; while at 6°.18 Reau. it only dissolved 5.6549 grains. By reducing these results to English weights and measures, we have at 66° of Fahrenheit, 7.498 grains in a cubic foot, and at 45½° Fahrenheit, 3.830 grains: a wonderful accordance with our theoretical determinations.

Mr. Anderson, in his highly interesting treatise upon Hygrometry, published in the " Edinburgh Encyclopædia," has also given us the results of his experiments, to determine the same point by a method less liable, perhaps, to objection. His manner of operating consisted in causing a large volume of air, saturated with moisture, to pass slowly in a stream through a sufficient quantity of sulphuric acid, or dry muriate of lime, cut off from all communication with the atmosphere; and then observing the increase of weight which these substances acquired in consequence of the air being transmitted through them. The weight of a cubic foot of steam, at different temperatures hence derived, is compared in the following table with those derived from calculation.

Temp.	Grs. by Expt.	Calculated.
49	4.085	4.407
59	5.679	6.046
77	9.828	10.387
83	11.660	12.354

Considering the nature of the experiment, and the complication of the calculations, this is again a very close agreement.

The manner of using the table will, perhaps, be best understood from an example. Let the temperature of the atmosphere be 70°, and the point of condensation, as found by the hygrometer, 55°; the pressure of the vapour, under these circumstances, is immediately found opposite to the degree of its constituent heat 55° = 0.476. To find its weight, we proceed thus:—Supposing that the temperature of the air had not differed from that of the dew-point, its weight would have been found upon the same line as its pressure = 5.342 grains. But its bulk is expanded by the excess of atmospheric heat; we must, therefore, seek in the fourth column for the degree of expansion at 55° = 1.0479, and at 70° = 1.0791, and apply the correction thus:—

Bulk at 70°		Bulk at 55°		Grs.		Grs.
1.0791	:	1.0579	::	5.342	:	5.175

which is the weight required.

Again,—the dryness of the atmosphere, under the above conditions, may be conveniently expressed as 15°, in terms of the thermometric scale: but it may be desirable also to know what it would be upon the natural scale of the hygrometer. This is readily ascertained by dividing the elasticity of vapour at the temperature of the dew-point, by the elasticity at the temperature of the air: the quotient will express the proportion of moisture actually existing, to the quantity which would be required for saturation; for, calling the term of saturation 1.000, as the elasticity of vapour at the temperature of the air is to the elasticity of vapour at the temperature of the dew-point, so is the term of saturation to the actual degree of moisture,—thus,

$$\frac{\text{Elas. at } 55^\circ}{\text{Elas. at } 70^\circ} \quad \frac{.476}{.770} = .168$$

The relation of this mode of expressing the degree of moisture to that of denoting the degree of dryness, by the thermometric scale, will be elucidated by selecting a different example. Let the temperature of the air be 47°, and the dew-point 32°; the dryness represented by the former expression will be 15°, as before, but by the latter the degree of moisture will be reduced to .593. In keeping a register, however, of observations, in which the final results are not calculated, the dew-point itself should always be recorded. By neglect of this the observations of the hygrometer published in the Philosophical Transactions are rendered nearly useless; as the mere record of the degrees of moisture on the hygrometric scale are of comparatively little interest, and the results of the hygrometer are thereby reduced almost to a par with those of a common hygroscope.

Thus, by two simple observations, and very easy calculations, we ascertain, with precision, the following points of the utmost interest to meteorology.

Temperature of the air	70°
Dew-point	55°
Degree of dryness on the thermometric scale	15°
Degree of moisture on the hygrometric scale	618
Elasticity of the vapour	.476 ins.
Weight of vapour in a cubic foot . . .	5.175 grs.

The state of the atmosphere, assumed above, would constitute fine weather; and one of two things, or a modification of both, must happen, before any precipitation of water could take place: either the temperature of the air must fall below 55°; or the quantity of vapour must increase to 8.392 grains in the cubic foot, the maximum quantity which could exist at 70°; or the point of condensation may become intermediate, by a corresponding rise and fall of the two.

In the first case, the precipitation would probably be only slight and transitory, such as mist or fog: in the second case, it would assume the form of hard rain and storms: while, in the third, some conjecture might be formed of its probable duration and quantity, according as one or other of its causes prevailed.

But the hygrometer can be made to measure not only the quantity and force of vapour existing at any time in the air, but may be applied at the same time to indicate the force and quantity of evaporation. Mr. Dalton, in the course of that important train of investigation to which I have before had occasion to refer, ascertained that the quantity of water, evaporated in a given time, bore an exact proportion to the force of vapour* at the same temperature. The atmosphere obstructs its diffusion, which would otherwise be instantaneous, as *in vacuo;* but this obstruction is overcome with a celerity proportioned to the force of the vapour. The retardation, however, does not arise from the weight of the air, for that would prevent any vapour from rising under 212°; but, as Mr. Dalton observes, is caused by the *vis inertiæ* of the particles of air, and is similar to that which a stream of water meets with in descending amongst pebbles. In ascertaining this point at ordinary atmospheric temperatures, regard must be had to the force of vapour already existing in the air. For instance, if water of 57° were the subject, the force of vapour of that temperature is $\frac{1}{60}$th of the force at 212°; and one might expect the quantity of evaporation to be $\frac{1}{60}$th also; but if it should happen that an aqueous atmosphere to that amount does already exist, the evaporation, instead of being $\frac{1}{60}$th of that from boiling water, would be nothing at all. On the other hand, if the aqueous atmosphere were less than that, suppose half of it, then the effective evaporating force would be $\frac{1}{120}$th of that from boiling water; in short, the evaporating force must be universally equal to that of the temperature of the water diminished by that already existing in the atmosphere. But the air, by its mechanical action, has another influence upon the rate of evaporation. When calm and still, it merely obstructs the process; but when in motion, it increases its effect in direct proportion to its velocity, by removing the vapour as it forms. Mr. Dalton fixes the extremes

* See the Essay upon Evaporation, p. 489.

that are likely to occur in ordinary circumstances at 120 and 189 grains per minute, from a vessel of six inches diameter, at a temperature of 212°. Upon these *data* the following Table was constructed, in which Mr. Dalton's results have been accommodated to the progression of elasticity adopted in Table I., from Dr. Ure, by the slight alteration of moving the temperature back two degrees.

TABLE II.—Showing the Force of Vapour, and the full evaporating Force of every Degree of Temperature, from 18° to 85°; expressed in Grains of Water that would be raised per Minute from a Vessel of Six Inches in Diameter, supposing there were no Vapour already in the Atmosphere.

Tem.	Force of Vapour.	Evap. Force in Grains.			Tem.	Force of Vapour.	Evap. Force in Grains.		
212°	30.000	120 gr.	154 gr.	189 gr.	212°	30 000	120 gr.	154 gr.	189 gr.
18	.131	0.52	0.67	0.82	52	.428	1.71	2.20	2.69
19	.135	0.54	0.69	0.85	53	.444	1.77	2.28	2.78
20	.149	0.56	0.71	0.88	54	.460	1.83	2.35	2.88
21	.146	0.58	0.73	0.91	55	.476	1.90	2.43	2.98
22	.152	0.60	0.77	0.94	56	.492	1.96	2.52	3.08
23	.158	0.62	0.79	0.97	57	.508	2.03	2.61	3 19
24	.164	0.65	0.82	1.02	58	.526	2.10	2.70	3.30
25	.170	0.67	0.86	1.05	59	.543	2.17	2.79	3.41
26	.176	0.70	0.90	1.10	60	.560	2.24	2.88	3.52
27	.182	0.72	0.93	1.13	61	.577	2.31	2.98	3.63
28	.188	0.74	0.95	1.17	62	.594	2.39	3.07	3.76
29	.194	0.77	0.99	1.21	63	.615	2.46	3.16	3.37
30	.200	0.80	1.03	1.26	64	.636	2.54	3.27	3.99
31	.208	0.83	1.07	1.30	65	.657	2.62	3.37	4.12
32	.216	0 86	1.11	1.35	66	.678	2.70	3.47	4.24
33	.224	0.90	1.14	1.39	67	.699	2.79	3.59	4.38
34	.232	0.92	1.18	1.45	68	.722	2.88	3.70	4.53
35	.240	0.95	1.22	1.49	69	.745	2.98	3.83	4.68
36	.248	0.98	1.26	1.54	70	.770	3.08	3.96	4.84
37	.256	1.02	1.31	1.60	71	.796	3.18	4.09	5.00
38	.264	1.05	1.35	1.65	72	.822	3.29	4.23	5.17
39	.272	1.09	1.40	1.71	73	.849	3.40	4.37	5.34
40	.280	1.13	1.45	1.78	74	.877	3.52	4.52	5.53
41	.292	1.18	1.51	1.85	75	.906	3.65	4.68	5.72
42	.304	1.22	1.57	1.92	76	.936	3.76	4.83	5.91
43	.316	1.26	1.62	1.99	77	.966	3.88	4.99	6.10
44	.328	1.31	1.68	2.06	78	.997	4.00	5.14	6.29
45	.340	1.36	1.75	2.13	79	1.028	4.16	5.35	6.54
46	.352	1.40	1.80	2.20	80	1.060	4.28	5.50	6.73
47	.364	1.45	1.86	2.28	81	1.093	4.40	5.66	6.91
48	.376	1.50	1.92	2.36	82	1.127	4.56	5 86	7.17
49	.388	1.55	1.99	2.44	83	1.162	4.68	6.07	7.46
50	.400	1.60	2.06	2.51	84	1.198	4.80	6.28	7.75
51	.414	1.66	2.13	2.61	85	1.235	4.92	6.49	8.04

The first column contains the degrees of temperature; the second the corresponding force of vapour; the third the amount of evaporation per minute from a vessel of six inches' diameter in calm weather; the fourth, the amount in a moderate breeze; and the fifth, in a high wind.

The use of this table, as applied to the hygrometer, is this:—Let it be required to know the force of evaporation at the existing state of the atmosphere: find the point of condensation by the instrument, as before directed; subtract the grains opposite that temperature, either in the third, fourth, or fifth columns, according to the state of the wind, from the grains opposite to the temperature of the air in the same column, and the remainder will be the quantity evaporated in a minute from a vessel of six inches' diameter, under the given circumstances. For example;—Let the point of condensation be 55°, the temperature of the air 70°, with a moderate breeze. The number opposite to 55° in the fourth column is 2.43, and that opposite to 70° is 3.96: the difference, 1.53 grain, is the evaporation per minute.

But it is, perhaps, simpler, and more convenient in many cases, to estimate the depth of the water evaporated in a day, and Dr. Young has shown how this may be done, from Mr. Dalton's *data*. It happens that the column of mercury equivalent to the elasticity of the vapour, expresses, accurately enough, the mean evaporation in 24 hours. Mr. Dalton's experiment gives 45 grains per minute, from a disc of 3¼ inches. Now $45 \times 60 \times 24 =$ 64800 grains, or 256.6 cubic inches, which would make a cylinder 30.9 inches in height, on a base $3\frac{1}{4}$ inches in diameter; and this differs only $\frac{1}{33}$ from the height of the column of mercury. We may, therefore, assume that the mean daily evaporation is equal to the tabular number expressing the elasticity of the vapour; sometimes exceeding it, or falling short of it about one-fourth; and we may readily allow for the effect of the moisture of the atmosphere, by deducting the number corresponding to the temperature of deposition. Thus, supposing the mean temperature, of 24 hours, to be 60°, and that of the dew-point 50, the evaporation will be equal to $.560 - .400 = .160$ inch.

It is evident that these estimates can be but mere approximations; for till we can obtain some accurate measure of the velocity of the wind, they must be liable to great uncertainty. They are, however, as much to be relied upon as the registers of the evaporating gauge in common use, whose only proper application can be to furnish a rough estimate of the state of atmospheric saturation, and the point of deposition. The notion that they afford the absolute measure of the quantity of water raised into the air is absurd, for the instrument can only give the amount of evaporation from the shallow body of water in the place where it has been fixed. The conditions which modify the process vary almost *ad infinitum*. They vary on the land and on the water; they vary in sun-shine and in the shade; they vary as land is more or less clothed with vegetation, or as water is more or less deep. The evaporating gauge, so far from representing the circumstance of those bodies which yield the great body of vapour on the earth's surface, probably does not correspond, in all essential particulars, with a dozen puddles in the course of the year; and the pains which are often taken to make the results tally with those of the rain gauge, or to compare the two, are wholly misdirected. The results of the hygrometer, as applied above, accommodate themselves more easily to the ever-varying conditions of the problem; and from these we can infer the effect of each combination of circumstances, and the capacity of the air for moisture modified by the velocity of the winds.

The next application of the hygrometer is not of inferior importance to any of those which we have been considering: I allude to its application to the correction of barometrical measurements. Ever since the celebrated and important experiment of Torricelli, the attention of some of the greatest philosophers has been drawn in succession to the interesting problem of the mensuration of heights by means of the barometer. The most laborious experiments have been undertaken for the improvement of the practical part of the operation, and the utmost refinements of mathematical calculation have been employed in the perfecting of its theory. To the former, M. de Luc, General Roy, and Sir George Shuckburgh have pre-eminently contributed; while the powerful minds of Halley, Newton, Playfair, and Laplace, have been applied to the latter. But one *desideratum* in Physics has stopped the progress of each at nearly the same point; a desideratum which all have felt, and all in succession have pointed out. I allude to the deficiency of means to measure the quantity and effects of aqueous vapour in the atmosphere. The relation of the air's density and elasticity, the effects of heat upon the relative weights of mercury and air, the diminution of gravity in ascending from the surface of the earth, its variation in different latitudes, and the disturbance of centrifugal force, have been appreciated and allowed for; but all the corrections, excepting the two first, are exceeded in value by that which has hitherto been only the subject of conjecture; namely, the correction for moisture. Some of the latter calculations have, indeed, assumed an appearance of considerable accuracy; but while the more important problem remains unsolved, such appearance is only illusory; and it may even be doubted whether the state of physical science is ever likely to be such as to render the introduction of the refinements which they exhibit, practically advantageous. The importance, however, of the problem, the solution of which I am now about to attempt, has, on the contrary, been universally admitted.

M. de Luc, in his valuable and laborious "Researches upon the Modifications of the Atmosphere," thus adverts to the knowledge which it is necessary to obtain of the effects of vapour in the air for the perfecting the mensuration of heights by means of the barometer.

"Voilà donc un nouveau champ ouvert aux expériences. Il s'agit de déterminer quel changement on doit faire à la hauteur trouvée par les logarithmes, quand l'air est plus ou moins chargé de vapeurs qu'un certain point fixe et de vapeurs échauffées plus ou moins qu'un certain dégré. Il me semble que pour découvrir cette loi, il faudroit pouvoir joindre l'observation d'un hygromètre comparable à celle du baromètre et thermomètre; car le point essentiel consiste à connoître s'il y a des vapeurs dans la colonne d'air qui est interceptée par les deux stations et quelle est leur quantité; puisque, si les vapeurs qui font baisser le baromètre sont plus élevées que cette colonnne, elles ne changent point la loi générale qui sert de fondement au calcul.

" Lorsqu'on aura obtenu ce premier point, il sera facile de connoître par l'expérience, 1°, Si les vapeurs influent de la même manière quelque soit la densité de l'air produite par la pression supérieure, et par conséquent, quelque soit la hauteur du mercure dans le baromètre. 2°, Quel rapport il y a entre la quantité des vapeurs exprimée par les degrés de l'hygromètre, et la diminution d'elasticité de l'air par une température donnée; ou plus directement, quelle partie proportionelle il faut déduire de la hauteur trouvée par le calcul, ou ajouter à cette hauteur, pour chaque dégré de l'hygromètre, quand l'air est à cette température. 3°, Enfin, quelle modification doit éprouver ce rapport lorsque la chaleur est plus ou moins grande que le point fixe, auquel la force expansive des vapeurs est egale à celle de l'air.

" Je conviens que tout cela présente bien des soins et des peines au premier coup-d'œil; mais j'ai éprouvé plus d'une fois que les difficultés connues s'applanissent beaucoup quand on les affronte avec courage."—Tome iii., p. 288.

General Roy, in commenting upon his experiments upon the different expansions of dry and moist air, for the elucidation of the same subject, says:—

"I am aware it will be alleged, that the proportion of moisture admitted into the manometer in these experiments, is greater than what can ever take place in nature; and, therefore, in order to be able to judge of the degrees of expansion the medium suffers in its more or less dense, and more or less moist, states, that not only air near the surface of the earth, but likewise that found at the top of some very high mountains should have been made use of. I grant all this; but, on the other hand, it must be remembered, that those experiments are very recently finished; that a good hygrometer, (if such can ever be obtained,) a great deal of leisure time, and the vicinity of high mountains, were all necessary for the carrying of such a scheme into execution. It is for these reasons, and in hopes that other people will sooner or later investigate this matter still further, not only by experiments made on the expansion of air taken at different heights above the level of the sea in middle latitudes, but likewise on that appertaining to the humid and dry regions of the atmosphere towards the equator and poles, that I have been induced to hasten the communication of this paper. In the mean time, having proved, beyond the possibility of a doubt, that a wonderful difference doth exist between the elastic force of dry and moist air, I may be allowed hereafter to reason by analogy on the probable effects this will produce in measuring heights by the barometer."—*Phil. Trans.* vol. lxvii. p. 714.

M. Laplace, who has applied the prodigious powers of his science to the perfecting the barometric *formula*, and has availed himself of all the accuracy of the modern method of experiment, was forced to leave the hygrometric state of the air in the catalogue of inevitable errors, contenting himself with an approximate correction:—

"Les corrections," says he, "relatives à la latitude, et a la variation de la pesanteur, sont très-petites ; mais commes elles sont certaines, il est utile de les employer pour ne laisser subsister dans le calcul que les erreurs inévitables des observations, et celles qui résultent des attractions inconnues des montagnes, *de l'état hygrométrique de l'air, auquel il serait nécessaire d'avoir égard ;* et enfin de l'hypothèse adoptée sur la loi de la diminution de la chaleur. On tiendrait compte, en partie, de l'état hygrométrique de l'air en augmentant un peu le coefficient 0.00375 de $\frac{t+t'}{2}$ dans la formule précédente ; car la vapeur aqueuse est plus légère que l'air, et l'accroissement de témperature en accroit la quantité, toutes choses égales d'ailleurs."—*Mécanique Céleste.* Tom. iv., p. 292.

The late lamented Mr. Playfair, in an elaborate paper upon the same subject, published in the Philosophical Transactions of Edinburgh, (vol. i. 1778,) thus enforces the same argument: "There is another cause of error, which, had the effects of it been sufficiently known, ought, no doubt, to have entered into this investigation. Moisture, when chemically united to air, or dissolved in it, so as to form part of the same homogeneous and invisible fluid, appears to have a powerful effect to increase the elasticity of the air and its expansion, for every additional degree of heat which it receives. Though the judicious and accurate experiments of General Roy have ascertained this effect of humidity, and have even gone far to determine the law of its operation; yet, for want of a measure of the quantity of it contained at any given time in the air, it is impossible to make any application of this knowledge to the object under our consideration."

Lastly Mr. Leslie, in an article upon barometrical measurements in the *Supplement to the Encyclopedia Britannica,* concludes his detail of corrections with the same acknowledgment. "The humidity of the air also materially affects its elasticity, and the hygrometer should therefore be conjoined with the thermometer in correcting barometrical observations. But nothing satisfactory has yet been done with regard to that subject. The ordinary hygrometers, or rather hygroscopes, are mere toys, and their application to science is altogether hypothetical."

Impressed with the importance of the object, so clearly pointed out by a succession of the most able philosophers, I had no sooner succeeded in constructing an instrument which, upon unerring principles, would show the quantity of vapour contained at any given time in the air, than I turned my attention to render it available to the desired purpose; and I shall now endeavour to explain a method of observation and calculation which, I trust, will be found fully and strictly to solve this important problem.

The most simple way of considering the subject, in a general point of view, appears to me to be that which was, I believe, first suggested by Sir George Shuckburgh (*Phil. Trans.* vol. lxvii. p. 556.), namely, to make a comparison of the specific gravities of mercury and air at a fixed temperature, and under a given pressure, the foundation of the operation. In this manner we calculate the height of a column of air, compared with any given column of mercury of equal base, supposing it of equal density throughout. The calculation of the gradual diminution of density which takes place for equal ascents in the atmosphere, according to a geometrical progression, is made in the usual way, by means of logarithms. This latter calculation may be deemed invariable under all circumstances; the former includes all the adventitious circumstances, and all the effects of disturbing causes.

The well-known accuracy of MM. Biot and Arago, assisted by the nicety of modern instruments, has determined the relative specific gravities of dry air and mercury at a temperature of 32°, and under a pressure of 30.00 inches, to be as 1 to 10435. The height of a column of air, therefore, of equal density throughout, which would balance a column of mercury of 30 inches, under these conditions, would be very nearly 26090 feet, or 4348 fathoms. Now the modulus of the common system of logarithms being .4342945, the height of the homogeneous atmosphere in fathoms may be considered identical with its thousandth part; and the decrease of density for the height is therefore immediately found by taking out the logarithm of the given height*.

* See the first Essay, p. 12.

These proportions may be disturbed in two ways by the operation of heat. In the first place, its expansive power, acting upon the mercury, may dilate or contract its particles ; so that a column of 30 inches, being more or less dense, will require an equipoise of greater or less length, according as its temperature is below or above the standard at 32°. This effect has been most minutely appreciated, and its correction is applied with the utmost ease and precision. In the second place, the power of heat, acting upon the air, occasions a much more considerable dilation or contraction of its parts, and gives rise to much greater differences in the height of the equiponderant column. The expansion of air has been determined with precision by the experiments of M. Gay-Lussac, and from them we infer, with confidence, that it increases or diminishes $\frac{1}{480}$th part for every addition or subtraction of 1° of heat. In this situation, therefore, the operation stands : the column of mercury, which is the measure applied, is rendered an invariable standard of comparison, by being brought, by an easy calculation, to a known density ; and the altitude measured is in proportion to the specific gravity of the air.

But heat is not the only agent which alters the specific gravity of the air; the admixture of aqueous vapour, it is well known, produces very important changes in its density. It did not, as I have shown, escape the observation of General Roy, that air, in contact with water, expanded much more than dry air ; and, from well-conducted experiments, he ascertained that the expansion was greater for equal increments as the temperature rose. From the mean results which he obtained, the following increasing rates of expansion were derived:—1000 parts of air, in contact with water, and under a pressure of 32.18 inches, expanded for each degree

From 0 — 32	. . . .	2.22799
32 — 52	. . . . . .	2.58800
52 — 72	. . . . . .	2.97228
72 — 92	. . . . . .	3.63194
92 — 112	. . . . . .	4.91072
112 — 132	. . . . . .	6.86550
132 — 152	. . . . . .	9.89494
152 — 172	. . . . . .	12.04087
172 — 192	. . . . . .	17.88344
192 — 212	. . . . . .	19.22470

I am indebted to M. Gay-Lussac for the following clear way of explaining the expansion of a dry gas, by the admission of aqueous vapour, and formula for the calculation of its effects:—

Let us suppose that the gas in contact with the water, in an inextensible vessel, has an elastic force equal to $p + f$ (p being the pressure of the atmosphere, and f the force of the vapour.) If the vessel should now become extensible, it would dilate until the interior pressure became equal to the exterior; so that, as f is constant, the gas will expand, until its elasticity become equal to $p - f$, and the volumes being in inverse proportion to the compression. v, the volume of air before its mixture with the vapour, is to V, its volume after mixture, as $p - f : p$; that is to say,

$$v : V :: p - f : p.$$

So that $V = v\left(\frac{p}{p - f}\right)$; or if $v = 1$, $V = \frac{p}{p - f}$.

Thus, if it be required to know the expansion which would take place in dry air, by placing it in contact with water, and raising its temperature from 0° to 32°, the force of vapour at that temperature, by Dr. Ure's table, is 0.216 ; therefore

29.784 : 30.000 :: 1.00000 : 1.00725.

This is the expansion which would arise from vapour only: to this we must add the expansion which would take place from the rise in temperature. Now .00223 (the expansion, per degree, for the bulk at 0°) × 32 = 0.7802 ; which, added to 1.00725, makes the total expansion 1.08527.

But the expansion which vapour causes in air, is not precisely similar to that occasioned by heat ; for while it dilates its parts, it adds its own weight to the mixture. Let it be required to know the specific gravity of air at 32°, saturated with vapour, compared with dry air, at the same temperature. Call the latter 1.00000 : the quantity of expansion will be .00725 ; which, considered alone, would reduce the specific gravity to .9928. But the weight of a cubic foot of air, under the conditions above named, is 558.131 grains, and the weight of a cubic foot of vapour at 32°, is 2.539 grains ; which, the former being 1.00000, will be nearly .00455 ; and which, added to the .99280 before obtained, will give .99735 for specific gravity sought.

Upon this principle I have constructed the following table, by means of which the specific gravity of any mixture of atmospheric air and aqueous vapour from 0° to 90°, may readily be found with sufficient precision. I have made air, under a pressure of 30 inches of mercury and at the temperature of 32°, the standard of comparison. The first column contains the degrees of Fahrenheit's thermometer ; the second shows the quantity due to each degree of heat, to be subtracted, or added to the original bulk, according as the temperature is above or below the standard; the third exhibits the expansion of volume occasioned by vapour of the respective degrees of elasticity appropriate to the several degrees of heat, and is always to be added; the fourth is the correction to be applied for the weight of the vapour, and is constantly to be added to the specific gravity corrected for the expansion; and the fifth is the correct specific gravity, supposing the air saturated with moisture at the given temperature.

TABLE III.—*For finding the Specific Gravity of any Mixture of Air and Aqueous Vapour, at Mean Pressure, from* 0° *to* 90°.—*Dry Air at* 32° *Temp. and* 30 *Inches' Pressure, being* = 1.0000.

Temp.	Alteration of Volume from Heat.	Alteration of Volume from Vapour.	Increase of Density from Weight.	Correct Specific Gravity of Saturated Air.
0	−.06666	+.00227	+.00153	1.0703
1	−.06458	+.00237	+.00159	1.0679
2	−.06249	+.00247	+.00166	1.0655
3	−.06041	+.00257	+.00172	1.0631
4	−.05833	+.00267	+.00179	1.0607
5	−.05624	+.00277	+.00185	1.0583
6	−.05416	+.00287	+.00191	1.0559
7	−.05208	+.00298	+.00197	1.0536
8	−.04999	+.00308	+.00204	1.0512
9	−.04791	+.00318	+.00210	1.0489
10	−.04583	+.00328	+.00216	1.0466
11	−.04374	+.00344	+.00224	1.0442
12	−.04166	+.00358	+.00234	1.0419

TABLE III. *continued.*

Temp.	Alteration of Volume from Heat.	Alteration of Volume from Vapour.	Increase of Density from Weight.	Correct Specific Gravity of Saturated Air.
13	−.03958	+.00372	+.00243	1.0396
14	−.03749	+.00385	+.00251	1.0373
15	−.03541	+.00398	+.00260	1.0350
16	−.03333	+.00412	+.00268	1.0327
17	−.03124	+.00425	+.00276	1.0304
18	−.02916	+.00439	+.00284	1.0282
19	−.02708	+.00452	+.00292	1.0260
20	−.02500	+.00469	+.00302	1.0239
21	−.02291	+.00489	+.00314	1.0215
22	−.02083	+.00509	+.00327	1.0194
23	−.01874	+.00529	+.00339	1.0171
24	−.01666	+.00549	+.00351	1.0148
25	− 01458	+.00570	+.00363	1.0125
26	−.01249	+.00590	+.00375	1.0104
27	−.01041	+.00610	+.00387	1.0082
28	−.00833	+.00631	+.00399	1.0061
29	−.00624	+.00651	+.00411	1.0041
30	−.00416	+.00671	+.00423	1.0017
31	−.00208	+.00698	+.00439	.9995
32	.00000	+.00725	+.00454	.9973
33	+ 00208	+.00752	+.00471	.9952
34	+.00416	+.00779	+.00486	.9927
35	+.00624	+.00806	+.00502	.9909
36	+.00833	+.00834	+.00518	.9887
37	+.01041	+.00864	+.00533	.9867
38	+.01249	+.00889	+.00549	.9846

TABLE III. *continued.*

Temp.	Alteration of Volume from Heat.	Alteration of Volume from Vapour.	Increase of Density from Weight.	Correct Specific Gravity of Saturated Air.
39	+.01458	+.00915	+.00564	.9824
40	+.01666	+.00942	+.00580	.9804
41	+.01874	+.00983	+.00604	.9783
42	+.02083	+.01024	+.00627	.9761
43	+.02291	+.01064	+.00650	.9741
44	+.02499	+.01105	+.00674	.9720
45	+.02708	+.01146	+.00697	.9699
46	+.02916	+.01187	+.00720	.9679
47	+.03124	+.01228	+.00743	.9658
48	+.03333	+.01269	+.00766	.9636
49	+.03541	+.01310	+.00789	.9616
50	+.03749	+.01351	+.00803	.9596
51	+.03958	+.01399	+.00839	.9575
52	+.04166	+.01447	+.00864	.9555
53	+.04374	+.01502	+.00896	.9535
54	+.04583	+.01557	+.00926	.9514
55	+.04791	+.01612	+.00957	.9494
56	+.04999	+.01667	+.00987	.9474
57	+.05208	+.01723	+.01017	.9453
58	+.05416	+.01784	+.01051	.9433
59	+.05624	+.01843	+.01083	.9414
60	+.05833	+.01902	+.01114	.9394
61	+.06041	+.01961	+.01146	.9374
62	+.06249	+.02020	+.01178	.9354
63	+.06458	+.02093	+.01217	.9334
64	+.06666	+.02166	+.01256	.9314

TABLE III. *continued.*

Temp.	Alteration of Volume from Heat.	Alteration of Volume from Vapour.	Increase of Density from Weight.	Correct Specific Gravity of Saturated Air.
65	+ 06874	+.02239	+.01295	.92 75
66	+.07083	+ 02312	+.01334	.92
67	+.07291	+.02386	+.01372	.92
68	+.07499	+.02466	+.01415	.9235
69	+.07708	+.02546	+.01457	.9216
70	+.07916	+.02634	+.01503	.9196
71	+.08124	+.02725	+.01551	.9177
72	+.08333	+.02817	+.01598	.9157
73	+.08541	+.02912	+.01648	.9137
74	+.08749	+.03011	+.01699	.9117
75	+.08958	+.03114	+.01752	.9098
76	+.09166	+.03221	+.01810	.9079
77	+.09374	+.03327	+.01861	.9059
78	+.09583	+.03438	+.01916	.9039
79	+.09791	+.03548	+.01973	.9020
80	+.09999	+.03663	+.02030	.9001
81	+.10208	+.03781	+.02090	.8982
82	+.10416	+.03903	+.02150	.8962
83	+.10624	+.04029	+.02213	.8943
84	+.10833	+.04159	+.02277	.8924
85	+.11041	+.04293	+.02343	.8905
86	+.11249	+.04431	+.02411	.8885'
87	+.11458	+.04573	+.02486	.8867
88	+.11666	+.04716	+.02549	.8847
89	+.11874	+.04858	+.02618	.8828
90	+.12083	+.05005	+.02688	.8809

To find the specific gravity of any mixture of air and aqueous vapour by means of this table, we must proceed as follows:—Note the temperature and the point of condensation by the hygrometer; if they coincide, that is to say, if the air be in a state of saturation, we shall find the specific gravity required in the fifth column opposite to the proper degree of heat in the first column. If the point of condensation be below the temperature, we must look for the amount of the alteration of volume due to the heat in the second column, and for the expansion due to the vapour in the third column. Add these together, if they have like signs; or subtract one from the other if they have different signs. As the volume corrected by this quantity is to the original volume, so is the standard specific gravity to the specific gravity as affected by the expansion or contraction. To this must be added the increase of weight due to the vapour in the fourth column, and the result will be the correct specific gravity sought.

For example:—If you wish to know the specific gravity of a mixture of air and vapour of the temperature of 60°, and of which the dew-point is 40°, we find in the second column opposite to 60° the number +.05833, and in the third column opposite to 40° we have + .00942: the sum of which is + .06775; therefore

$$1.06775 : 1 :: 1 : 93659.$$

In the fourth column opposite to 40° we find

$$+.00580, \text{ and } .93659 + .00580 = .94239;$$

which is the correct specific gravity under the assumed circumstances.

The application of this Table to barometrical

mensurations is sufficiently simple. For this purpose, with the usual operations at the upper and lower stations, must be combined simultaneous observations of the dew-point, by means of the hygrometer; and the approximate height, deduced in the common way, may, at once, be corrected for temperature and moisture, by the specific gravity of the air so obtained. As the specific gravity of the air at the time of the experiment is to 1.00000, the standard, so will the approximate height be to the real height.

If I am correct in my previous view of the constitution of the atmosphere, this is at once the most simple and correct method of applying the correction, both for temperature and vapour. By frequent observations of the hygrometer, in ascending a mountain, some estimate may be formed of the depth of the different beds of vapour which are mingled with the atmosphere, and a nearer approximation may be made to the mean temperature, by as frequently consulting the thermometer, than by assuming that regular gradation which two observations, at the bottom and top, suppose. If, upon ascending a height, I were to find that the dew-point was stationary during two-thirds of the ascent, and that during the remaining third it had fallen ten degrees, I should calculate the specific gravity of the atmospheric column, upon the supposition that two-thirds of it were affected by the expansion due to the elastic force of vapour of the higher degree, and one-third of the lower degree; and in the same manner I would calculate the expansion of the heat, by forming an estimate of the depth of the strata in which the thermometer indicated a steady temperature.

I shall now suppose a case, in which all the proper observations have been made, for the purpose of showing more distinctly the manner in which I propose to apply the table of correction.

Barom. at lower station		29.528	Temp. of mercury	58°
upper ditto		28.161	. . . . . .	51
Deduct for temp. of	29.528			
mercury	.084			
		29.444	Logarithm	.4689378
	28.161			
Ditto ditto	.060			
		28.101	Logarithm	.4487063
			Approximate height in fathoms	202.315
				× 6
			Ditto ditto in feet ..	1213.890

Temp. of air at lower station	55	Dew-point	40°	
Ditto ditto upper ditto	$51\frac{1}{2}$	Ditto	38	
Mean	$53\frac{1}{4}$			

Expansion of air at 53° per Table		.04374
Expansion for vapour at 40°	.00942	
at 38°	.00889	
Mean		.00915
Total expansion		.05289
105289 : 100000 :: 100000 :		94976
Increase of density for vapour at 40°. .	.00580	
at 38°. .	.00549	
Mean		.00564
Correct specific gravity. .		95540

	S. G. of air.		Standard.		Approx. height.		Correct height.
Then	.9554	:	10000	::	1214	:	1270

Or the correction may at once be made upon the comparative sp. grav. of mercury and air, and the height of the homogeneous atmosphere.

	S. G. of air.		Standard.		S. G. of mercury.		
For	.9554	:	10000	::	10435	:	10922

and 10922 × 30 = 327660 inches, or 27305 feet.

Correct height of the homogeneous atmosphere of the specific gravity found by the observations.

	Modulus of Logarithm.		Diff. of Logarithm.		Homogen. Atmosph.		Correct Height.
Then	.4342945	:	202315	::	27305	:	1272.

The hygrometer may be applied also to artificial atmospheres, and experiments upon confined air. Plate 1, Fig. 2, represents a receiver prepared for this purpose. A hole is drilled in its side, through which the tube proceeding from the ball within it, containing the thermometer, is passed, and welded with the tube proceeding from the other ball on its exterior, by means of a lamp; the stem is secured in the side of the glass with cement, the ether boiled, and the capillary opening closed, as before directed. The external ball is then to be covered with muslin: by this arrangement the evaporation from the latter produces a corresponding degree of cold upon the internal ball, which will measure the quantity of vapour included, by the precipitation, which may readily be marked. In delicate experiments a lighted taper, in a glass lantern, placed behind the bulb of the instrument, renders the deposition more easily visible, and ensures accuracy.

The hygrometric properties of any substance may thus be readily measured, by placing it under the receiver, and marking the absorption of the vapour.

Before I enter into the detail of experiments which I have made with the hygrometer under consideration, the results of which I cannot help indulging a hope may be found to be interesting to science, I feel myself called upon, rather unwillingly, to say something of the merits of the instrument, in comparison with others which have been intended to answer the same purposes. I was induced, at first, to hope that the universally acknowledged precision of the principle upon which its indications were founded, would have ensured for it a general adoption : and I hastened to communicate my observations in a more incomplete state than I should otherwise have done, from a conviction that the great value of the invention must be derived from the number, extent, and comparison of experiments performed with it by different observers, in different situations. Time, however, is necessary to overcome the force of prejudice, and some allowance must be made for the predilections of habit. I ought not, perhaps, to be surprised that continental philosophers have been slow in adopting the invention, when I look to the obstacles which have opposed its progress in this country. It was many years before the leaders of the scientific world even heard of its existence.

Professor Leslie, in an article upon meteorology, in the *Supplement to the Encyclopædia Britannica*, published three years after my first paper on the hygrometer, in one solitary remark upon the experiment of Le Roi, and the dew-point, observes, "*Could this method have been easily and nicely reduced to practice,* it might certainly have furnished an accurate estimate of the hygrometer, and state of the atmosphere."

I am happy to record this important testimony to the correctness of the principle, and have only to regret that the humble fame of the individual has been insufficient to attract the Professor's attention to means, which may surely be deemed *easy* and *nice*, of attaining an end, of the importance of which he is so thoroughly impressed. I am, doubtless, in courtesy, bound to suppose that Mr. Leslie would not be inclined to defend his encyclopædiacal knowledge at the expense of his candour.

It would never have occurrred to me to enter into a comparison of my hygrometer with the hygroscopic contrivances which have been hitherto in use; for I conceived that the vagueness and fallacy of their indications, their gradual and necessary deterioration, their liability to derangement, and accidental injury, had been universally admitted; and I thought that an instrument to measure the portion of humidity which a given portion of air holds, or is capable of sustaining, had been an acknowledged desideratum in physics. But here again I was deceived. I have lately learnt, with no small degree of surprise, that there are some, even amongst those whose rank in science give weight and importance to their opinions, who prefer the observations of a hair, a gut, or the beard of an oat. The only reason assigned for this preference is the time which is occupied in taking an observation with my hygrometer, while mere inspection is sufficient to ascertain the indications of the others. To this I reply, that one accurate observation upon fixed and certain principles is worth a thousand uncertain approximations; and that, therefore, if it were true that it requires much time for its management, the infallibility of the result should have ensured its adoption. But this waste of time must be a mere gratuitous assumption of those who never have fairly tried the experiment, for I speak from three years' experience, during which time I have used the instrument at least three times a day, when I assert that it requires less time to observe, in a proper manner, with the hygrometer, than with the barometer. To this we may also add, setting aside the uncertainty of the calculations, the time which is required to reduce the observations of the other instruments by *formulaic* processes to the dew-point, which must be performed before they can be applicable to any accurate purpose; a point at which we arrive at once with perfect precision by mere inspection.

The Editors of the *Bibliothèque Universelle*, of Geneva, in their number for March, 1820, have given an account of my invention, and have detailed their reasons for preferring the hygrometer of their illustrious countryman De Saussure. In replying to their observations, I shall, it is to be presumed, answer the strongest statement that can be made in favour of hygroscopic substances.

In the first place, however, I am proud to record so strong a testimony to the accuracy of the combination as the following:—"On peut ne pas adopter toutes les théories de l'auteur, ni partager sa prédilection pour l'appareil qui fait l'objet principal de son mémoire; mais *on ne peut disconvenir que cet appareil, tel qu'il est construit par M. Newman, fonctionne admirablement.*" The learned editors will pardon me, if I endeavour, by removing their objections, or rather their predilections, to make them absolutely partake of my preference for the instrument.

"Il est à présumer," say they, "que l'auteur ne faisant mention nulle part dans son mémoire de l'hygromètre à cheveu, du feu De Saussure, n'en avait aucune connoissance; fait assez étrange, vû la réputation qu'a acquise et que mérite à fort juste titre cet instrument pour toutes les recherches délicates. Il est pour le moins aussi sensible que celui de l'auteur; et pour la commodité du transport et de l'usage soit à l'air libre, soit en vases clos, l'hygromètre à cheveu l'emporte beaucoup. Il faut toujours faire une éxperience avec celui de l'auteur lorsqu'on veut connoître l'état hygromètrique de l'air; il faut une provision d'éther, etc. Avec celui de De Saussure, au contraire, il suffit de le regarder; en observant aussi le thermomètre dont les indications doivent toujours marcher pareillement à celles de l'hygromètre, ainsi que l'a prescrit soigneusement l'auteur dans son *Essai sur l'Hygrométrie*, l'un des fruits les plus remarquables de sa sagacité et de son génie."

It would, indeed, have been strange, had the presumption been correct, that I was totally unacquainted with the instrument invented by that indefatigable philosopher. Long have I been an humble admirer of his sagacity and genius, and to no work have I been more indebted for useful instruction on the subject of which it treats, than to the Essay above referred to. My reason for not making mention of the hair-hygrometer of De Saussure, was, as I have before stated, the conviction on my mind of the general admission of the inadequacy of any application of organic substances to the required accuracy of the purpose. I had selected one, as the best contrivance of this nature, to elucidate this point by contemporaneous observations with my own instrument; and the editors of the *Bibliothèque Universelle*, themselves, in recording my opinion, " on verra combien ses indications sont vagues et peu concluantes," add, " nous ne sommes pas très éloignés de cette opinion." Now, I must own, that I am quite at a loss to conceive any objection that can apply to the whalebone, that does not equally affect the hair as an accurate measure of vapour. But I shall prefer supporting this conclusion by the authority of others, rather than by any arguments of my own; especially, as I think, that I can produce authority, which the candour of the editors themselves will allow to be conclusive.

And let us hear the *Bibliothèque Universelle* itself upon this very subject. First, as to the proper construction of the instrument.—" Il en est peu qui exigent autant que l'hygromètre de l'adresse et des connoissances dans l'artiste qui l'entrepend; il devroit être physicien, mécanicien de tête et de main, et même un peu chimiste. La facilité de se procurer la substance hygrométrique qui fait l'ame de l'instrument à tourné à piège: on a cru que partout où l'on pouvoit se procurer des cheveux on fabriqueroit aisément les hygromètres; à la bonne heure, s'il s'agit d'hygromètres quelconques; mais on n'en obtient de réguliers, comparables, durables, que de la main d'un artiste expérimenté." Next, as to its permanency and the consequent reliance that may be placed upon its indications. " Dans ceux qui ont longtems éprouvé les inclémences de l'air, le cheveu acquiert un plus grande susceptibilité d'extension, et pour conserver à l'instrument une marche bien uniforme il seroit à-propos de changer le cheveu tous les deux ans."—*Bib. Univ.* Avril, 1819.

Baron de Humboldt, the celebrated philosopher and traveller, who is equally distinguished by his accuracy of observation, and by his philosophic generalizations, and who has had opportunities of making observations upon this subject which no other person ever yet enjoyed, and no other ever was more competent to appreciate, thus speaks of hygrometers in general, and of De Saussure's and De Luc's in particular *.

" We know, by very accurate experiments, the

* De Humboldt's Travels, translated by Helen Maria Williams. Vol. ii. p. 84, *et seq.*

capacities of saturation of the air at different degrees of the thermometer; but the relations which exist between the progressive lengthening of a hygroscopical body, and the quantities of vapour contained in a given space, have not been appreciated with the same degree of certainty. These considerations have induced me to publish the indications of the hair and whalebone hygrometers just as they were observed, marking the degree shown by the thermometers connected with these two instruments.

"As the fiftieth degree of the whalebone hygrometer corresponds to the eighty-sixth degree of the hair hygrometer, I made use of the first at sea and in the plains, while the second was generally reserved for the dry air of the Cordilleras. The hair, *below the sixty-fifth degree* of Saussure's instrument, indicates, by great variations, *the smallest changes of dryness*, and has, besides, the advantage of putting itself more rapidly into a state of equilibrium with the ambient air. De Luc's hygrometer acts, on the contrary, with extreme slowness; and on the summit of mountains, as I have often experienced to my great regret, we are often uncertain whether we have not ceased our observations before the instrument had ceased its movement. On the other hand, this hygrometer, furnished with a spring, has the advantages of being strong, *marking with great exactness, in very moist air*, the least increment of the quantity of vapour in solution, and acting in all positions; while Saussure's hygrometer must be suspended, and is often deranged by the wind, which raises the counterpoise of the index. I have thought that it might prove useful to travellers to mention in this place the results of an experience of several years." "*Notwithstanding the doubts which have been raised in these latter times respecting the accuracy with which hair or whalebone hygrometers indicate the quantity of vapours mingled in the atmospheric air ;* it must be admitted, that even in the present state of our knowledge, these instruments are highly interesting to a naturalist, who can transport them from the temperate to the torrid zone, from the northern to the southern hemisphere, from the low regions of the air which rests on the sea, to the snowy tops of the Cordilleras."

"I have never been able to reduce the hair or whalebone to the degree of extreme siccity for want of a portable apparatus, which I regret not having made before my departure. I advise travellers to provide themselves with a narrow jar containing caustic potash, quick-lime, or muriate of lime, and closed with a screw, by a plate on which the hygrometer may be fixed. This small apparatus would be of easy conveyance, *if care were taken to keep it always in a perpendicular position.* As, under the tropics, Saussure's hygrometer generally keeps above 83°, a frequent verification of the single point of humidity is most commonly sufficient to give confidence to the observer. Besides, in order to know on which side the error lies, we should remember that old hygrometers, if not corrected, have a tendency to indicate too great dryness."

Mr. Leslie, in his *Essay upon the relations of air to heat and moisture,* makes the following remarks upon the same subject. "But these substances (*viz.*, hygroscopic substances), especially the harder kinds of them, unless they be extremely thin, receive their impressions very slowly, and hence they cannot mark with any precision the fleeting and momentary state of the ambient medium." "The expansion of the thin cross-sections of box, or other hard wood, the elongation of the human hair, or a slice of whale-bone, and the untwisting of the wild-oat, of cat-gut, of a cord or linen thread, and of a species of grass brought from India, have, at different times, being used with various success. But the instruments so formed are either extremely dull in their motions, or, if they acquire greater sensibility from the attenuation of their substance, they are likewise, rendered the more subject to accidental injury and derangement, and all of them appear to lose, in time, insensibly, their tone and proper action."

But it is to the Essay of M. de Saussure himself, that I appeal with the most confidence, for the confirmation of this opinion. It is replete with acknowledgments of the obvious defects of instruments constructed upon this principle; defects which it was impossible that a mind like his could overlook or attempt to conceal: and it proves fully that his sagacity and genius were tasked to the utmost, to diminish the sources of uncertainty which it was out of his power wholly to remove. Any person, who had not seen the minute instructions given by this able philosopher for the construction of his hygrometer, would be surprised at the nicety required in its adjustment. The mere preparation of the hair is a process of great delicacy and uncertainty. It is previously exposed to an alkaline lixivium, upon the due strength and regulated application of which depends its most valuable properties; and hairs which have been unequally exposed to this action are no longer fit for comparison with one another. "Les cheveux n'ont une marche parallèle que quand ils sont également lessivés." So that it would be impossible for an artist in London, although he were "*physicien, mécanicien de tête et de main, et même un peu chimiste,*" with the most scrupulous attention to the directions contained in this Essay, to construct an instrument which should range with one made in Geneva, unless he had the means of actual comparison.

But after all the care which the ingenuity of such a philosopher could devise (and none but such a philosopher could be competent to take such precautions), thus guardedly and candidly does the inventor speak of the best instruments. "Quant à la comparabilité des hygromètres construits avec cette substance je puis dire que deux ou plusieurs de ces instruments, faits avec des cheveux semblablement préparés, gradués sur les mêmes principes, et exposés ensuite aux mêmes variations d'humidité et du sécheresse, ont des marches *que l'on peut nommer* parallèles. Je ne dirai cependant pas qu'ils indiquent toujours tous le même degré, mais que leurs écarts vont rarement audelà de deux degrés. Si après que deux hygromètres auront séjournés, pendant long-temps dans un air très sec, par exemple, au quarantième degré de ma division, on en porte un dans un air encore plus sec, qui le fasse venir, je suppose, à trente, et que pendant ce temps-là, l'autre ait été porté dans un air un peu moins sec, par exemple à cinquante degrés ; qu'ensuite on les replace tous les deux dans l'air où ils étoient d'abord, *ils ne reviendront ni l'un ni l'autre à quarante ;* celui qui vient de l'air le moins sec restera à quarante-deux ou quarante-trois ; et celui qui vient de l'air le plus sec ne montera qu'à trente-sept ou trente-huit." "Cet Hygromètre a l'inconvénient de ne pas revenir bien exactement au même point lorsqu'on l'agite un peu fortement, ou qu'on le transporte d'un lieu dans un autre, parceque le poids de trois grains qui tient la lame d'argent tendue, ne peut pas la ployer assez exactement pour la forcer à se coller toujours avec la même précision contre l'arbre autour du quel elle se roule : or on ne peut pas augmenter sensiblement le poids sans des inconvéniens plus grands encore. D'ailleurs si le cheveu est trop long, le vent, lorsqu'on observe en plein air, a trop de prise sur lui, et communique ainsi à l'aiguille des oscillations incommodes."

The relation of the degrees of this hygrometer, to the actual quantity of vapour in the air, is moreover very far from having been determined ; "C'est ce que j'ai tenté de faire," says the inventor, "pour mon hygromètre ; mais on verra que ce travail difficile est encore bien loin de sa perfection."

When we add to these admissions the disturbing influence of heat, which is so great, that the mere approach of the hand causes a sensible movement towards dryness; the adhesion of dust and spiders' webs ; the choaking of the pivot of the wheel; and the possibility of friction from the index; we shall have some notion of the sources of error in this instrument, which the great philosopher, its inventor, himself, has pointed out and laboured to modify.

It is thus that I reply, or rather it is thus that universal experience replies, to the "pour le moins aussi sensible," of the editors of the *Bibliothèque Universelle.* As to the "Commodité du transport et de l usage," I must remark, that the whole of the new apparatus packs in a box, which may very conveniently be carried in the pocket; and although each observation with it may, in strictness, be called an experiment, yet that infinitely less time is required to make this experiment, than would be necessary to assure an observer, with either the hair or whale-bone hygrometer, that "the instrument had ceased its movement." The inconvenience of carrying a supply of ether, may, I think, fairly be set against that of an apparatus for rectifying the instruments described by De Humboldt, and which he considers necessary to give confidence in their indications.

But upon this point I shall avail myself of direct evidence of the most unexceptionable nature.

Mr. Caldcleugh, in his "Observations in Brazil, and on the Equator," *(Jour. Royal Inst.* vol. xiv. p. 46,) remarks, "When I commenced using the instrument, I was almost afraid to touch it, from its apparent delicacy, but was soon convinced, from the many rude shocks it underwent, that it was stronger than I had imagined; more than common carelessness, indeed, is required to break it. I may be permitted to add, that I think no traveller will find any inconvenience from carrying this hygrometer or its accompaniment, a small stock of ether; the latter I usually placed among my linen."

Captain Sabine, also, after twelve months' experience between the tropics, during which time he daily made numerous observations, thus bears testimony to the same fact:

"I have great pleasure in remarking, that I found much less difficulty than I had anticipated in getting corresponding observations made with the hygrometer, on the correctness of which I could depend; the ingenuity in the principle of this instrument, and the simplicity of its application, together with the decisive nature of the results which it gives, independent of the labour, and, at best, the uncertainty of *formulaic* deduction, form its great advantage over the methods by evaporation, or the indications of hygroscopic substances: these particulars excite an interest in its trial, in persons to whom it was previously unknown, which is probably the reason that the distrust, which is almost always in the first instance expressed, of precision in the observation itself, is found to give way in practice so much sooner than might be supposed. It may be useful, also, to travellers in warm climates, to add a remark from my own experience, that in ascending elevations, or in journeying inland over rough roads, the ether carries perfectly well in a bottle in the waistcoat-pocket with a common cork capped with leather; and that the expenditure of ether altogether will probably fall much short of the estimate, as, with ordinary care, very little will be wasted." (*Journal Royal Inst.* vol. 15, p. 71.)

The quantity of ether expended by Captain Sabine during the year, fell something short of a pint.

The instrument, which I have presumed so strongly to recommend, shall be still further judged by the very competent authority to which I have been referred. M. de Saussure sums up, in his Essay, the qualities which a perfect hygrometer ought to possess; allowing, candidly, that his own falls very short of the perfection which he proposes. All I would ask is, if the one which I have invented fulfil all the conditions laid down as follows, that for the good of science it may be adopted as a standard by experimental philosophers.

"Un hygromètre seroit parfait: premièrement si les variations ètoient assez étendues pour rendre sensibles les plus petites différences d'humidité, et de sécheresse.

2. "Si elles ètoient assez promptes pour suivre pas-à-pas toutes celles de l'air, et pour indiquer toujours exactement son état actuel.

3. "Si l'instrument ètoit toujours d'accord avec lui-même, c'est-à-dire, qu'au retour du même état de l'air il se retrouvât toujours au même degré.

4. "S'il ètoit comparable, c'est-à-dire, si plusieurs hygromètres construits séparément sur les mêmes principes indiquoient toujours le même degré dans les mêmes circonstances.

5. "S'il n'ètoit affecté que par l'humidité ou la sécheresse proprement dites.

6. "En-fin si ces mêmes variations ètoient pro-

portionnelles à celles de l'air, en-sorte que dans des circonstances pareilles, un nombre double ou triple de degrés indiquât constamment un quantité double ou triple de vapeurs."

But there is yet another method of estimating the quantity of moisture at any time existing in the air, with which my own has been brought into comparison, and which is not liable to the same class of objections as those which we have been just considering; I mean that of a comparison between the temperatures of a moist and a dry thermometer. Dr. Hutton was the first who conceived the idea of applying such an observation to the purposes of hygrometry "I used to amuse myself," says he, "in walking in the fields, by observing the temperature of the air with the thermometer, and trying its dryness by the evaporation of water. The method I pursued was this: I had a thermometer included within a glass tube, hermetically sealed: this I held in a proper situation until it acquired the temperature of the atmosphere, and then I dipped it into a little water, also cooled to the same temperature. I then exposed my thermometer, with its glass-case thus wetted, to the evaporation of the atmosphere, by holding the ball of the thermometer, or end of the tube in which the ball was included, towards the current of the air; I examined how much the evaporation from that glass tube cooled the ball of the thermometer which was included."

Now this simple observation, it is probable, may furnish the necessary *data* for solving the problem in all its particulars: but if it do, it is by means of abstruse calculations, and many delicate corrections, upon the nature of which philosophers are by no means agreed.

Mr. Leslie has founded upon it an instrument, upon which he has expended much ingenuity and research: it consists of an air-thermometer, one of the balls of which is covered with muslin, and kept moist; but, in departing from the simplicity of the original experiment, he has multiplied the sources of error, both in construction and observation. He has, moreover, substituted an arbitrary scale for that of the common thermometer; and his hygrometer possesses a deceptive sensibility, which is liable to be affected by more causes than those which can be taken into account.

The observation, however, in its most simple and unexceptionable form, is by no means so easy to make with accuracy, as might, at first sight, appear. It is almost impossible to take the heat of the air to any degree of nicety, without the observation being affected by the power of radiation; and, if radiant caloric be allowed to interfere, the conditions of the calculation fail. The temperature of evaporation is no longer that constant quantity which it is supposed to be, if dependant only upon the temperature of air, and is liable to fluctuations with every change of place, and every breath of wind. The density of air must be also taken into the account, and it is allowed that the cold produced by evaporation from the moistened bulb of a thermometer, must depend, in some measure, upon the height of the barometer.

It would be foreign to my purpose here to enter into the theory of the experiment, which would itself be a work of much length: it is sufficient to observe that such men as Mr. Leslie *, Mr. Anderson †, Mr. Ivory ‡, and M. Gay-Lussac §, after having bestowed immense labour upon the subject, all differ from one another most essentially in their formulæ, corrections, and final results; and the *hygrometer by evaporation*, has not been able to save Mr. Leslie from the conclusion that hydrogen gas " must, in similar circumstances, hold in solution seven times as much moisture as the atmospheric medium;" or Mr. Colebrooke ||, from the probable error of placing the dew-point on the coast of Africa, below 0° of Fahrenheit's scale. I cannot better dismiss the subject than in the words of M. Gay-Lussac,—

" En genéral on peut parvenir à connaître l'état hygrométrique de l'air, d'après le froid produit par l'évaporation; mais comme ce froid est variable avec la pression de l'air, sa température, son degré d'humidité, il faudrait des tables très-étendues pour le déterminer avec exactitude. J'avais voulu entreprendre ce travail, en répétant mes expériences sur le froid produit par l'évaporation, et en faisant de nouvelles; mais j'ai été rebuté par sa longueur, et le défaut de données suffisamment exactes, et surtout par la considération que l'ingénieuse procédé de Le Roy était susceptible d'une application plus facile et que, dans l'état actuel de la Physique, il étoit de beaucoup préférable."—*Ann. de Chim.* tom. 21, p. 91.

* Supp. Ency. Brit., Art. Meteorology.

† Edin. Ency., Art. Hygrometry.

‡ Phil. Mag., vol. lx., p. 81.

§ Ann. de Chimie, tom. xxi., p. 82.

|| Journal of the Royal Institution, vol. xxvii. p. 115 et seq.

The observations of the hygrometer with which I have been chiefly engaged, relate principally to meteorology; and I conceive that it will be better to reserve such remarks as I have to make upon them for another Essay, than to enter upon their discussion in this place. I shall, therefore, conclude this paper by detailing a few experiments only, which may either throw some light upon the action of the instrument, or may tend to facilitate its use.

Exp. 1.—With the thermometer at 60°, I found the point of condensation to be 50°: I then took a receiver, fitted with a hygrometer, and ground to the plate of an air-pump, whose capacity was fifty-six cubic inches. The condensation was produced very visibly under the glass at the same temperature. Now the quantity of vapour in a cubic foot of air, under the above conditions, was only 4.445 grains; therefore the quantity actually included in the receiver, could only be 0.144 grains; which will serve to prove the extreme delicacy of the instrument, as it distinctly indicated so small a quantity. The receiver was then, without changing its contents, slid over a vessel containing water. In an hour and a half, the external temperature remaining the same, the precipitation took place at 57°. At the expiration of another hour and a half, the affusion of ether upon the exterior ball, caused instantaneous condensation upon the interior one, shewing that saturation, at the existing temperature, had taken place. The bell-glass was now slid from the water, and placed over a glass containing a few drops of sulphuric acid. After remaining a quarter of an hour in this situation, a depression of temperature of 30° produced no mist upon the instrument.

Exp. 2.—The receiver was placed upon the plate of the air-pump with some water under it; the air was then exhausted as perfectly as possible. The barometer stood at 29.79 inches, the thermometer at 62°; the gauge of the pump at 29.20; to the latter should be added the pressure of the included vapour at 62°=.59 inch, which would make the gauge and the barometer exactly correspond. When ether was dropped upon the exterior ball, precipitation was instantaneous. Air was now admitted gradually, till the gauge fell to 14 inches: the point of condensation was not altered, neither was it affected by restoring the equilibrium completely.

Exp. 3.—Temperature 64°; point of condensation 61°. The air in the receiver was rarefied till a copious cloud was formed; the gauge then stood at 8.1 inches, and the point of condensation had fallen to 54°. When the glass had risen to 60°, the air was suddenly restored, and a copious dew was formed upon it; the exhaustion was next carried on, till the cloud which was formed had totally disappeared, and the gauge stood at 24.2 inches. No precipitation took place at a temperature of 34°; the air was gradually re-admitted, and the deposition took place with the hygrometer at 36°, and the gauge at 15 inches.

Exp. 4.—The receiver was filled with oxygen in contact with water, and afterwards with hydrogen; but the point of condensation was the same as when filled with common air, under the same circumstances. This, with Experiment 2, fully coincide with Mr. Dalton's view of the theory of mixed elastic fluids, and prove, indeed, that the gases are as *vacua* with regard to vapour; and that, where they happen to be mixed together, they exist as independent atmospheres.

Exp. 5.—Having absorbed all the vapour contained in the receiver by means of sulphuric acid, I placed it over some spirits of wine; after remaining some time in this situation, a few drops of ether upon the hygrometer produced an instant precipitation. The experiment was also made with ether, in the place of the spirits of wine, with the same results.

Exp. 6.—The temperature of a room being 45°, I found the point of condensation in it to be 39°. A fire was lighted in it, the door and windows carefully shut, and no one was allowed to enter: the thermometer rose to 55°, but the point of condensation remained the same. A party of eight persons afterwards occupied the room for several hours, and the fire was kept up: the temperature increased to 58°, and the point of condensation rose to 52°.

I must now refer to the subjoined journals for further exemplifications of the application of the hygrometer, and to the Essays upon the Constitution of the Atmosphere, and upon the Climate of London, for such general theoretical conclusions as I have thought myself entitled to draw from my own extended observations, and the kindly-communicated observations of my friends.

Whatever opinion may be formed of my reasoning upon these subjects, let the hygrometer be judged upon its own merits alone; and if it shall be found to be liable to no errors of construction, and no deterioration, from use or age; if its indications shall prove to be infallible, and strictly comparable, under all circumstances; and if, moreover, it be easy to observe, and its observations applicable without the trouble and uncertainty of *formulaic* calculations, I shall still hope, that, for the good of science, it may be generally adopted.

AN ESSAY

UPON THE

RADIATION OF HEAT IN THE ATMOSPHERE.

The radiation of heat is a subject of the utmost interest and importance, and the difficulties which surround it have exercised the ingenuity and industry of some of the greatest philosophers of modern times. Count Rumford, Professor Leslie, Sir Humphry Davy, Professor Prevost, Dr. Delaroche, and M.M. Du Long and Petit, have particularly distinguished themselves by their experiments and reasonings upon it; and the latter gentlemen, more especially, have demonstrated some of the laws of the distribution of heat, with mathematical precision.

With regard to the influence of this power upon the œconomy of nature, however, but little is at present known; and the elegant and demonstrative Essay of the late Dr. Wells upon the formation of dew, stands almost alone in exhibiting its important connexion with the welfare of the vegetable kingdom. His successful labours were directed to one particular branch of, what would appear to be, a very extended inquiry; for there can be little doubt that radiant caloric must have a direct and very important influence upon many of the processes of vegetation.

It is with a view of exciting some attention to a

subject which appears to me to be so well worthy of elucidation, and to suggest some experiments, which, to render them beneficial, require much perseverance and extensive co-operation, that I venture to bring forward some observations of my own, which, I am sensible, are in a very imperfect state; but to which I have devoted much attention during the last three years. I hesitate the less to do so, as I am enabled, by the kindness of my friend Captain Sabine, (whose zeal for science prompted him, amidst the laborious operations connected with more important objects, to devote much of his leisure time to the study of atmospheric phenomena,) to give them additional interest, by combining with them his experiments in tropical climates.

It has often struck me with surprise, that, in the numerous meteorological registers which are published in different parts of the world, no one has ever thought of including observations upon the intensity of the solar rays at different seasons of the year, and in different situations. It is well known to the agriculturist and the gardener, that without the direct influence of the sun, whatever may be the temperature of the air, the fruits of the earth seldom come to perfection. What, then, is the force of this important agent? what the modifications to which it is subject? and how is its energy spent, when screened by concrete vapours from the surface of the earth? Does its influence increase with the temperature of the air from the pole to the equator? or is the rapid vegetation of the arctic regions, during the short summer of those climates, dependant upon any compensating energy of its operation?

Before I attempt to answer these questions, I will propose another, which, many will be surprised to find, cannot be met with an immediate solution; which is, the maximum degree of heat to which a plant, or the parts of a plant, are subjected by exposure to a mid-day sun at midsummer in this climate?

Many persons have, at different times, exposed naked thermometers to the direct light of the sun, and marked their rise; but such trials have never been persevered in, or registered with any exactness. Nor were the means employed calculated to resolve the problem with any precision. Few of the rays would impinge directly upon the bulb of the instrument so placed, and all, but the direct rays, would be reflected from it. The results would necessarily vary with size and shape, and no two thermometers would, probably, agree in their indications.

There are, no doubt, in all the plants of the vegetable kingdom parts which are calculated to absorb all the radiant heat which strikes upon them; and therefore it is desirable to know, with a reference to this subject alone, the utmost amount of temperature which radiant matter is capable of producing.

My meteorological register includes a column for observations upon this point. They are complete from November 1820, to the end of December 1821, and from the beginning of May 1822, to the end of August of the same year. They were made by means of a register thermometer of large range, having its bulb covered with black wool, and placed upon a south border of garden-mould, with a full exposure to the sun. The thermometer did not rest upon the earth, but was supported about an inch above it. The arrangement was by no means unobjectionable, but the irregularities, to which it was liable, would, it was hoped, be, in a great measure, balanced by the multitude of the observations. The maximum heat of the sun's rays during the day was thus measured and entered in the journal. The following Table presents us with the average intensity of the solar radiation for every month in the year, or the mean greatest height of the black thermometer above the surrounding medium, together with the utmost intensity observed in the same periods. The first column exhibits the month, the second the mean maximum temperature of the air, the third the average effect, and the fourth the maximum energy of the sun's light.

TABLE I. *Shewing the mean maximum Temperature of the Air, with the mean and maximum Power of the Sun, for every Month of the Year.*

	Mean Maxim. of the Air.	Mean Maxim. Force of Solar Radiation	Maximum Force of Solar Radiation
January . . .	39.6	4.4	12.
February . . .	42.4	10.1	36.
March	50.1	16.	49.
April	57.7	28.1	47.
May	62.9	30.5	57.
June	69.4	39.9	65.
July	69.2	25.8	55.
August	70.1	33.1	59.
September . .	65.6	32.7	54.
October . . .	55.7	27.5	43.
November . .	47.5	6.7	24.
December . .	43.2	5.4	12.

Hence it appears that, as it might have been predicted, the power of solar radiation follows the course of the sun's declination. The maximum intensity and effect occur in June, while the greatest mean temperature of the atmosphere does not take place till July. This arrangement, no doubt, has an important influence upon the processes of fructification in the vegetable kingdom. Agriculturists are well aware of the advantage of direct solar heat in the flowering of wheat, and other corn-crops; an advantage which is never compensated by any elevation of temperature under a clouded sky. A table, similar to that above given, founded upon the experience of several years, would furnish a very valuable standard of comparison, and the causes of fruitful and unfruitful seasons would, no doubt, be found to be intimately connected with the particulars of which it would be composed. For example, it will be seen in the register, that in the very fruitful year of 1822, the force of the sun's radiation in May was 7°, and in June 5°, above the corresponding months of the year 1821, in which the crops of corn were universally blighted and mildewed. The discordances above exhibited would also be found to vanish in a more extended average, and a more regular progression would be elicited from the balance of disturbing causes.

I have also been at some pains to ascertain the progression of radiation from the sun from its rising to the meridian, and from the meridian to its setting. The following are the details of a series of observations, made for this purpose in the month of June, 1822. The day was perfectly calm and cloudless, and the atmosphere so clear that the disc of the moon was visible throughout the day. The dew-point, by the hygrometer, was stationary at 57°, and only a few light *cirri* were discernible in the south-east quarter of the heavens.

TABLE II. *Shewing the Progress of Solar Radiation from Morning to Evening.*

Time.	Thermometer. In Sun.	Thermometer. In Shade.	Thermometer. Difference.
A. M. 9	93	68	25
9½	103	69	34
10	111	70½	40½
10½	119	71	48
11	124	71½	52½
11½	125	72½	52½
12	129	73	56
P. M. 0½	132	74	58
1	141	74½	66½
1½	140	75	65
2	143	75½	67½
2½	138	76	62
3	138	76½	61½
3½	132	77	55
4	124	76	48
4½	123	77	46
5	112	76	36
5½	106	75	31
6	100	73	27
Means	124¾	73½	51¼

The mean results of five series of experiments, conducted with every possible precaution, are contained in the following table, shewing the power of the sun's radiation from $9\frac{1}{2}$ A. M. to $6\frac{1}{2}$ P. M., in the month of June.

TABLE III. *Shewing the Progress of the Solar Radiation from Morning to Evening in June, upon an average of five Experiments.*

Time.	Force of Sun's Rays.
$9\frac{1}{2}$ A.M.	32
$10\frac{1}{2}$	46
$11\frac{1}{2}$	55
$12\frac{1}{2}$	63
$1\frac{1}{2}$ P.M.	65
$2\frac{1}{2}$	63
$3\frac{1}{2}$	58
$4\frac{1}{2}$	49
$5\frac{1}{2}$	35
$7\frac{1}{2}$	29

Some important questions now present themselves, which the present state of our knowledge will not, I fear, enable us to answer satisfactorily. It may, however, be of some use to point them out as objects of future investigation. As the mean effect of the sun's radiation upon the surface of the earth falls so much short of the impression which it is capable of producing, in what way is its energy spent? Is it absorbed or dissipated in mid air? How is the mean temperature affected by it? How does it modify the ascending gradation of temperature in the atmosphere? What is its influence upon the concrete vapour of the clouds? Is it not the source of partial and very unequal expansions and contractions in the aërial medium?

Before I proceed to detail some experiments which are connected (but slightly, I fear,) with these interesting considerations, I shall avail myself of Captain Sabine's kindness, to institute a comparison of the radiant power of the sun in different latitudes.

The first series of experiments to which I shall refer, were made in March, 1822, at Sierra Leone. The general state of the atmosphere was as follows:—

"The day commenced as usual, calm and clear; between a quarter and half-past ten the sea-breeze sprung up from the N.W., freshened at noon from the W.N.W., accompanied by a diffused haze. At one, P.M., cleared, the wind still freshening. At two, some very light clouds in the zenith, which clearing away before three, it became hot and oppressive in the sun, the sea-breeze gradually declining towards evening, and the land-wind setting in at half-past nine."

Observations were made with the thermometers now described.

"No. I. The mean of two thermometers, one with a silvered, the other with a blackened bulb, (differing never more than two or three tenths of a degree), freely suspended in the thorough draft in a store-house with open doors and windows every way, and with a veranda around open at the sides.

"No. II. A thermometer with a blackened bulb, suspended freely between the pillars of a transit instrument (the instrument itself being removed), about one and a half feet above the earthen parapet of the fort, and several feet above the general level of the ground, in a fair exposure to the sun and wind.

"No. III. A similar thermometer to the preceding in the same exposure, but *in vacuo* in a glass case.

"No. IV. A thermometer in the same glass case with the preceding, but having its bulb enclosed in a double case of polished silver, not in contact with the glass or bulb.

"No. V. A differential thermometer *in vacuo*; the sentient ball coloured dark, and the other enclosed in a double case of polished silver. The graduation of this instrument was on the millesimal scale; *i. e.*, the interval between the boiling and freezing of water, divided into 1000°. The three last were only exposed at the time of observation, being removed intermediately into the house, when the equilibrium in the bulbs of No. V. was gradually restored at the temperature indicated by No. I. These thermometers were registered, when the effect of the exposure had reached the maximum.

TABLE IV. *Experiments upon Solar Radiation at Sierra Leone.*

March the 2d.									
Time.	No. 1.	No. 2.	Difference.	No. 3.	No. 4.	Difference.	No. 5.	OBSERVATIONS.
A. M. 10	79.3	95.	15.7	110			70	
11	80.	93.	13.	109			78	
12	80.2	91.5	11.3	105			82	Haze.
P. M. 1	81.1	88.	6.9	109	102	7.	81	More clear, and wind freshening.
2	80.9	85.	4.1	109.5	102	7.5	64	Light clouds.
3	83.4	91.	7.6	118	107	11.	70	Clear.
4	82.9	90.	7.1	116	108	8.	53	Wind dying away.
5	81.4	83.5	2.1	100	98	2.		
$5\frac{1}{2}$	80.	81.5	1.5	86	86	0.		

The same thermometers were employed on the 4th of March, the variations of the weather being as nearly similar as can be described—the usual weather, in short, of the season.

TABLE V.—*Experiments upon Solar Radiation at Sierra Leone.*

March 4.							
Time.	No. 1.	No. 2.	Differ.	No. 3.	No. 4.	D iffe.	No. 5.
A.M. 9	80	95	15	110	102	8	69
10	80.5	93.	12.5	110	105	5	79
11	80.	94	14.	110	105	5	86
12	80.2	98.5	18.3	115	108	7	93
P.M. 1	80.8	96.	15.2	118	108	10	88
2	81.	97.	16.	118	110	8	77
3	83	90.5	7.5	113	105	8	69
4	82.5	89.5	7.–	109.5	102	7.5	59

The first, and most striking, result of these observations, is the very small comparative energy of the solar rays. All the means adopted to measure their effect concur in this conclusion. The utmost difference between a blackened thermometer in the sun and another in the shade was only 18.3°; and *in vacuo*, of one prepared to repel the radiant heat, and of another to absorb it, 11°. It is obvious that the whole difference between the first and third thermometer cannot be ascribed to radiation, for the latter, although placed in a rarefied medium, was still surrounded by attenuated air and aqueous vapour, the latter of which appeared in a pretty copious precipitation of moisture upon the sides of the glass. This atmosphere, however rare, was liable to become considerably heated under confinement. But even the maximum of this difference is only 38°.

The next series of experiments, which were made at Bahia, on the coast of Brazil, come into more immediate comparison with my own, and agree in the conclusion of the diminution of the force of radiation from a tropical sun. A mercurial register thermometer, having its bulb blackened and covered with black wool, was fully exposed to the sun on grass, and compared with a thermometer in the shade: the following Table exhibits the results.

TABLE VI.—*Experiments upon Solar Radiation at Bahia.*

	Sun.	Shade.	Difference.
July 24	114	82	32
25	123	82	41
26	124	83	41
27	123	83	40
28	95	78	17
29	115	78	37
30	127	80	47

Here the maximum effect was only 47°, with a nearly vertical sun; while the same influence, in our temperate climate in June, *in a medium not much less heated*, was 65°.

Captain Sabine instituted a third set of experiments, upon the same point, during his stay in the Island of Jamaica.

The thermometer, with black wool, was exposed to the sun on the vegetation by which Port Royal is surrounded. It is a tongue of sand, projecting a considerable distance into the sea, and overrun by the Tibullus Maximus, which was at the time in flower. The ball of the thermometer was in contact with the vegetation, and supported by it about ten inches off the ground.

TABLE VII.—*Experiments upon Solar Radiation at Jamaica.*

	Sun.	Shade.	Difference.
Aug. 25	122	86	36
26	123	87	36
27	122	86	36
28	122	86	36
29	123	86.5	36.5
30	123	86.5	36.5

A naked mercurial thermometer, suspended freely across between the upper branches of a stunted dead acacia, and expcsed to the sun, near the other thermometer, about four and a half feet above the ground, and not in contact with the tree, carefully observed at intervals of the fore and after-noon, from the 25th to the 30th of October, was never seen to rise higher than 92°. This point it usually attained at ten A.M., before the sea-breeze set in fell as the breeze commenced, but attained about the same height in the afternoon, although the breeze had freshened intermediately.

The indications of the differential thermometer, before described *in vacuo*, were as follow:

TABLE VIII. *Experiments upon Solar Radiation, at Jamaica.*

Date.	Hour.	Differential Thermometer.	OBSERVATIONS.
Oct. 24	12	88	Strong breeze—perfectly clear.
	P. M. 1½	88	Ditto ditto
26	A. M. 9½	82	No breeze—very clear.
	12	88	Fresh sea breeze—in 2 minutes.
29	12	90	Clear—little breeze—in 2 minutes.
	P. M. 2	86	Ditto—more breeze—in 2 minutes.
30	A.M. 10	88	Very clear—no breeze—in 2 min.
	P. M. ½	91	Very strong sea breeze—in 2 min.
	3½	74	Very strong breeze—in 2 minutes.
Nov. 3	A. M. 8	68	Calm and clear.
	9	82	Ditto ditto
7	A. M. 7	48	Ditto ditto

The smallness of the effect is no less striking from these results than from the last.

In looking over the interesting personal narrative of M. de Humboldt, (in which the inquirer, upon almost any subject, is sure to meet with valuable information) I find ample confirmation of these observations. At Cumana he remarks, "I have often endeavoured to measure *the power of the sun*, by two thermometers of mercury perfectly equal, one of which remained exposed to the sun, while the other was placed in the shade. The difference resulting from the absorption of the rays in the ball of the instrument *never* exceeded 3°. 7 (6°. 6 Fahr.) Sometimes it did not even rise higher than one or two degrees.—*Humboldt's Travels*, by H. M. Williams. Vol. ii. p. 58.

A fourth set of Captain Sabine's experiments, in the mountains of Jamaica, present a comparison of the greatest interest. The observations were made on the 31st of October, at Mr. Chisholm's house, situated on the summit of the Port Royal ridge, 4,000 feet above the sea. The woolled thermometer was laid upon the grass-plat about 100 yards from the house, and fairly exposed to the sun. In the forenoon, in intervals of the breeze, when the sky was clear, it rose above 130°; the thermometer in the shade, at the time being, 73°. The difference of 57° exhibits a much greater intensity of action than any that had been obtained at the level of the sea. The following observations of the differential thermometer *in vacuo*, in the same situation, confirm the conclusion.

TABLE IX. *Experiments upon Solar Radiation, upon the Mountains of Jamaica.*

Date.	Time.	Differential Thermometer.		OBSERVATIONS.
Oct$^{r.}$ 31.	A. M. 9	74	Maximum produced in	1½ Minute.
	10	100		1½ Minute.
	11	84		1½ Minute, light clouds.
	12	100		1½ Minute, very clear.

As it appears, I think, incontrovertible, from a comparison of Captain Sabine's experiments with my own, that the force of the sun's direct radiation decreases in approaching the equator, I became anxious to ascertain, if possible, whether an analogous contrary effect were observable in advancing towards the pole. In looking over the journal of the late Expedition for the discovery of a north-west passage, I found some observations, which tended much to establish this curious fact. At page 157 of that interesting narrative, Captain Parry observes—"On the 16th, (March), there being little wind, the weather was again pleasant and comfortable, though the thermometer remained very low. While it continued nearly calm, we observed the following differences in the temperature of the air in the shade, and in the sun; the latter were, however, noted by a thermometer placed under the ship's stern, which situation was a warm one, for the reasons before assigned." The difference of warmth in this situation had been before ascertained not to exceed 2° to 5°.

TABLE X. *Experiments upon Solar Radiation at Melville Island.*

Date.	Time.	Sun.	Shade.	Difference.
March 16	A. M. 9	+ 24	− 24	48
	10	+ 27	− 23	50
	11	+ 28½	− 22	50½
	12	+ 29	− 21	50
	P. M. 3	+ 19	− 13	32

Again, on the 25th of March, the thermometer in the sun being placed *at a distance from the ship*, and the weather very fine and calm.

TABLE XI. *Experiments upon Solar Radiation, at Melville Island.*

Date.	Time.	Sun.	Shade.	Difference.
March 25	12	+ 30	− 25	55
	P.M. 1	+ 17	− 22	39
	2	+ 25	− 22	47
	3	+ 21	− 22	43

Here it is seen that the sun had power to raise a thermometer, *which had not been prepared to receive its greatest impression*, 55° in the month of March, at Melville Island; the maximum effect in the vicinity of London, in the same month, *upon a thermometer covered with black wool*, being only 49°.

In Captain Scoresby's "Account of the Arctic Regions," there are also many remarks which powerfully confirm the same opinion. "The force of the sun's rays," he observes, "is sometimes remarkable. Where they fall upon the snow-clad surface of the ice or land, they are, in a great measure, reflected, without producing any material elevation of temperature; but when they impinge on the black exterior of a ship, *the pitch on one side occasionally becomes fluid, while ice is rapidly generated at the other;* or, while a thermometer placed against the black paint work, on which the sun shines, indicates a temperature of 80° or 90°, or, *even more*, on the opposite side of the ship, a cold of 20° is sometimes found to prevail. This remarkable force of the sun's rays is accompanied with a corresponding intensity of light."—Vol. I. p. 378.

To ascertain with precision the temperature denoted in the above extract, by the pitch becoming fluid, which appears to me to furnish the best measure of the force of the sun's rays, I tried the following experiment. I covered the bulb of a thermometer with pitch to the thickness of about $\frac{1}{10}$ of an inch, and suffered it to remain till it had become quite hard. I then held it at some distance from a fire, and noted the following points. At 100° Fahr., it began to soften. At 110°, it might be moulded into any form—and, from 120° to 136°, it rapidly approached fluidity, and, at the latter temperature, dropped off the ball. The degree denoted cannot, therefore, be placed lower than 120°; and if ice were forming at the same time in the shade, the force of the sun's radiation could not be less than 90°.

In the account of Captain Scoresby's last voyage to Greenland, a direct experiment in latitude 80° 19′, confirms the same conclusion.

"The sun broke through the clouds, and produced a powerful effect upon the temperature. At two, A. M., the thermometer was 3° or 4° below zero. At eight, A. M., it was + 6°, and at ten, A. M., about + 14° in the shade. But the genial influence of the sun was still more striking. In a sheltered air, it produced the feeling of warmth; the black paint work of the side of the ship, on which the sun shone, was heated to a temperature of 90° or 100°, and the pitch about the bends became fluid. Thus, while on one side, was uncommon warmth, on the opposite side, was intense freezing." Journal of a Voyage to the Northern Whale Fishery, p. 34. The radiating force of the sun must, therefore, have been 80° in the month of April.

With respect to the greater energy of the solar rays, upon the summit of a mountain, than upon a plain, I find that Monsieur de Saussure made some decisive experiments, which establish the same fact. In his "Voyages dans les Alpes," the following observations occur—"Je cherchais à Géneve un verre ardent assez petit pour qu'il n'eut précisément que la force nécessaire pour allumer de l'amadou. Je portai en suite le même verre et le même amadou sur le haut de Salève et je le vis là produire le même effet que dans la plaine *et même avec plus de promptitude.*" Tome II. p. 363.

"Sur la cime du Cramont (777 French toises = 4967 English feet above the plain) un thermomètre appliqué sur le liège noirci, exposé directement aux rayons du soleil pendant un heure précise, c'est à dire, depuis 2^h 12′ jusques à 3^h 12′, (le 16 Juillet 1774,) étoit monté à 21°, (79° Fahr.,) et un autre thermomètre, à boule nue, exposé en plein air aux rayons du soleil, à 4 pieds au-dessus du gazon, ne se soutenoit qu'à 5 degrés (43° Fahr.)—le lendemain, de retour à Courmayeur, où j'eus le bonheur d'avoir un tems clair, parfaitement semblable à celui de la veille, je choisis un prairie découverte dans laquelle j'établis mon appareil: le thermomètre placé sur le liège noirci montât dans un heure précise à 27°, (93° Fahr.,) et celui qui étoit en plein air à 19°, (75 Fahr.) Tome II. p. 365.

The same accurate observer also found, by comparative trials, that the chemical energy of the solar light, as well as its heating power, was much greater upon the summit of the Col du Géant, than on the plain of Geneva.

From these facts, then, I conclude, that the power of solar radiation in the atmosphere, increases from the equator to the poles, and from below, upwards. The obstruction, which the air offers to the passage of the rays, is not alone dependant upon its density at the surface of the earth, for most of the experiments, which establish the difference between the lower and the higher latitudes, were made under nearly equal heights of the barometer.

For the same reason, the difference cannot be ascribed to any change in the cooling power of the medium, for MM. Dulong, and Petit, have established, from experiment, that the velocity of cooling, in any gas, where it is solely owing to contact, remains the same, if the density and the temperature of the gas change in such a way that the elasticity remains constant. Part of the difference, upon the summit of the mountain, may be traced to the diminution of elasticity, but no such cause operated (or, if it did, in a degree too small for appreciation) in the experiments upon the plains.

May not the phenomena be owing to the differences in the thicknesses of the strata through which the radiant matter has to pass? The retardation of its passage, at the equator, may be dependant upon the inflation of the atmosphere over that zone, both from the centrifugal force of the motion of rotation, and from the expansion occasioned by the never-failing heat: its acceleration at the poles, to the comparative thinness of the aërial stratum, dependant upon cold, and a state of rest. As we ascend in the atmosphere, we obviously diminish the height of the stratum above us, and therefore the effect upon mountains may be referred to the same explanation.

It is well known that a considerable portion of the light of the sun is always detained and absorbed in its passage through the atmosphere; and it has been calculated, that a vertical ray, shot through the clearest air, would lose more than a fifth part of its intensity. This absorption, supposing the aërial covering of the earth to be every where equal, would be in proportion to the obliquity of the rays; for, as their course receded from the perpendicular, they would have to encounter a greater thickness of the aërial fluid. But if the form of the atmosphere be that of a greatly-oblate spheroid, the thickness of the equatorial stratum may readily be conceived to counterbalance or exceed the obliquity of the course by which the rays would penetrate to the flattened regions of the pole. Now the inequality of the temperature of the earth must evidently impress such a figure upon the elastic atmosphere; and supposing the pressure to be every where equal, and the heat to increase from the pole to the equator, from 0° to 80°, the equatorial axis would be to the polar, as 6 to 5.

Fig. 1.

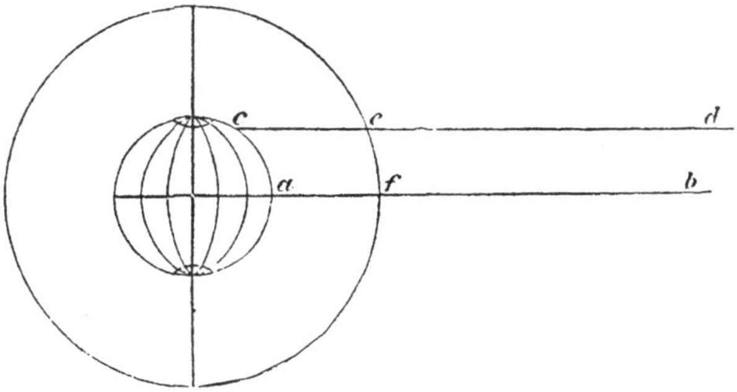

Fig. 2.

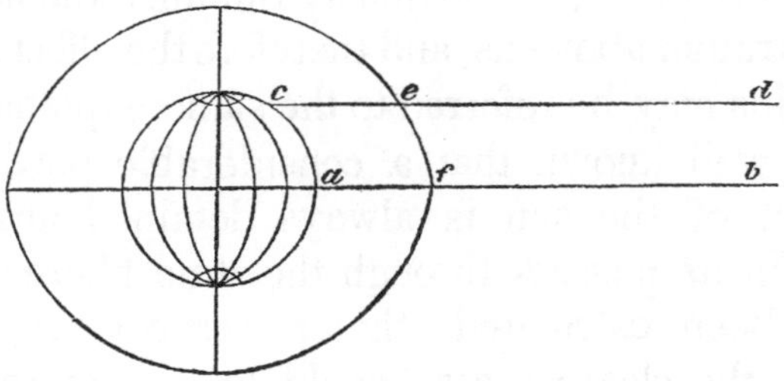

Let Fig. 1 represent a sphere, surrounded by an atmosphere every where equi-distant from the centre; and Fig. 2, the same sphere, with an atmosphere gradually expanding from the poles to the equator. Of the parallel rays, *a b*, *c d*, falling upon the surface of the sphere, the power of *c d*, in the first example, is diminished in proportion to the distance *c e* to *a f;* while, in the second example, the distances being equal, the power is undiminished. It may be objected, that the expansion caused by heat, does not, in fact, increase the quantity of matter through which the rays are obliged to pass, although it extends that matter through a larger space; and that, therefore, the proof is wanting of such diffusion increasing the obstruction.

The cooling power, however, of air, as I have before stated, has been proved to be in proportion to its elasticity; and, therefore, it is reasonable to suppose that the difficulty with which heat passes through it, is in the same proportion. The heat of the sun thus sets limits to its own energy; and by an admirable adjustment, the force of radiation is tempered in those regions where its full perpendicular action would be destructive to vegetable existence, and fully developed in climates where its utmost force is required, to counterbalance the obliquity of its course. If the experimental results could be obtained with sufficient precision, they would form the *data* of some curious and instructive calculations.

These suggestions, however, are offered with much diffidence, as I am well aware that the subject stands greatly in need of further elucidation. One great object which this arrangement answers, in the economy of nature, is too obvious to be passed over; I mean the additional stimulus which is thus afforded to vegetation in the polar regions, during the short but cheering visit of the sun to their inclement skies. The rapidity with which the earth, when it is uncovered of ice and snow, becomes covered with verdure at the first return of summer, has been often noticed; a rapidity which is totally unequalled at the departure of winter in more temperate climates. Most of the plants spring up, flower, and afford seed in the course of a month or six weeks.

Having traced some of the modifications to which radiant caloric is subject in its passage from the sun to the earth, and having shewn the importance of a further developement both of the cause and its effects, I shall now endeavour to collect and combine some particulars with respect to the radiation of heat from the surface of the earth into space; a process, in which the welfare of the vegetable kingdom is no less concerned than in that which we have just been considering. My journal of observations contains a column, also, of the results of experiment upon this division of the subject; they are complete for nearly the whole of the three years. They were obtained by exposing, upon short grass, a register thermometer to an open aspect of the sky, having its bulb covered with black wool. The lowest depression, during the night, was entered in the register. The theory and practice of this experiment have been so clearly elucidated in Dr. Wells's Essay, that I can have nothing to add upon the subject: my only aim is to carry a little further the observation of a principle, which, from limited experience, but with a masterly hand, he has shewn to be of such vast importance. The following table exhibits the mean effect of radiation for every month, deduced from the averages of the three years, together with its greatest observed intensity in the same intervals. The first column shews the month, the second the minimum temperature of the air, the third the mean effect, and the fourth the maximum force of radiation:—

TABLE XII. *Shewing the Mean Minimum Temperature of the Air, with the Mean and Maximum Force of Terrestrial Radiation for every Month in the Year.*

	Mean Min. of the Air.	Mean Depression from Radiation.	Max. Depression from Radiation.
January . . .	32.6	3.5	10
February . . .	33.7	4.7	10
March	37.7	5.5	10
April	42.2	6.2	14
May	45.1	4.2	13
June	48.1	5.2	17
July	52.2	3.6	13
August	52.9	5 2	12
September . .	50.1	5.4	13
October . . .	42.1	4.8	11
November . .	38.3	3.6	10
December . .	35.4	3.5	11

In the last column we may observe an approximation to the law of radiation, established by the experiments of MM. Dulong and Petit; namely, that the velocity of cooling *in vacuo* (or force of radiation) increases as the terms of a geometrical progression for excess of temperature in arithmetical progression. The power of radiation, as exhibited in the table, has evidently a tendency to increase with the heat, although the effect is masked by too many disturbing causes to have enabled us to discover the law of its progression. The amount of effect denoted in the third column is principally dependant upon the clearness of the atmosphere, and it affords no bad estimate of the comparative brightness of the different months. April appears to be the clearest month of the year, and the cloudy state of July in the midst of summer is very remarkable.

From the particulars of the diary it will be found, that vegetation is liable to be affected at night, from the influence of radiation, by a temperature below the freezing point of water ten months in the year; and that even in the two months, July and August, the only exceptions, the radiant thermometer sometimes falls to 35°.

The comparative experiments of Captain Sabine between the tropics are as follows:—

At Bahia he exposed upon grass to the aspect of the sky, an alcohol thermometer, registering the extreme cold, and having its bulb covered with black wool. The following is the comparison between its indications and those of a register thermometer placed under shelter.

TABLE XIII. *Experiments upon Terrestrial Radiation at Bahia.*

	Temp. of Air.	Temp. of Radiation.	Difference.	Observations.
July 24	68	63.5	4.5	Dew.
25	68	63.5	4.5	Ditto
26	72	62.5	9.5	Ditto
27	70	61.	9.	Ditto
28	64	60.5	3.5	
29	67	59.5	7.5	
30	65	64	1	

The register of cold was the same, whether the thermometer was placed on a grass-plat or on a thick bed of *Rotboëlia*, or on thick tufts of *Poa*.

At Jamaica, the radiating thermometer was placed in the manner before described, in contact with the vegetation, and supported by it about ten inches above the ground. The following are the results:—

TABLE XIV. *Experiments upon Terrestrial Radiation at Jamaica.*

	Temp. of Air.	Temp. of Radiation.	Difference.
October 25	76	72	4
26	76	69	7
27	76	65	11
28	76	66	10
29	76.5	65	11.5
30	76	65	11
November 3	76	67	9

On the mountains, at 4000 feet above the level of the sea, the thermometer, laid upon grass, afforded the following comparison:

TABLE XV. *Experiments upon Terrestrial Radiation upon the Mountains of Jamaica.*

Date.	Time.	Temp. of Air.	Temp. of Rad.	Difference.	OBSERVATIONS.
Oct. 31	P. M. 10	65	51	14	Clear and calm.
Nov. 1	A. M. 5	63	45	18	Ditto ditto.
	P. M. 11	64	51	13	Clear & gentle breezes
2	A. M. 5	64	55	9	Ditto ditto.

From all these experiments taken together, it would appear that the same cause which obstructs the passage of radiant heat in the atmosphere from the sun, opposes also its transmission from the earth into space. The force of radiation for the given temperature is less between the tropics, than at the latitude of London; and it obviously increases as we ascend above the surface of the earth.

I have sought, unsuccessfully, for facts which might, tend to throw any light upon the power of terrestrial radiation in the arctic regions; but it is to be hoped that this subject, as well as that of the solar power in the same latitudes, will shortly be determined by the indefatigable activity of Captain Sabine. The intense cold which was found to prevail, during calm weather, in Melville Island, so much beyond the amount of previous calculation, is a strong argument in favour of an increased effect.

In Captain Scoresby's Journal, the following account of the freezing of the sea while the temperature of the air was considerably above the point of congelation, must evidently be attributed to radiation of a very powerful degree.

"In cloudy weather, no freezing of the sea, I believe, ever occurs, when the temperature is above 29°; but in clear calm weather, the sea, in the interstices of the ice, generally freezes on the decline of the sun towards the meridian below the pole, though the temperature be 32°, or higher. In the instance now alluded to, the freezing commenced when the temperature was 36°, being 7½° or 8° above the freezing point of sea-water. About 2 A.M. the thermometer in the air fell to 33°, by which time the bay-ice was of such consistence that the headway of the ship, under a light breeze, was sometime stopped by it." *Scoresby's Journal*, p. 291.

I shall now proceed to detail some further experiments which I have made at different times, the results of which may not prove unimportant to the general subject of radiation. The apparatus which I employed was a concave reflector of copper, plated with silver, of a parabolic form; its diameter was 6 inches, and the length of its focus 1¼ inches. Through a collar in its side a thermometer could be passed, and its bulb fixed in the focus, the scale being kept on the outside. This reflector was placed upon a foot with a ball and socket joint, that enabled it to turn in any direction.

Dr. Wollaston was the first to expose a concave metallic mirror, turned upwards to the free air, with a thermometer placed in its focus; and proved the lowering of its temperature after its being thus exposed for a short time. This experiment he made before the publication of Dr. Wells's Essay, but it does not appear that he pursued the subject further. Mr. Leslie, some years afterwards, adapted the differential thermometer to this idea, and contrived an instrument which he has called *an Æthrioscope.* This was nothing more than a metallic reflector, with one of the balls of a differential thermometer placed in the focus, and the other out of it. He confirmed Dr. Wollaston's result, and the thermometer fell, upon being exposed to a clear sky. The effect he found to depend upon the clearness of the atmosphere.

Mr. Leslie, in his description of the construction and uses of his Æthrioscope, has, unfortunately I think, indulged in a brilliancy of imagination and figurativeness of language, which have greatly obscured his meaning. He ascribes, for instance, the action of the instrument to "cold pulses showered entire from the heavens." He speaks of "the higher strata of the atmosphere darting cold pulses downwards, and the lower strata projecting equal pulses of heat upwards."

The Æthrioscope, he says, "extends its sensation through indefinite space, and reveals the condition of the remotest atmosphere." Nay, more, he expects that "when constructed with greater delicacy, it may, perhaps, scent the distant winds, and detect the actual temperature of different portions of the heavens." With far humbler views I have made considerable use of Dr. Wollaston's apparatus, which, for reasons which I shall not now stop to discuss, I very much prefer to Mr. Leslie's. The standard thermometer, to which all my observations refer, when not otherwise expressed, had its bulb covered with black wool.

My first object was to ascertain the force of radiation from a thermometer, so guarded from the influence of surrounding bodies, compared with another, exposed, as I have before described, upon grass. The following Table exhibits the results. The first column shews the lowest temperature of the air during the night, the second the lowest temperature of a thermometer on grass, and the third that of the thermometer in the reflector.

TABLE XVI. *Comparison of the Force of Radiation in a Reflector and on the Grass.*

	Temp. of Air.	Temp. of Grass.	Temp. in Reflector.	OBSERVATIONS.	
	42	34	30	Very fine and clear.	
	47	39	35	Ditto	ditto.
	52	44	42	Ditto	ditto.
	44	35	32	Ditto	ditto.
	44	36	34	Ditto	ditto.
	54	48	45	Ditto	ditto.
	58	52	52	Dull.	
	57	51	49	Very fine—moon hazy.	
	56	51	50	Light clouds.	
	51	41	41	Very fine and clear.	
	45	35	35	Ditto	ditto.
	50	42	41	Ditto	ditto.
Mean	50	42.3	40.5		

The average difference is not quite two degrees.

I consider this as much the most accurate method of measuring the force of terrestrial radiation: at the same time it is gratifying to find that the means which I had adopted, before this idea had occurred to me, were not, upon a mean of observations liable to any very considerable error. The radiant thermometer is so completely insulated by the reflector, from the counter-radiation of surrounding bodies, that it may be applied with equal effect in any situation where the aspect of the sky is very limited. Even in the streets of London, where the radiation of an exposed thermometer is nearly neutralized, and the utmost effect never exceeds two or three degrees, that of the thermometer, guarded by the reflector, is wholly unimpeded. Experiments that are thus made, in whatever situation, are strictly comparable, provided they are screened from any strong action of the wind.

Being thus in possession of the means of cutting off the access of radiant matter from any source, and of directing it to any required object, I was anxious to ascertain the force with which it was given off while the sun was above the horizon, compared with what it was in the absence of that luminary. Under the most favourable circumstances, when the air was calm and the atmosphere clear, I never could obtain an effect of more than five or six degrees with the thermometer covered with black wool. It then occurred to me, to try the influence of colour in modifying the results. I had another reflector made exactly similar to the former, and their power, upon trial, was found to be precisely equal; that is to say, the radiating thermometer fell to an equal amount in each of them. I now covered the ball of a thermometer with white wool, and placed it in the focus of one reflector, and the thermometer with black wool in the focus of the other. I selected a cloudless day for the experiment, and placed the two instruments, side by side, in the shade of a tree, inclining them at equal angles towards the clear eastern sky. The following Table includes the results.

TABLE XVII. *Comparison of the Force of Radiation from Black and White Wool.*

May 16.	Radiation from Black Wool.	Radiation from White Wool.	Temp. of Air.	OBSERVATIONS.
P.M. 3½	58	53	63	Atmos. cloudless.
4	58	53	63	During the experiment the reflectors were changed.
8	44	43	54	
11	36	36	47	
During Night	35	35	45	

The amount of radiation, therefore, from the white wool, was equal, during the time the sun was high in the heavens, to what it was during the night; while it was one-half less from the black wool. During the absence of the sun, the radiating power of the two was equal.

The experiment was repeated under varying circumstances; and in examining the results, as included in the following Table, it will be necessary to attend particularly to the collateral circumstances.

TABLE XVIII. *Comparison of the Force of Radiation in Black and White Wool.*

May 17.	Radiation from Black Wool.	Radiation from White Wool.	Temp. of Air.	OBSERVATIONS.
P.M. 1½	64	59	65	Overcast, with cumulo-stratus.
2	68	60	65	Clearing—reflectors turned to clearing space.
2½	73	62	65	Faint sun-shine.
3½	74	62	66	Strong sun-shine—reflectors turned, so that the shadows of the bulbs just appeared on the metal.
4	76	64	68	
11	51	51	55	Lightly overcast.
Night.	42	42	51	Fine.

The power of radiation was nearly neutralized in the black wool, while the sky was over-cast, but in the white wool was only reduced to about one-half. As the sky cleared, the reflectors being turned towards the sun's place, the black thermometer rose above the temperature of the air, and the white thermometer still gave off more heat than it received. In full sun-shine, the reflectors being just turned out of the direct rays, the black thermometer rose 8° above the temperature of the air, and the white thermometer fell 4° degrees below it. In estimating these effects it must be remembered, that the action of the reflector, in receiving and transmitting heat, is different. In the former case, we have an exaggerated action; the heat which falls upon the surface of the speculum is thrown upon the thermometer in a concentrated form. In the latter case, the heat, which radiates from the thermometer in the focus, falls upon the concave metal, and is reflected into space in parallel lines. The effect is, therefore, only slightly augmented from the larger aspect of the sky. Whenever the reflector, with the black-wooled thermometer, is turned, while the sun is above the horizon, towards a cloud, the mercury rises above the temperature of the air, excepting in the winter months; and a distinct effect is produced even from the quarter most distant from the sun. The concrete vapour seems to disperse the radiant matter, and to act upon it in much the same way as ground glass upon transmitted light. I subjoin some experiments which illustrate this point, and shew the effect of two similar thermometers in similar reflectors, directed to different quarters of the heavens.

TABLE XIX. *Effects of Radiation under different Aspects of the Sky.*

Date.	Hour.	Position of the Reflector.	Temp. of Black Wool.	Temp. of Air.	OBSERVATIONS.
July 2	12	Horizontal Inclined 30°	76 79	63	Sky overcast—cumulo-stratus—sun's place not visible—and brisk wind from S.W.
	1	Horizontal Inclined 30°	83 83	63	Sun's place just visible, but no shadows.
	2	Horizontal Inclined 30°	86 86	63	Ditto ditto.
	2½	Horizontal Inclined E. 30°	83 82	63	Sun's place not visible.
	3	Horizontal Inclined N. 30°	79 79	63	Ditto ditto.
		Inclined N. 30° S. 30°	76 79	63	Ditto ditto.
		Inclined N.E. 30° S.W. 30°	85 88	63	Ditto ditto.
	3½	Inclined N.E. 30° S.W. 30°	96 100	63	
		Inclined W. E.	55 55	61	Just before sun-set.
	11 P.M.	Inclined N. S.	57 57	61	
	Night.	Inclined N. E.	40 40	48	Fine.
July 4	All Night.	Horizontal Vertical	41 43	51	Very fine.

No effect is produced by any cloud after the sun has sunk below the horizon; and in an overcast night the action of radiation is perfectly neutralized. I have appended to this essay a series of observations made in London, at different hours of the night and day, from January to April, 1822.

As the power in different bodies of absorbing heat, and the power of emitting it, do not appear to be equal under every circumstance, as has been demonstrated in the case of the black and the white wool, it becomes a curious inquiry to ascertain the relation of various substances to these effects. I regret that I have not had leisure to pursue this branch of the subject with the attention which it deserves. I shall subjoin the results of two or three experiments, to shew that much curious information might be expected from the investigation. The standard of comparison was, in all cases, the black-wooled thermometer, and the substance compared was stuck upon a thermometer, in a similar reflector, by its side.

TABLE XX. *Comparison of the Force of Radiation from different Substances.*

		Substance Compared.	Radiation.	Black Wool.	Temperature of Air.	OBSERVATIONS.
May 21	11 P.M.	Naked Alcohol Therm.	48	46	55	Very fine night.
	Night.	. . . . . .	40	38	48	
22	9 A.M.	. . . . . .	53	58	56	Haze.
23	8 P.M.	Garden Mould.	41	38	46	Very fine.
	11	. . . . . .	40	37	46	Ditto
	Night.	. . . . . .	38	35	45	Ditto
24	9 A.M.	. . . . . .	59	59	56	Ditto
	8 P.M.	Chalk.	46	43	53	Ditto
	11	. . . . . .	43	40	50	Ditto
	Night.	. . . . . .	35	33	44	Ditto
25	A.M. 11	. . . . . .	67	61	62	Lightly overcast.
18	P.M. 11	Leaf of the Rose-Campion.	46	45	57	Very fine.
19	P.M. 8	. . . . . .	51	50	61	Ditto
	11	. . . . . .	44	43	53	Ditto
	Night.	. . . . . .	39	38	48	Ditto
22	P.M. 11	Integument of the Flower of an Iris.	44	42	52	Ditto
	Night.	. . . . . .	38	37	47	Ditto

Whilst engaged in this course of experiment, it occurred to me that a favourable opportunity presented itself of determining a question which has at different times occasioned considerable controversy, and concerning which, many discordant statements have often been made: I mean, the radiation of heat from the body of the moon. Dr. Howard has lately published the following result of an experiment by means of a delicate differential thermometer, which seems to establish the reality of such an effect.

"Having blackened the upper ball of my differential thermometer, I placed it in the focus of a thirteen-inch reflecting mirror, which was opposed to the light of a bright full-moon. The liquid began immediately to sink, and in half a minute was depressed 8°, where it became stationary. On placing a screen between the mirror and the moon, it rose again to the same level, and was again depressed on removing this obstacle."—*Silliman's Journal*, vol. ii. p. 329.

Upon reading the above extract, it struck me that it did not clearly explain in which leg of the instrument the depression of the liquid took place; and that the effect, *as described*, might just as well be attributed to the radiation of heat from the blackened ball of the thermometer, as to radiation to it from the moon. To determine this doubt, I tried the following experiments:—

I selected an unexceptionable opportunity, 26th of December, 1822. The moon was in that part of her orbit when she is nearest to the earth, and was approaching to the full. The atmosphere was cloudless, and perfectly calm. The smallest writing was distinctly legible in the moon-light. At 9 P.M. the temperature of the air was 28°. I placed the black thermometer in the focus of the reflector, and directed it to a part of the sky at a distance from the moon. In a few minutes it fell to 20°, and was stationary. I then turned it immediately towards the moon, and caused the focus of light to fall upon the ball of the thermometer. It still remained stationary at 20°, and for half an hour, during which the rays were concentrated upon it, the mercury never moved.

At 11 P.M. the temperature of the air .	27°
reflector turned from the moon	19°
——— in the moon-beams .	19°

Dec. 28th, 7 P.M.

Moon full—atmosphere perfectly calm and clear.

Temperature of the air	24°
Reflector turned from the moon	15°
——— in the moon-beams	15°

At 11 P.M. the sky became lightly clouded, and the amount of radiation was only 2°.

Temperature of air	22°
Radiating thermometer	20°

Thus it appears that, so far from possessing the power of radiating heat to the surface of the earth, the moon does not even diminish the amount of radiation from the earth; and the lightest vapour is more efficacious in this respect than the concentrated influence of the lunar light.

Observations of a Black Radiating Thermometer in a Concave Reflector.

Turned to the North, Angle 30°.

Date.	Hour.	Temp of the Air.	Day.	Night	Difference	State of Reflector.	State of Weather.
1822 Jan. 13	p.m. 4	48	45	—	– 3	———	Fine, but misty
	11	45	—	37	– 8	———	Ditto ditto
	night	40	—	32	– 8	———	Very fine
14	a.m. 10	42	35	—	– 7	Spots of rain . .	Very fine and clear
	p.m. 4	42	35	—	– 7	———	Cloudless
	11	39	—	30	– 9	———	Light clouds
	night	40	—	36	– 4	———	Ditto ditto
15	a.m. 9	40	32	—	– 8	Bright	Very clear
	p.m. 11	31	—	21	–10	———	Ditto ditto
	night	30	—	20	–10	Bright	Ditto ditto
16	a.m. 9	31	22	—	– 9	———	Ditto ditto
	p.m. 4	31	21	—	–10	———	Ditto ditto
	11	29	—	19	–10	———	Ditto ditto
	night	29	—	19	–10	———	Ditto ditto
17	a.m. 9	33	33	—	0	———	Snow
	p.m. 11	33	—	26	– 7	———	Very fine
	night	32	—	25	– 7	———	Ditto ditto
18	a.m. 9	35	30	—	– 5	Bright	Ditto ditto misty
	p.m. 4	39	34	—	– 5	———	Ditto ditto ditto
	11	39	—	37	– 2	———	Overcast and dull
	night	38	—	33	– 5	———	Dull
19	a.m. 9	41	41	—	0	Tarnished	Dull and foggy
26	a.m. 9	40	32	—	– 8	———	Very fine
	p.m. 11	40	—	35	– 5	———	Dull
	night	32	—	22	–10	Dull and spotted	Very fine
27	a.m. 10	34	27	—	– 7	———	Ditto ditto
	p.m. 4	47	47	—	0	———	Overcast and dull
	11	40	—	40	0	———	Ditto ditto
	night	35	—	33	– 2	———	Dull
28	a.m. 9	45	43	—	– 2	Spotted with rain	Mild and misty
	p.m. 11	41	—	39	– 2	———	Overcast and dull
	night	35	—	27	– 8	———	Very fine but misty
29	a.m. 9	37	31	—	– 6	Blacks in the mirror	Ditto ditto
	p.m. 11	38	—	34	– 4	———	Lightly overcast
	night	32	—	22	–10	———	Very fine
30	a.m. 9	34	26	—	– 8	———	Ditto ditto
	p.m. 11	34	—	30	– 4	———	Fog

Observations, &c., continued.

Date.	Hour.	Temp of the Air.	Day.	Night	Difference	State of Reflector.	State of Weather.
Jan. 30	night	30	~	23	− 7	———	Fine
31	a.m. 9	35	33	~	− 2	Dew upon lower half	Light clouds
	p.m. 11	35	~	28	− 7	———	Drops of rain
	night	35	~	28	− 7	———	Fine
Feb. 1	a.m. 9	42	37	~	− 5	Spotted with rain	Very fine
	p.m. 11	38	~	29	− 9	———	Ditto ditto
2	a.m. 9	47	47	~	0	Tarnished & spotted	Small rain
	p.m. 11	49	~	45	− 4	———	Fine
	night	39	~	39	0	Full of rain	Rain
3	a.m. 10	44	39	~	− 5	———	Very fine
	p.m. 11	37	~	28	− 9	———	Ditto ditto
	night	33	~	25	− 8	———	Ditto ditto
4	a.m. 9	38	36	~	− 2	Moisture running off the bulb	Foggy
	p.m. 11	47	~	45	− 2	———	Overcast
	night	38	~	37	− 1	———	Stormy
5	a.m. 10	49	42	~	− 7	———	Fine
	p.m. 11	39	~	30	− 9	———	Very fine
	night	32	~	22	−10	Hoar-frost upon the bulb and stem	Ditto ditto
6	a.m. 10	35	27	~	− 8	———	Ditto ditto
	p.m. 11½	41	~	35	− 6	———	Dull and close
7	a.m. 9	45	43	~	− 2	Tarnished . . .	Ditto ditto
	p.m. 3	48	45	~	− 3	———	Ditto ditto
	11	47	~	43	− 4	———	Ditto ditto
	night	45	~	41	− 4	Full of rain . .	Ditto ditto
8	a.m. 9	48	45	~	− 3	———	Overcast but fine
	p.m. 11	45	~	42	− 3	———	Ditto ditto
	night	42	~	36	− 6	Spotted with rain	Ditto ditto
9	a.m. 10	47	47	1	0	———	Overcast and mild
	p.m. 11	49	~	47	− 2	———	Ditto ditto
	night	46	~	41	− 5	———	Ditto ditto
10	a.m. 10	47	45	~	− 2	———	Ditto ditto
	p.m. 11	47	~	45	− 2	———	Light rain
	night	41	~	33	− 8	Spotted with rain	Very fine
11	a.m. 10	44	41	~	− 3	———	Ditto ditto
12	p.m. 11	39	~	36	− 3	Stained . . .	Fog
	night	37	~	33	− 4	Some water . .	Ditto
13	a.m. 9	41	38	~	− 3	———	Fine but misty
	p.m. 4	44	38	~	− 6	———	Ditto ditto
	night	40	~	32	− 8	———	Very fine
Feb. 24	a.m. 10	46	48	~	+ 2	———	Lightly overcast

Observations, &c., continued.

Date.	Hour.	Temp of the Air.	Day.	Night	Difference	State of Reflector.	State of Weather.
Feb. 24	p.m. 11	48	~	46	− 2	———	Overcast and dull
	night	46	~	43	− 3	———	Ditto ditto
25	a.m. 10	51	55	~	+ 4	———	Lightly overcast
	p.m. 11	47	~	43	− 4	———	Ditto ditto
	night	44	~	39	− 5	Bright	Ditto ditto
26	a.m. 10	49	53	~	+ 4	———	Ditto ditto
	p.m. 4	51	54	~	+ 3	———	Overcast, rain
	11	45	~	41	− 4	———	Clearing
	night	36	~	26	−10	Spotted with rain	Very fine
27	a.m. 10	42	34	~	− 8	———	Ditto ditto
	p.m. 11	35	~	24	−11	———	Ditto ditto
	night	31	~	22	− 9	Hoar frost on bulb	Ditto ditto
28	a.m. 10	34	30	~	− 4	———	Ditto ditto, fog
	p.m. 5	43	37	~	− 6	———	Very fine
	11	35	~	26	− 9	———	Ditto ditto
	night	30	~	20	−10	Hoar frost on bulb	Ditto ditto
Mar. 1	a.m. 10	34	31	~	− 3	———	Ditto ditto, fog
	p.m. 11	39	~	33	− 6	———	Light clouds
	night	34	~	31	− 3	———	Ditto ditto
2	a.m. 10	46	47	~	+ 1	———	Overcast and mild
	p.m. 11	43	~	36	− 7	———	Very fine
11	p.m. 4	46	50	~	+ 4	———	Turned to a dense cloud
	5	44	40	~	− 4	———	Very fine
	11	37	~	28	− 9	———	Ditto ditto
14	a.m. 10	54	60	~	+ 6	———	Overcast and dull
19	a.m. 10	56	61	~	+ 5	———	Lightly overcast
	5	59	57	~	− 2	———	Ditto ditto
	11	54	~	51	− 3	———	Ditto ditto
	night	51	~	46	− 5	———	Fine
20	a.m. 10	54	58	~	+ 4	———	Lightly overcast
	p.m. 5	56	50	~	− 6	———	Very fine
	night	46	~	37	− 9	———	Ditto ditto
21	a.m. 10	51	52	~	+ 1	———	Overcast
	p.m. 5	56	51	~	− 5	———	Very fine
	11	51	~	49	− 2	———	Overcast
	night	41	~	31	−10	———	Very fine
27	a.m. 10	52	57	~	+ 5	———	Overcast
	p.m. 5	56	49	~	− 7	———	Very fine
	night	46	~	37	− 9	———	Ditto ditto
April 1	a.m. 10	42	44	~	+ 2	———	Dense clouds

Observations, &c., continued

Date.	Hour.	Temp of the Air.	Day.	Night	Difference	State of Reflector.	State of Weather.
April 1	p.m. 5	46	46	~	0	~~~~~~	Overcast and dull
	11	42	~	32	−10	~~~~~~	Very fine
	11½	43	~	39	− 4	~~~~~~	Lightly overcast
	night	39	~	31	− 8	~~~~~~	Fine
3	a.m. 10	50	59	~	+ 9	~~~~~~	Lightly overcast
	p.m. 1	54	60	~	+ 6	~~~~~~	Ditto ditto
	2	53	61	~	+ 8	~~~~~~	Ditto ditto
	3	53	50	~	− 3	~~~~~~	Clear
	5	50	42	~	− 8	~~~~~~	Very fine
	11	43	~	35	− 8	~~~~~~	Ditto ditto
	night	40	~	31	− 9	~~~~~~	Ditto ditto
4	a.m. 10	49	58	~	+ 9	~~~~~~	Lightly overcast
	10½	51	61	~	+10	~~~~~~	Ditto ditto
	11	51	62	~	+11	~~~~~~	Ditto ditto
	12	51	60	~	+ 9	~~~~~~	Ditto, drops of rain
	p.m. ½	52	64	~	+12	~~~~~~	Ditto ditto
	1	52	58	~	+ 6	~~~~~~	Ditto ditto
	3½	51	53	~	+ 2	~~~~~~	Clearing
	5	49	47	~	− 2	~~~~~~	Ditto
	11	47	~	42	− 5	~~~~~~	Overcast
	night	43	~	34	− 9	~~~~~~	Fine
5	a.m. 10	51	62	~	+11	~~~~~~	Lightly overcast
	p.m. 5	51	50	~	− 1	~~~~~~	Ditto ditto
	11	46	~	37	− 9	~~~~~~	Very fine
	night	43	~	33	−10	~~~~~~	Ditto ditto
6	a.m. 10	52	52	~	0	~~~~~~	Ditto ditto
	p.m. 5	49	49	~	0	~~~~~~	Ditto ditto
	11	41	~	31	−10	~~~~~~	Ditto ditto
	night	39	~	30	− 9	~~~~~~	Ditto ditto
7	a.m. 10	45	51	~	+ 6	~~~~~~	Overcast
	night	36	~	27	− 9	~~~~~~	Very fine
8	a.m. 10	45	46	~	+ 1	~~~~~~	Ditto ditto
	p.m. 5	39	39	~	0	~~~~~~	Hail showers
	night	35	~	30	− 5	~~~~~~	Fine
15	a.m. 10	57	67	~	+10	~~~~~~	Lightly overcast
16	p.m. 5	54	53	~	− 1	~~~~~~	Dull
	11	46	~	37	− 9	~~~~~~	Very fine
	night	41	~	34	− 7	~~~~~~	Foggy
28	a.m. 10	56	67	~	+11	~~~~~~	Very hazy
	night	46	~	40	− 6	~~~~~~	Fine

AN ESSAY

UPON THE

HORARY OSCILLATIONS OF THE BAROMETER.

THE barometric column was first observed to have a daily periodical vibration between the tropics, by the expedition under the command of the unfortunate Peyrouse. M. Lamanon, the naturalist, has given an account of these observations in the second volume of the voyage, at page 521. He states that from about the 11th degree of north latitude, he began to perceive a certain regularity of motion in the barometer, so that the mercury stood highest about the middle of the day, from which time it descended till the evening, and rose again during the night. As they approached the equator, the effect became more distinct, and on the 28th September (1785) a series of Experiments were begun in 1° 17′ north latitude, and continued for every hour, till the 1st of October. The following are the results of the observations on the 28th and 29th.

Sept. 28.	From 4 to 10 A.M., barometer	rose 0.19 inch.
	From 10 A.M., to 4 P.M.	fell 0.12
	From 4 to 10 P.M.	rose 0.09
Sept. 29.	From 10 (28th) to 4 P.M.	fell 0.13
	From 4 to 10 A.M.	rose 0.15
	From 10 A.M. to 4 P.M.	fell 0.13
	From 4 to 10 P.M.	rose 0.10

The observations on the 30th were to the same effect.

Hence it was inferred that there is a periodical flux and reflux of the atmosphere, at the equator, producing in the barometer a variation of about 0.12 inch (English,) corresponding according to M. Lamanon to a height in the atmosphere of about 100 feet. The latitude of the ship on the 28th, was 0° 50′ north; and 0° 11′ north, on the 29th.

In the year 1794, Dr. Balfour published, in the Asiatic Researches, an account of some observations made at Calcutta, which agreed in a remarkable manner in the same conclusion. During one whole month, he observed the barometer every half hour: the mercury constantly fell from ten at night to six in the morning; from six to ten in the morning it rose; from ten in the morning, to six at night, it fell again; and, lastly, rose from six to ten, at night. The *maximum* height was therefore at ten, P. M., and ten, A. M., and the *minimum* at six, P. M., and six, A. M. The oscillations sometimes amounted to 0.1 inch, but in general were considerably less.

The observations of M. de Humboldt, of a later date, confirm the existence of these semi-diurnal variations in the torrid zone, and extend them to the south of the equator. According to his results, the barometer generally falls from ten o'clock, A.M., till 4, P.M.; then rises again till ten, P.M. —again drops till four, A.M., and mounts till ten, A.M.

Captain Sabine, also, amongst his other numerous, laborious, and interesting, pursuits, turned his attention to this subject, while between the tropics, and has favoured me with the following results of his Experiments, to ascertain the amount of the horary oscillation at Sierra-Leone, St. Thomas', Trinidad and Jamaica:—

TABLE I. *Horary Oscillations of the Barometer, observed at Sierra Leone, February and March, 1822.*

OBSERVATIONS.

Hours	February 25 Bar.	February 25 Ther.	February 26 Bar.	February 26 Ther.	February 27 Bar.	February 27 Ther.	March 1 Bar.	March 1 Ther.	March 2 Bar.	March 2 Ther.
6 A.M.	29.861	75	—		29.850	76	29.876	73.5	—	
7	—		—		—		—		—	
7½	—		—		—		—		—	
8	29.864	79	—		29.874	77	—		29.918	77.5
8½	—		—		—		—		—	
9	29.885	80	29.872	80	—		—		—	
9½	—		—		—		—		—	
10	29.878	80.2	29.884	78	29.884	78	—		29.930	79.3
10½	—		—		—		29.929	78	—	
11	29.880	80.5	—		—		—		—	
11½	—		—		—		—		—	
12 noon	29.878	81.2	29.875	82	29.865	82	29.927	78.5	29.921	80.2
½ P.M.	—		—		—		—		—	
1	—		—		—		—		29.908	81.2
1½	—		—		—		—		—	
2	29.850	85	29.850	84	29.840	84	29.876	80.6	29.878	81
2½	—		—		—		—		—	
3	—		—		—		—		29.860	83
3½	29.830	84	—		—		—		—	
4	29.828	84	29.813	81	29.844	82	29.868	80	29.856	83
5	—		—		—		—		29.850	81.4
6	29.825	82.5	29.813	79	29.846	77	—		29.861	79.6
7	—		—		—		—		—	
8½	29.868	80	—		29.880	76	—		—	
9	—		—		—		—		—	
10	—		29.865	77	29.900	79	29.926	79	—	

Hours	March 20 Bar.	March 20 Ther.	March 21 Bar.	March 21 Ther.	March 22 Bar.	March 22 Ther.	March 28 Bar.	March 28 Ther.	March 29 Bar.	March 29 Ther.	March 31 Bar.	March 31 Ther.
6 A.M.	—		—		—		—		—		—	
7	29.903		—		29.818	80.2	29.820	80	—		—	
7½	29.912		—		29.820	80.3	—		—		—	
8	29.920		—		29.822	80.7	29.825	81	—		—	
8½	29.915		—		29.818	81	—		—		—	
9	29.910		29.875	81.2	29.824	81.3	29.825	82	29.808	81	—	
9½	29.908		29.876	82.5	29.828	81.5	—		29.812	81.2	29.860	
10	29.908		29.872	83.5	29.828	81.8	29.825	83	29.815	81	29.865	
10½	29.920		29.874	83	29.830	82	—		29.815	81.2	29.860	
11	29.912		29.876	82.8	29.828	82.5	29.825	84	29.815	81.5	29.860	
11½	29.900		29.876	82.8	—		—		29.812	81.3	29.850	
12 noon	29.860		29.876	82.8	—		29.795	84	29.810	81.6	29.830	
½ P.M.	29.848		29.872	83	29.820	83	—		29.785	81.8	29.825	
1	29.846		29.872	83	—		29.795	83	29.784	82	29.815	
1½	—		29.850	83	29.800	83	—		—		29.815	
2	—		29.828	83	—		29.780	82	29.750	82.5	29.800	
2½	—		29.830	84	29.776	82.8	—		29.750	82.5	—	
3	—		29.830	83	29.774	82.6	—		29.750	82.8	—	
3½	—		29.810	83	29.760	83.5	—		29.748	82.5	—	
4	—		29.808	81	—		—		29.740	82	—	
5	—		—		29.772	84	—		29.737	80	29.790	
6	—		—		29.780		—		—		—	
7	—		—		—		—		—		—	
8½	—		29.812	80	29.790	82.2	—		29.760	81.7	29.890*	
9	—		29.850	80	29.808	83	—		29.780	80.7	29.902*	
10	—		29.870	80	29.814	82.5	—		29.800	80	29.930* Preceding a Tornado	

Differences from the Maximum of each Day.

Hours	F. 25	F. 26	F. 27	M. 1	M. 2	M. 20	M. 21	M. 22	M. 28	M. 29	M. 31	Average.
6 A.M.	.034	—	.034	.053	—	—	—	—	—	—	—	.040 at 6 A.M.
7	—	—	—	—	—	.017	—	.012	.005	—	—	.011 — 7
7½	—	—	—	—	—	.008	—	.010	—	—	—	.009 — 7½
8	.021	—	.010	—	.012	0	—	.008	0	—	—	.009 — 8
8½	—	—	—	—	—	.005	—	.012	—	—	—	.009 — 8½
9	0	.012	—	—	—	.010	.001	.006	0	.007	—	.005 — 9
9½	—	—	—	—	—	.012	0	.002	—	.004	.005	.004 — 9½
10	.007	0	0	—	0	.012	.004	.002	0	0	0	.002 — 10
10½	—	—	—	0	—	0	.002	0	—	0	.005	.001 — 10½
11	.005	—	—	—	—	.008	0	.002	0	0	.005	.003 — 11
11½	—	—	—	—	—	.020	0	—	—	.003	.015	.009 — 11½
12 noon	.007	.009	.019	.002	.009	.060	0	—	.030	.005	.035	.018 — noon
½ P.M.	—	—	—	—	—	.072	.004	.010	—	.030	.040	.031 — 0½ P.M.
1	—	—	—	—	.022	.074	.004	—	.030	.031	.050	.034 — 1
1½	—	—	—	—	—	—	.026	.030	—	—	.050	.035 — 1½
2	.035	.034	.044	.053	.052	—	.048	—	.045	.065	.065	.049 — 2
2½	—	—	—	—	—	—	.044	.054	—	.065	—	.054 — 2½
3	—	—	—	—	.070	—	.044	.056	—	.065	—	.059 — 3
3½	.055	—	—	—	—	—	.066	.070	—	.067	—	.064 — 3½
4	.060	.071	.040	.061	.074	—	.068	—	—	.075	—	.064 — 4
5	—	—	—	—	.080	—	—	.058	—	.078	.075	.073 — 5
6	.063	.071	.038	—	.069	—	—	.050	—	—	—	.068 — 6
7	—	—	—	—	—	—	.064	—	—	.055	—	.059 — 7
8½	.027	—	.004	—	—	—	.024	.040	—	.035	—	.025 — 8½
9	—	—	—	—	—	—	.006	.022	—	.015	—	.015 — 9
10	—	.019	+.016	.003	—	—	—	.016	—	—	—	.011 — 10

The Barometer was suspended in a room 190 feet above the level of the sea; the average of the maximum (in the eleven days) was 29.877 at 80.4, or 30.037 at the same temperature at the level of the sea; whence 30 inches at a temperature of 80, may be considered the corresponding *mean* height of the Barometer.

[*To face Page* 252.

The material originally positioned here is too large for reproduction in this reissue. A PDF can be downloaded from the web address given on page iv of this book, by clicking on 'Resources Available'.

TABLE II. *Barometrical Observations in Fort Charles, Port Royal, Jamaica; Latitude 17° 56′ N. the Cistern being Six Feet above the Level of the Sea.*

	Height of the Barometer. Thermometer 81° to 85°.									Differences from the Maximum at the several hours of the day.										
Hours.	Oct. 22	23	24	25	26	27	28	29	30	Oct 22	23	24	25	26	27	28	29	30	Mean	
8 A.M.	——	——	——	——	30.053	——	——	——	30.081	——	——	——	——	.025	——	——	——	.019	.022	
9	——	30.050	——	——	30.068	——	——	——	30.093	——	.020	——	——	.010	——	——	——	.007	.012	Mean maximum 30.066
10	30.020	30.070	30.070	30.089	30.078	30.070	30.045	30.070	30.100	0	0	0	0	0	0	0	0	0	0	Mean minimum 30.0075
11	30.016	30.065	30.060	——	——	——	——	——	——	.004	.005	.010	——	——	——	——	——	——	.006	Mean . . 30.039
12	30.014	——	——	——	——	30.050	——	30.062	——	.006	——	——	——	——	.020	——	.008	——	.011	Being rather higher than the mean of the 24 hours.
1 P.M.	30.008	30.041	30.045	30.060	——	——	30.028	30.028	——	.012	.029	.025	.024	——	——	.017	.042	——	.025	
2	——	30.025	30.042	——	——	——	——	30.018	30.010	——	.045	.028	——	——	——	——	.052	.030	.039	
3	29.970	——	30.028	——	30.020	30.018	30.000	——	30.038	.050	——	.042	——	.058	.052	.045	——	.062	.052	
4	29.967	30.005	30.028	——	30.012	30.005	29.990	30.002	30.033	.053	.055	.042	——	.066	.065	.055	.068	.067	.059	
5	29.964	30.005	——	30.028	——	——	30.001	——	——	.056	.055	——	.056	——	——	.041	——	——	.052	

In comparing this Table with the similar observations at Sierra Leone, I remark that the hours of the maximum and minimum correspond; but that the amount of the differences from the maximum at the several hours of the day are, throughout, less at Jamaica than at Sierra Leone, in Latitude 17° 56′, than in Latitude 8° 29′. Does the amount of the atmospherical tides diminish progressively in receding from the equator to the tropics?

The following maxima and minima were observed at Trinidad, in Latitude N. 10° 39′. The barometer was placed in the Protestant Church, 18 feet above the surface of the sea.

TABLE III.

Sept. 34	10 A.M. 30.028 and at $4\frac{1}{2}$ P.M. 29.972	Diff. .056	.063	Ther. 83° to 85°.
25	10 A.M. 30.032 and at $4\frac{3}{4}$ P.M. 29.970	Diff. .062		
26	10 A.M. 30.053 and at $4\frac{1}{3}$ P.M. 29.966	Diff. .067		
29	$10\frac{1}{4}$ A.M. 30.028 and at 4.50 P.M. 29.962	Diff. .066		
Oct. 1	$10\frac{1}{4}$ 30.033 and at $4\frac{1}{3}$ P.M. 29.970	Diff. .063		

The following maxima and minima observed at St. Thomas's in lat. N. 0° 24′.

TABLE IV.

Date.	Maximum.	Minimum.	Diff.	
May 27	30.077	30.000	.077	Mean height of the barometer in the day 30.066. Thermometer 81° to 86°, 21 ft. above the level of thesea.
28	30.070	30.000	.070	
29	30.074	30.000	.074	
30	30.100	30.028	.072	
June 1	30.150	30.074	.076	
3	30.114	30.040	.074	
Corrected May 28	30.140	30.065	.075	
Means . .	30.103	30.029	.074	

These all appear to indicate a progressive decrement, though it must be allowed the materials for generalization are scanty.

The material originally positioned here is too large for reproduction in this reissue. A PDF can be downloaded from the web address given on page iv of this book, by clicking on 'Resources Available'.

It is not, however, alone in tropical latitudes, that these horary motions of the barometer may be detected: the absence of those disturbing causes which affect the atmosphere in temperate climates, and produce the much more considerable but irregular fluctuations of the mercurial column, render them more prominent in those situations; but by a system of averages, which balances the irregularities, the regular movement is elicited, even when most concealed. M. Ramond found at Clermont-Ferrand, in latitude 45° 47″, that a mean of ten days sufficiently neutralized the irregular oscillations, and the periodical motions were distinctly exhibited in intervals of that length. The hours of the fluctuations were, as nearly as possible, coincident with those at the equator; but the effect was considerably less, and did not amount to more than 0.039 inch. The monthly means of the observations made at the observatory at Paris, present the same result, with a still further reduction of the effect; the average of six years' observations being 028 inch: and, lastly, my own meteorological journal exhibits the horary movements with great regularity, but only to the average extent of 0.015 inch.

Thus, there can be no doubt that the suggestion of Captain Sabine is correct, that "the amount of the atmospherical tides diminishes progressively from the equator to the tropics," and, further, that it continues at a diminishing rate, at least as far as the fifty-second degree of latitude. The following Table presents a synoptic view of the mean results.

TABLE V. *Shewing the Mean Periodical Movement of the Barometer at different Latitudes.*

Names of Places.	North Latitude.	Mean Periodical Movement of the Barometer.
St. Thomas'	0° 24′	0.074 Inch.
Sierra Leone	8° 29′	0.073
Trinidad	10° 39′	0.063
Jamaica	17° 56′	0.058
Clermont-Ferrand . .	45° 47′	0.039
Paris	48° 50′	0.028
London	51.31	0.015

There can no longer, therefore, be any hesitation in admitting that, while the irregular movements of the atmosphere and the general range of the barometer increase, in going from the equator towards the poles, there is a regular concomitant fluctuation, which augments, as we proceed from high latitudes towards the equator. This phenomenon presents an universally acknowledged difficulty, and is, as yet, one of the unresolved problems of Meteorological Science. Attempts have been made, by some, to explain it upon the supposition of a tide produced as in the waters of the sea; but the regularity of its horary recurrence is obviously inconsistent with the notion of lunar influence. Others have imagined it to be dependant upon the alternations of land and sea breezes; but observations made in mid-ocean, totally disprove the fact.

I must solicit indulgence for the following attempt to solve the difficulty.

In drawing up my essay upon the constitution of the atmosphere, this, amongst the other phenomena, occupied much of my attention: but I have only slightly alluded to it amongst the meteorological facts in the third part. I thought it better, as it presented matter of some debate, and as its explanation was not essentially necessary to the general investigation, to reserve its consideration to a separate paper. Those who have had the patience to peruse my first pages, will, probably, most readily comprehend the following ideas.

Let us suppose that in the atmosphere surrounding the earth, a circulation is kept up between the poles and the equator; and that the cold dense air of the former regions flows in a lower current to the latter, while the elastic air of the latter is returned in an upper current to the former. There can be no difficulty in imagining further that, as long as these currents are maintained with regular velocities, a barometer, at all intermediate stations, might exhibit an equal pressure of the aërial columns: for as much air would flow from their summits, as would be returned to their bases. A general alteration of temperature which equally pervaded both currents, would produce no alteration in the weight of a vertical section, comprising both; nor would a partial alteration, equally diffused through the upper and under sections of any one column. The velocities of the currents would be partially altered, but the higher and the lower would still compensate each other. But an alteration of temperature, which affected the upper and lower currents unequally, would produce partial expansions and contractions, which would effect an unequal distribution of the ponderable matter. If the lower stratum of any perpendicular section were expanded by heat, while the upper were unaffected, the outgoing current of that section would be increased, while the in-coming current would be checked; and the balance of the two being disturbed, the total weight would be diminished: and, on the other hand, a local decrease of temperature would produce the analogous contrary effect. Now, the alternations of heat and cold, produced by the changes of day and night, may be regarded, in a general way, as equally affecting both the main currents of the atmosphere, and as equally pervading the whole length of the aërial columns. The heating surface being below, the warm particles quickly ascend, and are immediately replaced by the colder particles from above; and, by this vertical circulation, the diffusion of the heat is very rapid. But a minuter examination will satisfy us that, though rapid, this action is not in effect instantaneous; and the lower stratum, which is in contact with the heating surface, must, in the act of receiving heat, have its temperature disproportionately augmented.

The exchange of particles between the upper and lower strata, must occupy some time, however small the interval: the consequence must be, that the barometer will measure by its fall the amount of this inequality. So, on the other hand, in the process of cooling, in the absence of the sun, experiment has proved, that the lower strata of the air become more rapidly affected by radiation than the upper, and the total increase of weight from this cause will be shewn by the rise of the mercurial column.

Let us endeavour to trace this effect a little more minutely along any given meridian, beginning at the equator.

At this station, the only circumstance which we have to appreciate is, the irregularity of the lateral expansion or contraction. As the earth acquires warmth from the sun, the barometer falls; but the check thus communicated to the in-coming currents from the poles, must be felt along the whole line of their course; and their due velocity being opposed, without any adequate compensation in the upper currents, the barometer, from this cause, would have a tendency to rise at all latitudes between the equator and the pole. Assuming then an intermediate station upon the same meridian, we should have the same effect produced by the unequal expansion of the lower current of the atmosphere, but opposed now by the impulse communicated from the equator. The fall of the barometer would only then represent the balance of the two effects, and must be less than at the equator. The further we proceed towards the pole, the more must the revulsive action accumulate, and the less must the balance of the two become, till, at some neutral point, they are exactly equal. Beyond this point, again, the former action would exceed the latter, and the barometer would rise in the higher latitudes while it was falling in the lower.

The following Figure may possibly tend to illustrate these propositions.

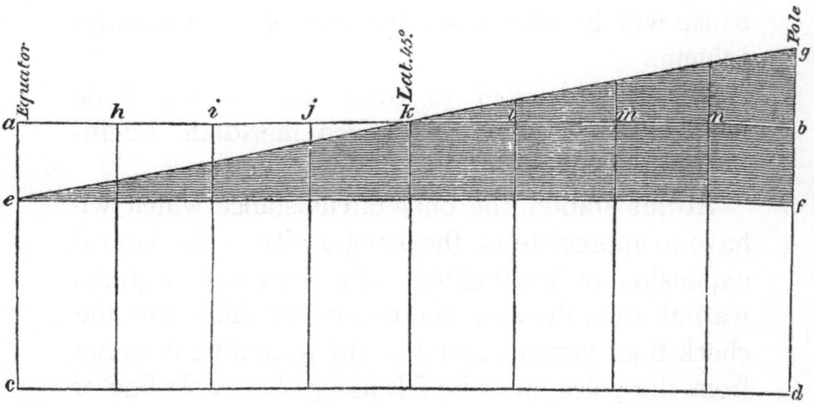

Let the parallelogram *a b c d* represent the lower current of air flowing from the pole to the equator in its undisturbed state, and the perpendiculars *h i j k*, *&c.*, different degrees of latitude. Let the lesser parallelogram *a b e f* be the equal diminution of weight which would arise from the partial expansion by the increase of daily temperature, and the triangle *e g f* the gradually increasing density arising from the retardation of the current. The rise and fall of the barometer on either side of the neutral point *k* would then be represented by the portions of the perpendiculars included between the hypothenuse *e g* and the side of the parallelogram *a b*.

The results of observations in different latitudes, included in the preceding Table, obviously coincide with such a gradual progress towards a neutral point; but we have as yet no experiments to prove the corresponding opposite effect beyond this limit. Whilst considering this subject, it occurred to me, that Captain Parry's observations at Melville Island, might possibly afford some light upon this interesting point. Upon consulting, however, the meteorological register, as published in his Journal, I was disappointed to find, that it only recorded the maximum and minimum height of the barometer in the twenty-four hours without mentioning the periods of their recurrence. I happened, very fortunately, to discourse with Captain Sabine upon the subject, and he assured me, that the observations were made and entered four or six times a day with the utmost regularity, and very obligingly offered to apply to the Admiralty for liberty to inspect the manuscript. His application was immediately complied with, and I was favoured with the loan of the original registers.

I found, upon inspection, that the Journal had been kept with the greatest precision, and the height of the barometer had been entered, during part of the time, at four regular periods, *viz.*, 6 A.M., noon, 6 P.M., and midnight; and the remainder of the time six times in the twenty-four hours, *viz.*, 4 A.M., 8 A.M., noon, 4 P.M., 8 P.M., and midnight. I immediately, with the utmost interest, undertook the arrangement of the observations to suit my purpose. I selected the twelvemonth from Sept. 1819, to August 1820, during which time the Hecla was constantly between latitudes 74° and 75°, and the greater part frozen up in Winter Harbour. The following Tables exhibit the results of my calculations. The first contains the monthly mean heights of the barometer and thermometer, taken four times in the day from September to February and part of March, and the second the monthly means taken six times in the day, from the latter part of March to August inclusive.

These Tables present a complete confirmation of the opinion which I had formed from theory.

In the first, including the winter half year, it will be observed, that the mean temperature scarcely varied between noon and midnight, the effect of the remote equatorial expansion was therefore unopposed, and the barometer constantly rose from 6 A.M. to 6 P.M., in coincidence with the fall in the lower latitudes. From 6 P.M. to 6 A.M. it as regularly fell.

In the second half of the year, while the sun was above the horizon, the daily variations of temperature were considerable, and the effect less regular; but, nevertheless, the barometer constantly rose from noon to 8 P.M., and then descended to midnight.

I am enabled, by the publication of the account of the Expedition to the Rocky Mountains of America, under the command of Major Stephen Long, to subjoin from their Meteorological Journal, for the same year, a comparison of the motions of the barometer at three different periods of the day, almost upon the same meridian, and at a distance of 33° of latitude. The expedition took up their winter quarters at "Engineer Cantonment," in latitude 41°25′ N., and longitude 95°43′ W., almost the centre of the great North American Continent, and the following Table contains my calculations of the means of the observations made during their stay in this situation.

ON BOARD THE HECLA, between Latitude 74° and 75°.

TABLE VI. *Shewing the Mean Height of the Barometer and Thermometer at Four different Hours of the Day at Melville Island.*

1819		6 A. M. Temperature	6 A. M.	Noon.	Noon. Temperature		6 P. M.	6 P. M. Temperature	Midnight. Temperature	Midnight.	
September . .	. . . .	+21.5	−29.884	+29.906	+23.7	. . . .	+29.920	+22.7	+21.3	−29.890	
October. . . .	. . . .	− 4.	−29.777	+29.808	− 2.8	. . . .	+29.840	− 3.9	− 5.	−29.825	
November . .	. . . .	−21.	−29.935	+29.946	−20.1	. . . .	+29.946	−20.1	−21.2	−29.937	
December . .	. . . .	−23.	−29.874	−29.872	−21.	. . . .	+29.881	−21.1	−21.6	−29.893	
January . . .	. . . .	−30.3	−30.040	−30.036	−30.	. . . .	+30.068	−29.9	−30.4	−30.063	
February . . .	. . . .	−32.8	−29.741	+29.758	−30.8	. . . .	+29.782	−32.6	−33.5	−29.771	
March . . . . (10 Ds.)	. . . .	−19.1	−29.551	+29.561	−14.5	. . . .	+29.614	−18.5	−20.5	−29.571	
			208.802	208.887			209.051			208.950	
			29.8288	29.8410			29.8644			29.8500	Max. 6 P. M. . . 29.8644
			− .0212	+ .0122			+ .0234			− .0144	Min. 6 A. M. . . 29.8288
											Difference .0356

TABLE VII. *Shewing the Mean Height of the Barometer and Thermometer at Six different Hours of the Day at Melville Island.*

1819	4 A. M.	8 A. M. Temperature	8 A. M.	Noon.	Noon. Temperature	4 P. M.	8 P. M.	8 P. M. Temperature	Midnight. Temperature	Midnight.	
March . . . . (20 Ds.)	−29.894		−29.885	−29.880		+29.902	+29.906			+29.910	
April	−29.963	− 9.2	+29.976	−29.971	− 3.7	+29.973	+29.988	− 8.1	−12.8	−29.987	
May	+30.116	+15.	+30 119	−30.099	+20.3	30.099	+30.109	+18.2	+13.1	30.109	
June	+29.826	+36.3	+29.828	−29.821	+38.6	+29.823	+29.819	+36.5	+33.6	−29.817	
July	+29.668	+42.5	+29.675	−29.674	+45.	+29.663	+29.665	+42 7	+39.1	−29.660	
August	−29.733	+32.7	−29.727	−29.734	+35.5	+29.737	+29.738	+32.	+30.5	−29.735	
	179.200		179.210	179.179		179.197	179.225			179.218	
	29.8666		29.8683	29.8631		29.8661	29.8708			29.8696	Max. 8 P. M . . 29.8708
	− .0030		+ .0017	− .0052		+ .0030	+ .0047			− .0012	Min. noon . . . 29.8631
											Difference .0077

To face Page 260.

The material originally positioned here is too large for reproduction in this reissue. A PDF can be downloaded from the web address given on page iv of this book, by clicking on 'Resources Available'.

TABLE VIII. *Shewing the Mean Heights of the Barometer and Thermometer at three different Hours of the Day at the Rocky Mountains of North America. Lat. 41°.25′, long. 95°.43′.*

	Morning.		Noon.		Evening.	
	Temp.	Barometer.	Temp.	Barometer.	Temp.	Barometer.
1819.						
September	60.	+28.650	80.3	−28.634	72.1	−28.633
October	40.3	+28.812	61.8	. 28.730	55.6	−28.720
November	35.8	+28.705	48.	−28.607	46.	−28.604
December	20.2	+28.808	29.	−28 660	26.3	+28.703
1820.						
January	2.1	+28.966	16.9	28.966	10.8	−28.954
February	22.1	+28.618	36.5	−28.501	32.	+28.550
March	27.4	+28.902	41.8	−28.815	37.	+28.881
April	45.7	+28.465	65.4	−28.267	62.4	−28.261
May	55.2	+28.496	69.7	−28.309	65.8	+28.370
	Mean	28.713		28 609		28.630

Maximum, Morn.	28.713
Minimum, P.M.	28.609
Difference	.104

Notwithstanding the height of this latter station above the sea, we still find the same principle prevail, and it is satisfactory to discover amongst so many various circumstances, whose influence upon the results are at present unknown, that, in accordance with the theory, on the same hours of the same months, the barometer, upon nearly the same meridian, periodically rose in latitude 74°47′, and fell in latitude 41°25′.

It must be acknowledged that much remains to be done for the complete elucidation of this subject.—Much, that is not difficult of performance, but requires extensive co-operation, and some nicety of observation. Had the numerous meteorological registers, which have hitherto been published, been kept with that exactness and that attention to the accuracy of the instruments employed, which is so necessary in scientific pursuits, the *data* for these and for other highly important calculations would have been already abundant: but, notwithstanding the multitude of labourers employed in this interesting field, there is a want of unity, and especially a carelessness, in their exertions which render them totally unavailing to the nicer purposes of the science. Let the meteorologist inquire by what means it is, that astronomy, the sublimest of all the sciences, has attained to its present wonderful state of perfection; and he will find that it is by the most *microscopic* attention to the perfecting of its instruments of research, and by the most faithful precision of unremitting observations; and let him be assured that it is only by the same painful care to *minutiæ*, that his own favourite science will ever be raised to that standard of exactness of which there can be no doubt that it is susceptible, but from which it is to be lamented that it is at present so far removed.

AN ESSAY

UPON THE

CLIMATE OF LONDON.

AFTER the very interesting and laborious work of Mr. Howard, upon the climate of London, it may, at first sight, appear presumptuous in me to claim attention upon a subject which has been so ably and so extensively pre-occupied: but when it is considered that that able philosopher was unprovided with any sufficient means of measuring the quantity, or estimating the changes of the sea of vapour which necessarily permeates and pervades every part of the great aërial firmament; and when it is borne in mind, that one of the main springs of all the wonderful motions of the air, and the changes of the weather, is the slow and silent influence of the aqueous steam ; I shall be excused for attempting to elucidate, from experiments, a part of the subject so important, but, hitherto, so neglected.

Its connexion with the vegetable kingdom, and with all the most important processes of the agriculturist, must be evident to the most superficial observer ; and it is more than probable, that it will be found of equal importance to those who make a study of the complicated processes of the animal economy.

It has ever been a favourite speculation with philosophers to trace in the constitution of the atmosphere the origin of some of the diseases which affect the human race. The discovery of pneumatic chemistry, and the new means of questioning nature, which it put into their hands, seemed at first to promise a solution of this interesting problem; and hopes were entertained that the cause of epidemic and local complaints might be found in the varying elements of the compound air we breathe. The eudiometric processes which were immediately instituted and repeated in every part of the world, proved, however, the unvarying proportions of the permanent gases of which it is composed. It is not, therefore, irrational to suppose that an accurate method of estimating the varying quantity of aqueous vapour in the elastic medium which surrounds us, which is the only fluctuating ingredient of its composition, may lead to some useful hints upon this important subject. Certain it is, that some indications of this kind may be perceived even by the healthy, and those who are not conversant with the progress of disease. There are days on which even the most robust feel an oppression and languor, which are commonly and justly attributed to the weather; while on others they experience exhilaration of spirits, and an accession of muscular energy. The oppressive effect of close weather and sultry days, may probably be accounted for from the obstruction of the insensible perspiration of the body, which is prevented from exhaling itself into the atmosphere, already surcharged with moisture; while unimpeded transpiration from the pores, when the air is more free from aqueous vapour, adds new energy to all the vital functions. In bodies, debilitated by disease, indeed, the contrary effects may be produced: they may be unable, from weakness, to support the drain of free exhalation which is exhilarating to the healthy; and hence, probably arises the benefit of warm sea-breezes in cases of consumption, and diseases of the lungs. Observations upon climate, with a more particular regard to the hygrometric state of the atmosphere, may reasonably be expected, amongst other certain advantages, to throw some light upon the treatment of these complaints; and may, perhaps, teach us to construct an artificial atmosphere, of greater efficacy than any that has yet been recommended, in cases when the relief of local change may be impossible.

The foundation of the following attempt is a series of observations unremittingly continued three times in the day for three years; a period of time which, though it may not entitle me to say that it includes all the changes occasioned by the revolution of the seasons, is sufficiently long to furnish, by the system of averages, a very near approximation to the distinguishing characters of the climate. I have assumed the *data* as furnished by my own experience, and it is gratifying to find that, with the barometer and thermometer, my results agree very closely with those of Mr. Howard, who has founded his calculations upon a long series of years, and to whose conclusions, in cases of discrepancy, I willingly cede the preference which is due. The general accordance between us, however, upon these points, encourage me to presume that the indications of the hygrometer will not be found to differ very far from the real mean.

I shall proceed, first, to consider the general characters of the climate, as derivable from the averages of the three years together; and I shall then endeavour to institute a comparison of the separate years with the mean, and with each other.

I shall not, now, enter into any detailed account of the instruments employed, or the mode of placing and observing them; I shall reserve for another place, a few observations upon the precautions which ought to be taken in making meteorological observations, and shall only here assume credit for moderate care in these particulars: I must, however, premise, that the times of the day, denoted by morning, afternoon, and night, were, for the first, from eight to ten o'clock, A.M.; for the second, from half-past three to half-past five, P.M.; and, for the third, from ten to half-past eleven, P.M.

The mean pressure of the total atmosphere, denoted by the barometer, I find to be 29.881 inches. The mean of twenty years, deduced by Mr. Howard from the observations of the Royal Society, is 29.8655 inches. The mean temperature derived from the daily *maxima* and *minima* of the thermometer, is 49.5°, which corresponds even to the decimal place, with Mr. Howard's estimate. The mean dew point I consider 44.5°, it being also calculated from the daily maxima and minima. The elastic force of the vapour is, therefore, 0.334 inch, and a cubic foot of the air contains 3.789 grains of moisture. The degree of dryness is represented by 5° upon the thermometric scale, and the degree of moisture by 850, upon the hygrometric scale. The average quantity of rain is 22.199 inches, and the amount of evaporation, calculated from the hygrometer, 23.974 inches. The weight of water, raised from a circular surface of six inches diameter, is 0.31 grain per minute.

The accordance of this method of estimating the amount of evaporation with the results of actual measurement is gratifying, and proves most incontestibly the accuracy of the calculations upon which it is founded. From the mean of four years, the gauge being upon, or near, the ground, Mr. Howard found that the annual results averaged 21.46 inches. Of these, one year, the summer of which was hot and dry, afforded 25. inches, and when it is considered that two out of the three summers now in question were likewise distinguished by a high temperature, the coincidence becomes very striking.

The range of the barometer is from 30.82 inches to 28.12 inches; the range of the dew-point from 70° to 11°. The pressure of the vapour varies with it from 0.770 inch. to 0.103 inch. The maximum temperature of the air is 90°, the minimum 11°.—The force of radiation from the sun averages 23.3° in the day, and the force of radiation from the earth at night 4.6°: the highest temperature of the sun's rays, is 154°, and the lowest temperature on the surface of the earth 5°. The greatest degree of dryness is 29°, or the least degree of moisture upon the hygrometric scale 389. The time of the day influences in some degree all the mean results. One of the most constant effects is that produced upon the barometer. The mercurial column reaches its maximum height in the morning, declines to its minimum in the afternoon, and again rises at night. The average difference of these periods, as exhibited by the journal, are as follow:—Morning above night, +.005 inch.—Afternoon below morning, — .015 inch.—Night above the afternoon, +.010 inch. —The means of the monthly observations, present but one or two exceptions to the fall in the middle of the day, or to the rise from afternoon to night, but the rise from night to morning is not quite as constant.

With respect to the dew-point, it may be considered that the journal includes four daily observations; for the observation of the minimum temperature of the air, which constantly falls a few degrees below the term of precipitation taken in the day, must obviously be included. From morning to afternoon, it rises but 0.3 of a degree; from afternoon to night, it falls 0.9 of a degree, and below this again, the minimum temperature is 2.7°, and the mean is calculated from the latter, and the afternoon observation.

The temperature of the air varies in the twenty-four hours from 56.1°, its mean maximum, to 42.5° its mean minimum. The mean temperature of a climate, is generally regarded as made up of the average impression of the sun due to its latitude, upon the surface of the globe. The mean quantity of aqueous vapour must also be referable, finally, to the same principle. But there is another way of considering the subject more accurate in detail, though upon an average of years ending in the same conclusion: that is, to regard the mean temperature as made up of the temperature of different currents flowing from different points of the compass; and it will be necessary to my purpose to contemplate the atmosphere of vapour particularly, in this point of view. The medium dew-point 44.5° is therefore made up of the following proportions of the means from eight points of the wind—

87 North 40.1° – 133 north-east 40.7°.
80 East 42.3° — 111 south-east 45.6°.
70 South 48.7° – 225 south-west 48.6°.
215 West 44.8° — 174 north-west 41.3°.

Before I enter upon the consideration of the effect of the sun's progress in declination, and the succession of the seasons, I shall endeavour to point out the influence of the geographical situation of the island of Great Britain upon its aqueous atmosphere. The mean quantity of the vapour follows exactly the changes of the mean monthly temperature, that is to say, the dew-point rises and falls with the increase and the decrease of the heat. But the winds which transport the vapour may be divided into two classes; namely, the land winds which blow from off the great continent of Europe, and which comprise the north-east, the east and south-east; and the sea-winds which blow from the great oceans which surround it on every other side; namely, the north, north-west, west, south-west, and south. In the former, we may expect to find that the course of the mean temperature is exactly followed; for the sources of the vapour must be comparatively shallow streams, and reservoirs of water, whose temperature must soon adapt itself to that of the surrounding air. But in the unfathomable depths which supply the latter, the law by which the density of water is regulated, must, at particular seasons, maintain a temperature above the mean of the declining season; whilst at others, the increasing heat of the latter must outstrip the progress of the former. The following Table contains the dew-point of the several winds, divided into the two classes for every month in the year, beginning with the autumnal quarter.

TABLE I. *Shewing the Difference of the Dew-point in the Land and Sea-winds.*

	Land Winds. NE. E. SE.	Sea Winds. N. NW. W. SW. S.
September . .	53°	53°
October	45	46
November . .	41	42
December . .	31	37
January . . .	29	35
February . . .	31	35
March	34	38
April	45	42
May	47	44
June	54	54
July	52	55
August	56	57

And here the effect anticipated is clearly perceptible. The vapour of the land-winds, it will be seen, declines in force from September to January, in which month it reaches its minimum, and from that point gradually rises till it reaches its maximum in August; and this, it will be afterwards seen, is the exact progress of the mean temperature of the air. In the sea-winds the vapour follows the same course from September to November, and the balance is such, that the elastic force of both divisions is nearly the same. The north and south winds neutralize each other, and the north-west, west, and south-west, are equivalent to the north-east, east, and south-east. Having descended to about 40°, which is somewhere about the point of the greatest density in water, in November, the accordance proceeds no further. In December, the vapour from the land has descended six degrees below that from the sea, and the difference continues in January. In February the former rises two degrees, and the latter remains stationary. The difference of four degrees continues through March, and is diminished to three degrees in April and May. In June, they again attain their former equality. The reason of this is obvious; the temperature of 40°, being that of the greatest density, cannot be lowered till the whole mass of the waters has passed this term; and in the deep seas, this must necessarily be a process of some duration. The shallow waters, on the contrary, soon assume the temperature of the ambient air, and continue to decline with it in heat. Upon the return of spring the contrary effect is produced. The great deeps must again repass the fortieth degree before the superficial waters can take the higher temperature of the incumbent atmosphere. The consequences we should expect from this progression, would be an increase of humidity in December and January, and a rapid decrease in the four following months; an expectation which we shall find correct in our further investigation.

There is another law of the aqueous fluid, which we might also expect to have an influence upon the emission of its steam—the evolution, namely, of heat in the process of congelation and its absorption during the liquefaction of ice. The British isles are placed in such a position, as would induce us to suppose that, at particular seasons of the year, this influence might be perceptible in one direction more than in any other. We may bring this idea to the test, by comparing together the northerly and southerly winds, as is done in the following Table:—

TABLE II. *Shewing the Effect of the Ice in the North Seas upon the Dew-point.*

	Southerly. SW. S. SE.	Northerly. NE. N. NW.
September . .	58°	48°
October	51	41
November . .	47	37
December . .	42	32
January . . .	38	31
February . . .	36	31
March	42	32
April	47	40
May	51	41
June	58	50
July	58	50
August	60	54

Here we may observe, that the decline of the vapour from September to December is exactly equal in both classes, but from that time it ceases about the temperature of 32° in the northerly winds, and continues in the southerly to the month of February. In March, again, the temperature of the latter has increased from the minimum 6°, but in the former it still remains at 32°. In April, on the contrary, the increase in the northerly winds exceeds that of the southerly; and in May, they have again attained their original relative distances, and resume their parallel progression. It would be difficult, I think, to assign any other cause for this modification of the phenomena than the one which has just been suggested. The evolution of heat, in the process of freezing, stops the decline of temperature in the regions exposed to its influence, while it proceeds in those which are not exposed to the change ; and the absorption of heat, in the operation of thawing, prevents the accession of temperature, which is due to the returning influence of the sun. When this operation has ceased, the vapour quickly attains its former relative degree of force.

Wonderful adjustments these, to mitigate the rigours of a northern climate! They both operate from November to February, by the evolution of heat in the coldest season of the year; and at the same time, by an extra supply of vapour, decrease the degree of dryness, and prevent the consumption of heat which always attends the process of evaporation.

Let us now endeavour to trace the order from which, " while the earth remaineth, seed-time and harvest, and cold and heat, and summer and winter, and day and night, shall not cease."

In the month of January, the first month of the year, but which, in the most natural division of the seasons, constitutes the second month of the winter quarter, heat is at its minimum in all its particulars. The mean temperature is 36.1°, varying from 39.6°, the mean highest, to 32.6°, the mean lowest; the utmost range of the thermometer being from 52° to 11°. The average power of the sun is 4.4°, and the utmost intensity of its rays 12°. The cold, produced by radiation from the earth, is 3.5°, and the greatest effect 10°

The mean force of the vaporous atmosphere is also at its lowest point, 0.234 inch, the dew-point being 34 3°. The mean degree of dryness, calculated from the mean temperature and the mean dew-point, is 1.8°, and the state of the air's saturation 939. The average degree of greatest daily dryness is 3.5°, and that of least saturation, 878.

The quantity of rain in this month greatly exceeds the amount of evaporation, the former being 1.483 inch, and the latter, at its minimum, 0.413 inch.

The height of the barometer is 29.921, and its mean range 1.60 inch.

In the month of February the mean temperature increases to 38°, nearly two degrees. This accession takes place principally while the sun is above the horizon; the maximum temperature rising to 42.4°, nearly three degrees, while the minimum only advances about one degree to 33.7°. This difference is partly owing to the increased influence of radiation under a less clouded sky, which dissipates the accumulated heat; the temperature of the radiant thermometer averaging 29°, one-tenth of a degree lower even than in January. The greatest force of radiation is 10°, as before, but the average effect is increased to 4.7°. The power of the sun rises to 10.1°, and its greatest intensity to 36° The range of the diurnal temperature of the air is from 53° to 21°.

The dew-point advances to 34.9°, only 0.6 of a degree; the peculiar laws of the evaporating fluid keeping it back, as before explained. The force of the vapour is 0.239 inch. The consequence of this retardation is, that the mean degree of dryness advances to 3.1°, and the hygrometric state of the air falls to 905. The average degree of greatest dryness is 6.1°, and that of least saturation 816.

The quantity of rain is at its minimum, being 0.746 inch, which is very little more than 0.733 inch, the amount carried off by evaporation.

The mean pressure of the atmosphere is 30.067 inches, and the range of the barometer 1.36 inch.

With the month of March commences the spring quarter, the seed-time of the husbandman, when it is so important to the interests of agriculture that the superfluous moisture should be exhaled from the earth, which would prevent the proper preparation of the soil, and destroy the germinating principle of the grain. By a wise Providence, therefore, the temperature of this month advances six degrees, while the dew-point rises only four, checked by the same cause which began to restrain it in the last month. The mean temperature is 43.9°, and the point of precipitation 39° ; making the degree of dryness 4.9°, and reducing the moisture of the air to 831. The elasticity of the vapour is .272 inch. The evaporation is rather more than doubled, amounting to 1.488 inch, and exceeding the quantity of rain, which is 1.440 inch. The average degree of greatest dryness is 9.6, and that of least saturation 715.

It is still during the day that the heat accumulates most, the maximum rising to 50.1, and the minimum to 37.7, an increase of 7.9° in the former, and only 4°. in the latter. The temperature of the air ranges from 66° to 24°. The amount of radiation is 5.5°, an increase of nearly one degree, but its maximum effect is 10°, as before. The force of the sun's direct rays is 49°, and their mean maximum effect 16°.

The height of the barometer is 29.843, and its range 1.26 inch.

In April, the mean temperature of the air rises six degrees to 49.9°, and the constituent temperature of the vapour only 4.5° to 43.5°, making the amount of dryness 6.4°. The degree of moisture is consequently no more than 783. The mean of maximum dryness 12.8°, and the mean of minimum saturation 651. The elasticity of the vapour .322 inch. Evaporation is increased to 2.290 inches, and the quantity of rain does not exceed 1.786 inch. The power of radiation from the earth is raised to 14°, and its mean effect attains its highest amount of 6.2°. The power of the sun averages 28.1°, and the highest observed effect is 47°. The heat of the air ranges between 74° and 29°.

The mean height of the barometer for this month is 29.881 inches, and its average range 1.11 inch.

In May the temperature of the air still outstrips the advance of the vapour, and the atmosphere attains very nearly its state of greatest dryness. The mean of the former is 54°., that of the latter 46.1°. The state of saturation 769, the degree of dryness 7.9°, the mean minimum of the former 597, the mean maximum of the latter 15.6°. Elastic force of the vapour 354 inch. Evaporation amounts to 3.286 inches, and rain to 1.853 inch. The power of the sun is 57°, its mean greatest influence 30.5. The force of radiation, from the surface of the earth, is 13°, its nightly effect 4.2°. The reduction of this effect implies a rather more clouded state of the atmosphere than that of the preceding month. The mean maximum of the air is 62.9°, the minimum 45.1°: the range of the thermometer from 70° to 33°.

The height of the barometer is 29.898, its range 1.09 inch.

In June, the first month of the summer quarter, the advance of the dew-point and of the daily temperature are nearly equal: the former averages 50.7° the latter 58.7°. The degree of dryness is therefore 8°, and the state of the air's saturation 762.

The force of the vapour .410 inch.

The quantity of evaporation rises a little above

that of the last month, and amounts to 3.760, the maximum of the year, and the quantity of rain is 1.830 inch.

The energy of the sun's beams is at its height, and also its maximum effect: the former amounts to 65°, and the mean of the latter 39.9°. The temperature of the air does not attain its maximum till the two following months. This arrangement must have an extremely important influence upon the fructification of the vegetable kingdom, and the horticulturist and botanist would do well to attend more particularly, than has hitherto been done, to the different modifications of heat of radiation and heat of temperature. Experience has suggested many practical precautions and artifices evidently connected with this subject, and it is almost certain that a scientific attention to these particulars would tend much to the benefit of the art of gardening.

The force of radiation from the earth, I have once observed in this month to be 17°, the greatest effect that has ever come under my notice: its mean amount 5.2°.

As connected with the subject to which I have alluded above, it is worth while to notice that there are but two months in the year, in which vegetation, in particular situations, is not exposed to a temperature below the freezing point. These two months are July and August, and even in them the radiant thermometer descends to 35° and 34°. Thus, a plant might be so situated, in the month of June, as to undergo all the changes of heat from 154° to 30°.

The mean maximum dryness of the month is 16°:

the mean minimum saturation, 597. The maximum temperature of the air averages 69.4°, the minimum 48.1°, and the greatest difference between the two happens at this time. This difference is evidently chiefly dependant upon the power of the sun, and the time that it remains above the horizon; therefore, like the direct heat of the solar rays, it follows the progress of the sun's declination. The range of the thermometer is from 90° to 37°.

The mean pressure is 30.020 inches, and the mean variation 0.64 inch.

In July, the increase of vapour is rather greater than that of temperature, and both approach their maximum. The mean heat of the air is 61°, and that of the dew-point 54.5°. The force of the vapour 468 inch. The degree of dryness is 6.5°: the hygrometric degree, 811. Mean maximum dryness of the day 13.7. Mean minimum moisture 658. Evaporation decreases to 3.293 inches, and the rain attains its maximum quantity 2.516 inches.

The increase of the mean temperature here appears to be wholly derived from the night, for the mean maximum is only 69.2°, while the mean minimum has risen to 52.2°. This must be owing to the cooling power having been checked by a cloudy sky, and accordingly we find that the effect of radiation has fallen to 3.6°, while its greatest power is 13°.

The force of the sun's rays decrease to 52.5°, but it is probable that this is not their utmost power, as the following month exhibits an increase. Their average greatest effect is 25.8°. This decrease of the solar power does not immediately check the mean temperature, for the earth having become heated in the preceding months, acts as a warm body on the atmosphere, and gives out again the heat which it has received. Mr. Howard's explanation of the mean temperature always being about a month behind the sun's place in declination, is, no doubt, as correct as it is ingenious; namely, that "as the sun advances in north declination, the heat we derive from him increases, actually in proportion to his altitude, but not sensibly; because a part of it is required to heat the earth, and is lost there by absorption. As he declines southward in the autumn, the heat we receive actually grows less in proportion, but not sensibly; because we now receive back a certain quantity from the warm earth."

The greatest range of the temperature of the air for this month is from 76° to 42°.

The height of the barometer is 29.874 inches, and the mean range 0.79 inch.

The particulars of the month of August remain much the same as those of the month of July. The warm nights continue, and the heat of the day is undiminished. The mean temperature is 61.6°; the maximum of the day, 70.1°; and the minimum of the night, 52.9°. The range of the thermometer from 82° to 41°. The force of the sun's rays 59.5°, and their average maximum effect 33.1°. The power of radiation from the earth 12°, and its mean amount 5.2°.

The dew-point is 55.3°, and the elastic force of

the vapour .481. The degree of dryness 6.3°, and state of saturation 819. Mean maximum dryness 12.4°; mean minimum moisture 677.

Evaporation is the same as in the last month, 3.327 inches, but the rain is decreased nearly one-half; the amount being only 1.453 inch.

Mean height of the barometer 29.891 inches; mean range 0.73 inch.

In September, the first month of Autumn, the reduction of temperature begins to be sensibly felt; but, still, less in the night than during the day. The mean temperature declines to 57.8°; the maximum to 65.6°, and the minimum to 50.1°; the greatest range of the thermometer being between 76° and 36°.

The mean dew-point is 52.3°, and the elasticity of the vapour .432 inches; the dryness of the air 5.5°, and its state of saturation 827. Mean maximum dryness 11.1°; and mean minimum moisture 702. The precipitation and evaporation are again nearly upon a par, the former averaging 2.193 inches, the latter 2.620 inches. The power of the sun is but little decreased, its greatest energy being 54°, and its mean daily amount 32.7. Terrestrial radiation also remains nearly the same, rising to 13°, and averaging 5.4°. The height of the barometer is 29.931 inches, and its mean range 0.88 inch.

In October, the mean temperature falls nearly 9°, and does not exceed 48.9°; the maximum and minimum averaging, respectively, 55.7° and 42.1°. The dew-point declines almost in the same proportion

44.8°. The dryness is reduced to 4.1°, and moisture increases to 870. Evaporation decreases to 1.488 inch, while the rain continues in nearly the same quantity; the amount for the month being 2.073 inches. Now, that the fruits of the earth are laid up in store, this increase of wet is attended by no injurious effects; the remaining heat of the earth is preserved from a needless expenditure, and guarded from dissipation by an increasing canopy of clouds. The effect of radiation is reduced to 4.8°, and its greatest force to 11°. The power of the solar rays declines to 43°, and their mean effect to 27.5°. The greatest range of the air's temperature is from 68° to 27°. The mean elasticity of the vapour is .336 inch, the pressure of the whole atmosphere 29.774 inches, and the average range of the barometer 1.38 inch.

In the dark and dreary month of November, the atmosphere is nearly saturated with moisture. The temperature of the air is 42.9°, and the dew-point averages no lower than 40.5°; the dryness is, therefore, only 2.4°, and the dampness amounts to 910. The precipitations are augmented to 2.400 inches, and only 0.770 inch is carried off by evaporation. The maximum dryness of the days is but 4.7°, and the least degree of moisture 845. The effect of the sun's rays, whose greatest power is 23.5°, is scarcely 6.8°, and that of terrestrial radiation only 3.6°; its intensity being 10°

The mean highest point of daily temperature is 47.5°, and the mean lowest 38.3°; the utmost range of the thermometer being from 62° to 23°.

The mean elasticity of the vapour is .286 inch; the pressure of the whole atmosphere, 29.776 inches, and the range of the barometer 0.92 inch.

The month of December closes the year with nearly the same characters as those of the last month; mean temperature, 39.3°; mean maximum, 43.2°; mean minimum, 35.4°; greatest range, from 55° to 17°.

The greatest force of the sun's rays, 12.5°; their mean influence, 5.4°; power of terrestrial radiation, 11°; mean effect, 3.5°.

Temperature of the dew-point, 37.6°; degree of dryness, 1.7°; and state of saturation, 952; mean maximum dryness, 3.3°; mean minimum moisture, 888.

Amount of precipitation, 2.426 inches; of evaporation, 0.516 inch.

The elasticity of the vapour, .261 inch; pressure of the atmosphere, 29.693 inches, and range of the barometer, 1.13 inch.

I have not, in the preceding summary, noticed the prevalent winds of the several months, or distinguished the quality of the vapour transported from the different quarters of the compass. I have thought it better to separate this view of the subject from the preceding, and to present the results in a tabular form. The following Table exhibits the average number of days on which the different winds blow in each month of the year, together with the mean dew-point of the vapour which is wafted by them:—

TABLE III.,

Shewing the Dew-Point of Eight different Winds in each Month, and the average Number of Days on which each prevails.

	N.		N.E.		E.		S E.		S.		S.W.		W.		N.W.	
	No. of Days.	Dew Point. °	No. of Days.	Dew Point. °	No. of Days.	Dew Point. °	No. of Days.	Dew Point. °	No. of Days.	Dew Point. °	No. of Days.	Dew Point. °	No. of Days.	Dew Point. °	No. of Days.	Dew Point. °
January . . .	3½	—31.5	4½	—27.5	1½	—23.5	2¼	—34.5	1¾	—39.	6½	—42.5	6¼	—37.	4½	—32.
February . .	1¼	—30.0	4¾	—29.	2½	—32.	2¾	—34.5	2¼	—37.5	5	—39.5	5½	—39.	3¾	—34.
March . . .	2½	—31.5	4	—31.	—	—	2	—35.0	2¼	—47.0	9½	—44.5	6¼	—42.	4½	—35.
April	2½	—40.	3¼	—40.5	3	—45.	3½	—49.0	2¾	—47.	4	—45.	5½	—44.	5½	—42.
May	3	—42.	4	—40.5	4½	—45.5	4	—50.5	1	—54.	6¼	—49.5	5¼	—46 5	3	—41.
June	5	—49.5	6½	—49.5	2	—56.	4	—57.	1	—62.	3½	—56.	3	—52.	5	—50.5
July	2½	—50.	3	—49.	2	—50.5	4	—58.	2¼	—58.5	7	—59.	5	—56.	5¼	—53.
August . . .	1	—55.5	2¾	—53.	1½	—55.5	3	—60	2¼	—63.0	6	—58.5	11½	—55.	3	—53.
September . .	2	—45.	4	—50.	1	—52.	4	—56.	1	—61.	6	—58.	6	—54.	6	—49.5
October . . .	3	—38.5	3¼	—41.5	2	—45.5	3½	—49.	2¼	—53.5	5¾	—50.5	5	—46.5	6¼	—43.
November . .	3	—38.0	3	—37.	3	—40.	2	—46.	3	—48.0	6	—47.5	5	—42.	5	—35.5
December . .	1	—31.5	2¼	—29.	3½	—27.5	4	—38.	2	—45.5	8¼	—44	6	—40.	4	—35.
	30½	—40.	45¼	—39.5	26½	—43.	39	—47.	23¾	—51.	73¾	—49.	70¼	—47.	55¾	—42.

This Table is constructed from the observations of the morning, afternoon, and night, leaving out those of the minimum temperature; and therefore the total means differ slightly from those already given. This has been done for the sake of forming a standard of comparison, whereby to judge of the state of the weather from hygrometric observations. The mean monthly temperature of the dew-point affords a useful criterion for this purpose, but the average state of each wind is much more accurate; and when the Table shall have been improved by the results of a longer series of experiments, an almost infallible judgment may be formed from it of the probability of atmospheric changes. I shall return to this subject hereafter.

It will be observed in the Table, that the northerly winds and the southerly are in nearly equal proportions, but that the westerly are to the easterly nearly as two to one. These proportions are preserved in the several quarters of the year.

It is also worthy of remark, that the dew-point of the sea-winds, *viz.*, the S.W. W. and N.W. is 3° higher than that of the land-winds from the opposite quarters, *viz.*, N.E. E. and S.E.

Thus much we have learnt from the mean observations of the three years together: let us endeavour to advance a few steps further, by a comparison of the years, one with another, and with the mean.

To render this analysis more practically useful, I shall endeavour to connect the popular description of the weather with the scientific details; and I shall, more particularly, aim at elucidating the influence of atmospheric changes upon those branches of vegetation, with which are connected the wealth, happiness, and subsistence of the community. By connecting together the reports of the agriculturist, and the observations of the meteorologist, we may, in time, obtain some insight into the nature of the various blights which affect the different products of the soil, whether it be the mildew and smut of wheat, or the, so called, fly of the turnip or the hop. Something may be learnt from a comparison of a forward or a backward season, and the effect of weather upon the soil; some precautions may be suggested, and the grounds of anticipation strengthened. Knowledge, in short, in this, as in all other instances, will surely end in power; at all events, a running commentary, in the common language of the farmer, may draw the attention of those most interested in the subject, to details which might otherwise appear uninteresting or unintelligible.

The observations of the Journal commence with the first month of the autumn of 1819. The particulars do not differ essentially from the mean in any respect. The weather was fine, warm, and seasonable, and corn-harvest was completed under favourable auspices. With respect to the aggregate products of the earth, the season was one of the most plentiful that had occurred for some years. In the early part of the month rain was greatly wanted, and water for the cattle; the pastures were burnt up, and the stubbles bare. The showers in the latter part produced a very beneficial effect.

The results of October, taken together, present nothing very remarkable but the unequal distribution of the fine weather. The first part of the month was much above the mean in temperature, and the latter part as much below; and a considerable quantity of snow fell on the 21st of the month, which is very unusually early in the season. The thermometer ranged from 68° to 27°, which are its greatest limits for the month. The degree of dryness was altogether below the mean. Southerly and westerly winds prevailed till the middle of the month, and afterwards the north and north-easterly winds set in; and it was probably owing to their influence that the cold precipitations of the vapour took place. The ground was in a very favourable state for the sowing of wheat, and the turnips derived considerable improvement from the increase of moisture, and were deemed, upon the average, a fair crop. Grass and fodder were superabundant

The prevalence of catarrhal, rheumatic, and other inflammatory disorders, together with affections in the bowels, were ascribed by physicians to the extraordinary vicissitudes of the weather.

The month of November was colder than had been known for many years. The mean of the thermometer was 3.7° below the season, and it descended to its lowest limit for the month. It was likewise very damp, the degree of moisture being 941, while the average is not above 845. The amount of rain was not great, but that of evaporation very small. The precipitations fell chiefly in light showers, and seldom lasted the whole day.

This season of the year affords but little room for remarks upon the influence of the weather upon the important processes of vegetation. The young wheats, which looked generally promising, received a sudden check from the early frosts, and were not expected to benefit from the change to extremely mild and moist weather, which suddenly came on in the latter days of the month. Green food for cattle still abounded

With the month of December commenced, what may be deemed, a severe winter. The temperature averaged 5° below the mean, but its state of saturation did not differ much from its usual proportion, and the precipitations were altogether light. The mean state of the barometer was rather low, but its range uncommonly small for the time of year.

The long continuance of frost put an end to all the business of the field, but the young crops were well secured by a seasonable covering of snow. The turnip crops, however, received injury in proportion to the benefit derived to the wheat. The alternations between frost and thaw were particularly injurious to that useful root.

In January, 1820, the mean temperature was almost six degrees lower than usual, and the frost continued, with nearly unremitting rigour, till the 23d of the month. The barometer fluctuated very much, and its maximum was higher than had been observed for many years. The weather was altogether damp and unpleasant.

The wheats, and all arable lands, were considered to derive a full portion of the advantages which never fail to accrue from a frost of some duration, accompanied with a sufficient cover of snow. The young clovers, and crops of that description, suffered greatly from the rigour of the frost; and vast quantities of turnips were totally destroyed.

The vicissitudes of the weather, and extraordinary severity of the season, produced a more than ordinary number of those disorders which affect the pulmonary organs.

The month of February was still below the mean in temperature, and unusually damp. The amount of precipitations was large, and evaporation very trifling. In other respects, there was no variation worthy of remark.

March was more than usually dry, and there were scarcely any precipitations of rain and snow. The temperature continued lower than usual.

The lands were in a good and fertile state, and the sowing of the Lent corn, was carried on with expedition. The effect of the preceding frosts was to pulverize the soil, which turned up in most places like garden-mould. The latter sown wheats appeared weak, and dependant entirely upon the genial nature of the coming spring. The early wheats stout and healthy. All had, however, suffered in some degree from the sharp N.E. winds, which retarded vegetation, particularly of the grass. The green crops were affected severely.

The mean temperature, in April, nearly recovered its proper amount, and the weather was altogether dry and seasonable.

The sowing of the corn was never completed under more happy auspices. The season had been previously most favourable to manuring, and the heaviest clays worked admirably. Vegetation, which had suffered very much, recovered with the timely showers of this month, but many crops, upon loose and spongy soils, were injured past recovery. On the whole, however, the appearance of the country was favourable, and the shew for fruit abundant.

The month of May was genial and seasonable in every respect. The temperature rather above the mean. Rain was interspersed in just such proportion as seems most conducive to the welfare of vegetation. The season, however, was above the mean in dryness. Vegetation was sudden and rapid, and an appearance of luxuriance took place in all the productions of the soil, although the fruit blossoms suffered partially from the frost in the early days. The pulse, artificial grass crops, and hops, which were beginning to suffer severely from their peculiar blight-insects, were well washed by the showers, and speedily recovered the incipient damage.

June was extremely remarkable for being unusually cold and wet in its commencement, and, for the very extraordinary rise of temperature in its latter half. This rise was not accompanied by a proportionate increase in the quantity of vapour, and the weather was beautiful in the extreme. The temperature of the whole period was above the mean, and rain very abundant. The variation of the barometer was great, and its mean very high during the hot weather.

The cold and wet weather increased the bulk of vegetation, but without proportionally accelerating the fructifying process; but after the change, the wheats blossomed most beautifully. Hay-harvest was at its height, and an abundant crop most successfully secured.

That precarious plant, the hop, appeared strong and luxuriant.

July was above the average in heat, but a want of sun was experienced to ripen the corn, and a consequent fear of mildew was prevalent. The spring crops and hops were of good promise, and turnips looked very well. The heavy periodical rains of this season commenced on the 16th of the month, and were accompanied at intervals by violent thunder and lightning. Both the quantity of precipitation and evaporation exceeded the mean amount.

August—this month, so important to the agriculturist, was as favourable to the operations of harvest as could be wished: of full average temperature, with a sufficiency of direct radiation from the sun; without which the fruits of the earth scarcely ever arrive at perfection. The mean degree of dryness was greater than ordinary, but a full proportion of rain was not wanting. The weather was particularly propitious till the 19th, when a degree of *blighting* cold took place, which is marked in the journal by the radiating thermometer falling to 35° This state of check did not fortunately last long, and was succeeded by a mild temperature, well calculated to forward the labours of the field.

The appearance of the turnips was not so favourable as could have been wished—they had suffered from blight, or from what the farmers term the fly.

With September, 1820, we commence our second annual period, for which we now possess other terms of comparison, besides the mean, in the divisions of the year which we have already analyzed.

The temperature of the month was below the mean, and still more below the last year, the difference taking place entirely in the night. This may, probably, have arisen from the great effect of radiation, from a particularly clear state of the atmosphere. The register of the radiating thermometer is in accordance with this supposition; for the mean effect will be found to be considerably greater than usual, and nearly amounting to 7°. The injury which the hop-plant received about this time, may not improbably be ascribable to this cause: the flower was stinted and small, and mould very prevalent.

This, as well as the last, was one of those plentiful seasons which are not of frequent occurrence. Corn, pulse, and fruit of every kind, abounded. The only draw-back to these advantages was the blight occasioned by the various sudden changes of weather. The usual complaint of fine harvest weather—want of rain for grass and turnips,—was generally heard. The average quantity, however, was not deficient.

October—the temperature of the month was below the mean, and that of last year. Notwithstanding this, the dryness was above the mean, and there was very little rain. The consequence was, great injury to the turnip crops and grasses. The harvest was completed in perfection, by the housing of beans in good condition; and all orchard fruits were very abundant. In many places, the soil was too dry to be worked to advantage, and the sowing of corn was very much postponed till after Christmas.

November—the temperature was considerably higher than last year, but still a little below the mean; the dryness about the mean. The quantity of rain was very unusually small.

In consequence of this continuance of drought, the turnip crops were generally lost. The young wheats looked healthy and well.

December—the temperature a little above the mean, and six degrees higher than the last year. The quantity of rain was small, but in all other respects, the season did not vary from the average.

Owing to the openness of the weather, a considerable quantity of wheat was put into the ground in the early part of the month. The appearance of the crops was fine, and very luxuriant.

January, 1821, presents a complete contrast to the same month of the last year. The mean temperature was nearly 8° higher, and the degree of dryness much the same. Instead of being bound up in frost and snow, the country, every where, presented an open appearance, and the operations of husbandry were in universal forwardness. Early pulse and beans, had already been planted on the forward lands. Some of the wheat, and other green crops, experienced damage, on clay soils, during the frost for want of a covering of snow.

February—the temperature of the month about the same as last year, and below the mean. The barometer averaged considerably above the mean, and on the 6th, it attained the extraordinary height of 30.82 inches. Scarcely any rain or snow fell, and the season was contrasted with that of last year in being extremely dry.

The season was most propitious in all respects to the cultivation of the soil. The operations of husbandry proceeded almost uninterruptedly, and the favourable dry season of last year, which occurred in March, was anticipated in this by a month. The wheats, and winter crops in general, had a most promising appearance; so little rain fell during the winter quarter, that moisture was wanting to convert the abundance of straw into manure.

March was a complete contrast to the same month in the last year: very damp and rainy. Though the temperature was very little below the mean, it was cold and unpleasant to the feelings. The seed-time had, however, been anticipated in the preceding month, and the operations completed most favourably. The cold rains of the present month had not any very signal effect in forwarding vegetation, but the crops of every description had a healthful, if not a forward, appearance. The backwardness of the spring is, upon the whole, rather a favourable prognostic for the summer season.

April was of medium character, in most respects, but decidedly warm, its temperature being nearly two degrees above the mean. The warmth and moisture together, began to have their usual influence upon vegetation, which was not, hitherto, in a forward state. The wheat put on a luxuriant appearance; the pulse and Lent corn crops made a very fair shew; and all the operations of husbandry were in a forward state.

The month of May was unseasonably cold and dry, and constituted a very backward season. It was the more irksome to the human constitution as well as detrimental to vegetation, from being in occasional contrast, with a day or two of summer temperature. The blossoms of all fruit trees suffered from blight; for the cause of which we need, probably, look no further than the radiant thermometer, which we shall find was on five nights below the freezing point. A sufficiency, however, in most cases, remained for an abundant crop. The shew of grass, lucern, and clover, was great. The wheats appeared strong and healthy, and the spring crops had a thriving appearance. Hops were checked by the cold, but the vine was strong.

June.—The temperature of the month was nearly three and a half degrees below the mean, owing to the prevalence of north-east winds. All hopes vanished of good crops of fruit, and the hops suffered much. The wheats were fortunately backward, and escaped the injury which they would have received had they been in their usual progress, towards the flowering process. The spring corn was retarded in its growth from the same cause, and looked yellow and sickly. The hay-harvest, upon the whole, was good, and turnips, which had been in general early sown, were already beginning to spring up.—The season was rather one of uncomfortable sensation than of positive sickness.

July was deemed an average season, and the different crops upon the ground improved in their appearance. Want of solar heat was very much felt, and high winds had an injurious effect upon the flowering of the wheat; but the crops, generally, appeared in a thriving state. The hops were much mended, and the rains drew up the turnips to a size and substance which put them beyond the reach of serious injury.

The month of August, also, did not differ materially from the average. Harvest began about the middle of the month, but in some parts was very backward. Owing to the spring and early part of the summer being unfavourable, and the subsequent beating down of the corn by the rains, the wheats had received considerable damage. Mildew and smut are the diseases consequent upon such seasons as these. Barley, pulse, and tares, were expected to be full crops. Turnips and potatoes very promising.

September was warmer in this year than in the two last, but the degree of dryness very much less, and below the mean. The minimum state of humidity during the day, in the three, was as follows: 675, 673, and 759. The quantity of rain was about the usual quantity, but the amount of evaporation one fourth less.

As the two preceding years may be taken as examples of the weather which is most desirable at this season, the present one will furnish a specimen of those expensive and distressing harvests to which the uncertainty of the climate sometimes exposes the farmer. Far worse seasons, however, are by no means of very rare recurrence, in which cold, wet, and blighting weather are more constant. The present was relieved, throughout, by warm and genial alternations. The atmospheric diseases took place early, and the rains, which clouded the harvest, uncompensated by drying winds, completed the misfortune of the crop. The bulk, indeed, was great, but samples of good quality very scarce. There was a vast quantity of black and sprouting wheat, and of discoloured barley. On the other hand, the crops of peas and beans abounded, and the country never saw a finer turnip crop, or more luxuriant after-grass. Hop-picking began about the second week of the month, and the crop turned out large, though not of the first quality.

The warm weather still continued in October, but with an average degree of dryness. The crops of turnips covered the land completely, but were considered to be too luxuriant in their foliage. Grass was in great plenty. The worst fears of the farmer, with respect to the grain, were confirmed by its produce upon the barn-floor.

November differed from the mean, and from both the preceding years, in a very extraordinary way. The average temperature was 5° above the usual amount; and, although its dryness was in excess, the quantity of rain exceeded the mean quantity by one-half. The barometer, upon the whole, was not below the mean. All the low lands were flooded, and the sowing of wheat very much interrupted by the wet. The early sown, upon good lands, was very forward and luxuriant. A complaint was made, that the turnips had more foliage than bulb, and the potatoe crops were small, and of a blighted and inferior quality.

In December, the quantity of rain was very nearly double its usual amount The barometer averaged considerably below the mean, and descended lower than had been known for thirty-five years. Its range was from 30.27 inches to 28,12 inches. The temperature was still high for the season, and the weather continued, as in the last month, in an uninterrupted course of wind and rain; the former often approaching to a hurricane, and the latter inundating all the low grounds.

The water-sodden state of the soil, in many parts, prevented wheat-sowing, or fallowing the land at the regular season. The mild temperature pushed forward all the early-sown wheats to a height and luxuriance scarcely ever before witnessed. The grass, and every green production, increased in an equal proportion.

January, 1822.—This most extraordinary season still continued above the mean temperature, but the rain, as if exhausted in the preceding month, fell much below the usual quantity in this. There was not one day on which the frost lasted through the twenty-four hours.

The wheats, where they had not been flooded, were generally found to look well; but drawn up as they had been, by warm and moist weather, without the slightest check from frost, serious apprehensions were entertained lest they should be exhausted by excessive vegetation, and, ultimately, be more productive in straw than corn.

The month of February, still five degrees above the mean, ended a winter which has never been paralleled. The dryness, also, above the mean; evaporation great, and very little rain. There was no snow in Middlesex. Under these circumstances the earth became in the finest state for cultivation. The superfluous moisture was exhaled with far less damage from the floods than had been anticipated. The spring culture for every article was forward. The wheat looked finely, and grass was in profusion.

March was again 4° above the mean in temperature; dry, with a proper average proportion of rain. Wheat, taken generally, never looked better; the risk from winter frosts at an end, and nothing to be apprehended but those cold easterly winds, which are generally expected to succeed a mild winter. The appearance of the country, with respect to the lands and the crops, was universally favourable. Great evaporation took place, and the soil was every where getting into the finest state. The grass had a beautiful verdure, and all the spring crops were got in without hinderance from the weather. The shew for fruit, luxuriant and beautiful.

April did not differ much from the general average, but its temperature was low, with a more than mean quantity of rain. The frosts, in the early part of the month, were very injurious to the fruit-blossoms. The wheats suffered in colour and condition from the alternations of heat and cold. All the spring crops had been well got in, though on clay soils, with considerable trouble and expense.

A genial May, of the usual character of that month, visibly improved the crops upon cold, unsound land, which had, in course, suffered most. The appearance, throughout the country, was generally promising. The fallows and lands, for the spring crops, had been worked with much difficulty, owing to the want of the disintegrating influence of frost. All the green crops and grass were of good promise. Great complaints were made of the weak and blighted state of the hops, and fruits had suffered materially.

The month of June was unusually dry and hot; evaporation was enormous, and rain very short of the usual quantity. The amount of the former was 4.65 inches, and that of the latter only 1.11 inch.

The aspect of the country was considerably reduced in verdure and luxuriance, by almost constant drought and excessive solar heat. The amount of radiation from the earth was also great, and the radiating thermometer four times descended to the freezing point. There were several thunder showers, but the rain was not sufficient to moisten beyond the surface of the earth, or effectually nourish the vegetable roots. The autumnal wheats flowered in the most beautiful manner, and there was every prospect of an early harvest. The Lent corn crops suffered for lack of moisture. Hay-harvest was completed in the best possible manner, and the crops were good. The hops improved, and the apples promised well.

The timely showers of July revived the appearance of the Lent corn and pulse, which were generally injured by drought. The oldest person did not recollect either an earlier hay or corn harvest; or more successful ones, to this time, both with respect to weather, and quantity, and quality of produce. The turnip sowing, which had been attempted in the dry weather, was obliged to be repeated after the rains. The crop of potatoes middling.

The weather continued through August the most beautiful, and best adapted to getting in the harvest, and, indeed, to every agricultural purpose, that could have been chosen.

The wheat crop was safely housed early in the month, and the next article in rank for human subsistence, potatoes, were of equal promise with the wheat, both in regard to quantity and quality. Barley, oats, and beans, were only good in some favoured situations; and, in general, those crops were considerably below an average, though much improved by the showers which succeeded the long drought. Oats, particularly, suffered from smut in many parts. No crop received greater benefit from the rains and subsequent warm weather than the hops. Turnips were a failing crop, destroyed almost entirely by the fly.

The method adopted by Mr. Howard of expressing a series of results, by a curve, is so extremely useful and striking, and so greatly facilitates the comprehension of their different connexions, that I shall here proceed to connect my own observations with his, by a similar occular comparison of certain particulars, and endeavour to supply the void in his graphic illustrations, by tracing upon the same plan, some of the variations of the aqueous atmosphere.

Fig. 1, Plate I. presents a comparison of the monthly mean temperature of the air, and of the dew-point, throughout the year, founded upon the observations of the three years. The full line exhibits the progress of the former, the dotted line that of the latter; and the degree of dryness of the different periods, is accurately represented by the space included between the two. From this we clearly perceive how closely the constituent temperature of the vapour follows the mean temperature of the air; but we may, at the same time, remark that it is more speedily diminished by the fall of the latter, than it is increased by its rise. Hence arises the contrast between the dryness of the spring and summer months, and the dampness of the autumn and winter. The two horizontal lines are drawn upon the annual means, and form the standards of comparison for the several variations.

Fig. 2 presents the separate means of the three years, by the conjunction of which the last curves were constructed. The relation of the different lines will be sufficiently intelligible from inspection, without minute description. The character of the various seasons which we have just been analysing and comparing, will be found to be very prominently marked upon this diagram. The wet autumn of 1821, which proved so injurious to the harvest, is characterized by the near approximation of the two dotted lines in September; and the very cold winter of 1820, by the low descent of the continuous lines in the scale in January. The hot and seasonable summers of 1820 and 1822, and the blighting June of 1821, together with the very extraordinary winter of 1821-1822, are also well defined. The oscillations of the seasons, round the horizontal lines of the general means, exhibit many other points of interesting comparison, which will be sufficiently obvious to those who are inclined to take this method of studying the results of the observations. The general correspondence in the progress of the mean temperature, and that of the dew-point is sufficiently evident in the separate years; but there are some evident modifications of the general law, presented by these curves, which are well worthy of attentive consideration.

To render the general accordance between the mean temperature and the dew-point, still more evident, Fig. 3 has been constructed, in which the variations of the *daily* mean, are represented for forty-five days, in September and October, 1819: and to carry the analysis of this connexion to its greatest extent, Fig. 4 presents the observations four times in the day, for eleven days, including the daily *maxima* and *minima*. By inspection of the latter figure, it will be evident that the *maxima* of the temperature of the air have but little influence upon the force of the vapour, which appears to be governed chiefly by the daily minima; and a line connecting together the points of greatest depression would not differ very essentially from the curve of the dew-point; while a similar connexion of the points of greatest elevation would scarcely resemble it in any particular.

Plate II represents the comparative fluctuations in the elasticity of the air and vapour, and in their temperatures, taken at morning and evening, for a period of three months, in 1819. The former are laid down upon the same scale; and the two upper lines depict the real proportionate differences of the simple and compound pressures. An attentive consideration of these will shew that the undulations are generally in contrary directions, and a rise in the line of vapour is generally accompanied by a fall in the barometric curve. The lower lines depict the variations of the temperature: the continuous one that of the atmosphere; the dotted that of the point of condensation. In fine weather, it will be observed that these two are widely separated, while, during the time of aqueous precipitation, they coincide. In this diagram may be distinctly traced the much greater variations of the aqueous atmosphere at the upper part of the thermometric scale, than for equal differences at the lower; and, in short, all the connexions of temperature, moisture, and pressure, may be comprehended at one glance. The conv g-ing lines are intended to indicate the direction of the winds.

The preceding pages I would wish to be considered as an Appendix to Mr. Howard's valuable and laborious work; and, as such, I trust that they will not be found to be without interest. There are many points upon which I have refrained from touching at all, which have been discussed, with all the precision which the present state of our knowledge will permit, by that able and indefatigable observer; and for information upon these, I prefer a reference to the work itself, to an attempt to recapitulate his clear and comprehensive statements. The climate of London has altogether received a degree of elucidation, (to which I shall be proud to be considered to have contributed, in however small a degree,) which distinguishes it from any other on the globe, and which, it may be hoped, will excite some degree of emulation in other countries. But much remains to be done even in this field of meteorology, and it is probable that the most interesting and practically-useful points, are those which are still enveloped in the greatest obscurity. If the interest which has lately been excited on the subject, and the consequent spirit of observation be properly directed, there can be little doubt that the progress to perfection will be sure and rapid.

METEOROLOGICAL OBSERVATIONS

MADE AT

MADEIRA, SIERRA LEONE, JAMAICA, AND OTHER STATIONS, BETWEEN THE TROPICS,

By Captain EDWARD SABINE, R.A., F.R.S.

MY friend, Captain Sabine, before his departure upon his voyage to the south, in 1822, for the purpose of ascertaining the length of the second's pendulum, at different stations near the equator, with that unaffected zeal for science, for which he is so much distinguished, kindly undertook to try any experiments, or make any observations which I would point out as likely to be of interest to meteorology. I eagerly embraced so desirable an offer; and, not to be backward in promoting to the utmost of my ability such an important object, I sketched out some imperfect directions for his guidance. The results which he communicated to me upon his return, are of the highest interest and importance, and of particular value, from his well-known accuracy. I have already availed myself largely of their general authority in the preceding Essays, and shall now, with his liberal permission, proceed to detail, in his own words, the particulars of his notes. I shall here omit such experiments as relate particularly to the subjects of radiation, and the horary oscillations of the barometer, as they are to be found in the preceding pages. The observations commenced at the island of Madeira.

"The mountainous parts of the interior of Madeira have been rendered accessible to a greater distance than formerly, by roads of recent construction, passable at most seasons by mules, or by the small horses of the island, which vie with mules in the sureness of their footing. I availed myself of the opportunity which our short stay afforded, of making an excursion to the summit of the Pico Ruivo, the highest of the island, with a view to obtain a measurement of its height, and to make a first essay with a portable barometer having an iron cistern, on which Mr. Newman had bestowed much pains, to obviate the liability to the various errors to which these instruments are generally subject. The party consisted of Captain Clavering, of his Majesty's ship Pheasant, Mr. Whitelaw, surgeon of the Iphigenia, Mr. George Don, naturalist of the Horticultural Society, and two midshipmen of the frigate; we were accompanied by Mr. Blackburne, an English merchant resident at Madeira, who, having before ascended the Peak, was kind enough to undertake to conduct us, and by his local knowledge and authority over our Portuguese attendants and guides, as well as by his own enterprising spirit, enabled us finally to accomplish our purpose. Lieutenant Stokes, of the Iphigenia, was so kind as to remain on board the frigate throughout the day, to note the variations in temperature and density of the atmosphere, and the point of deposition indicated by the hygrometer. These were observed hourly by a chronometer, so as to be simultaneous with others which we should make at the heights at which we might find ourselves.

" We quitted Funchal before day-break, and proceeded about six miles along the coast to the westward to Camera de Loubos, from whence we commenced the ascent in a northerly direction. At eight we stopped to breakfast at the Jardim de Serra, a house which Mr. Veitch, the British consul-general, has built, at an elevation of nearly 2,800 feet. In approaching this height, the vegetation reminded us at every step of England; the people of the country, whom we met on their way to mass, impressed us favourably by their courteous demeanour towards each other, as well as to strangers; they were well, and even handsomely, clothed; the men able-bodied and good-looking, but the women, almost without exception, very plain.

" We found the temperature at Mr. Veitch's 16° less than at Funchal, being a much greater difference than we had expected as due to the elevation. An ascent of about half an hour from the Jardim opens the first sight of the Curral, which struck me, who am, however, but little accustomed to mountain scenery, as the most magnificent view I had ever seen; the Curral das Freiras, which means literally, I believe, the Sheepfold of the Nuns, is a ravine extending several miles in a north and south direction, and of considerable width, the sides extending four thousand feet in height, in character frequently precipitous, and where so, being in fine contrast with the deep green foliage of the trees, by which the sides are more generally clothed; these trees are principally laurels, amongst which we noticed the Nobilis, Indica, and Fœtens. The valley of the Curral is occupied by a small river, which descends from the high land of the interior with all the character of a mountain torrent. Our route led into the Curral, for the purpose of ascending its valley, but the descent being impracticable at the spot where the first view is obtained, the road continues to ascend, passing over an elevated ridge, on which there was much snow. In descending, on the Curral side of this ridge, and at some distance beneath its summit, is a copious spring, which collects in a shaded basin formed in the rock by the workmen by whom the road was made. The temperature of the water in this basin was 47.2°, that of the air 46°, and at Funchal 65°; its elevation 4454 feet.

"Whilst these observations were making, the summit of the Pico Ruivo, which was enveloped in clouds during the day, was visible for some minutes; and it may be worthy of notice, that this was the only period in which the proportion of moisture, in the upper air, to saturation, was observed to be less than at Funchal. The wind throughout the day was easterly and light, but with little of the unpleasant sensation which usually characterizes the *Leste*.

"The time pressing, we committed our horses to the Portuguese attendants, and descending ourselves on foot, more quickly than we should have done on horseback, although stopping occasionally in admiration of the splendid scenery on every side, which it was impossible to pass without notice, we crossed at noon the Ribeiro di Curral, on a tree which had fallen across the torrent, the horses fording it lower down, and pursued a road which led to the head of the valley. We there recommenced the ascent, and passing through districts of brooms and ferns, entered the snow at a somewhat lower elevation than on the heights near the coast. At two P.M. we reached the highest point attainable on horseback, by reason of the depth of snow, and of the frequent *quebradas*, or breaches, in the road, caused by the descent of torrents. It is a ridge 4380 feet above the sea, over which the road passes at the foot of the Pico das Torrinhas, which is inferior in height only to the Pico Ruivo. From hence, Mr. Whitelaw and myself proceeded on foot, the others of our party returning to the valley to await us. Entering a thick wood of evergreens, consisting of laurels, of the quercus ilex, and of the erica arborea, which attains a large size, and grows even at the summit of the mountains, we were soon enveloped in the clouds by which the Peak was hid from our sight; and, after an hour and a half's good walk through snow, which latterly exceeded two feet in depth, impeded occasionally by the *quebradas* which are passable only by the aid of roots and branches of trees, and not without danger, as a slip unrecovered would generally be fatal, we attained the summit. We experienced no other inconvenience than being wet by the rain, and a little cold, whilst we remained to make the necessary observations to ascertain the height; certainly none that need deter others from a similar undertaking at the same season of the year, when, should the weather be clear, they will be amply repaid. The Peak being nearly in the centre of the island, the view from it must be very splendid, though of this we were only able to form an imperfect judgment from the unfavourable circumstances of the weather. It is not otherwise interesting than as relates to its height and situation, being merely one of several pinnacles in an island of volcanic formation.

"It was dark before we had rejoined our party in the valley. We had then to reascend the opposite side of the Curral to that which we had descended in the morning, in order to gain a nearer road to Funchal than by the Jardim de Serra. This ascent was more precipitous than any we had yet traversed, and made those amongst us feel nervous who had not learned from habit to confide in the sure footing of the horses, inasmuch as, during the greater part of the way, a single false step would have precipitated the horse and rider many hundred feet into the valley beneath; the apprehensions of danger were, perhaps, augmented by the accompaniment of torch-light, and induced some of the party to trust to themselves rather than to the horses; we all, however, reached Funchal in safety by midnight.

"The barometer was found to answer extremely well, both in conveyance and in use. I am not aware of any objection to the iron cistern to counterbalance its many advantages over those of leather or of wood, the former of which are especially faulty in being affected by damp, whilst the certain freedom of the mercury from air and moisture in barometers of this construction, give them a decided preference over those which are filled on the spot, and which I cannot consider as otherwise than very uncertain. I regret extremely that I had not an opportunity of ascertaining its performance in the more important ascent of the Peak of Teneriffe, but our departure from England had been so long delayed by contrary and tempestuous winds, that we were only able to remain seven hours at Santa Cruz. We were told, indeed, that the Peak was inaccessible in the winter season, but we had heard the same at Funchal of the Pico Ruivo. I am aware that the difficulty in the two cases does not admit of comparison, but the true interpretation is, that neither is accessible without more exertion than travellers are ordinarily disposed to bestow. Had Sir Robert Mends felt at liberty to have remained at Teneriffe for three days, we should certainly have made the attempt; and as Captain Baudin succeeded in December, I trust we should not have failed in January. The precise determination of the height of this peak is yet to be accomplished, and appears worthy of being undertaken, were it only to submit barometric measurement to the test of a more exact comparison with the geometric method, (both conducted with the precision of which modern instruments are capable,) than has yet been effected. A residence of some days, at the proper season, near the summit of this remarkable Peak, which rises so abruptly, and to so great an elevation, from the middle of the basin of the Atlantic, might indeed be expected to produce many important meteorological and other results; and would certainly throw much light on the extent of variation, to which barometric measurement is liable, from varying circumstances connected with the atmosphere itself, independently of errors of instrument or observation, or of the formula by which a result is deduced; the limit within which this liability might be apprehended, would appear, by a comparison of the registry of the barometer at the top and at the bottom, continued for a sufficient time.

"We experienced a similar disappointment, and scarcely in an inferior degree, in passing hastily by Fuego, one of the Cape de Verds. I am not aware of any good account of this very remarkable island having been published, and am surprised that it has been so little visited. It rises in a cone almost from the water's edge, to an height much exceeding that of St. Antonio, which is estimated by Captain Horsburg at 7400 feet, and we had reason to conclude, from the angle which it subtended at different distances, justly estimated. The summit of Fuego was visible from the ship for two days, rising much above the clouds, and always clear; no smoke proceeded from it, although it is said to be generally burning. I cannot conceive a station more eligible for interesting experiments, connected with the relations of heat and moisture to the atmosphere.

"The detail of the observations, and the heights computed from them, is as follows:—

Observations made at Madeira, January 13, 1822, to determine the Elevation of several Stations in the ascent to the Pico Ruivo.

STATIONS.	Observations at the Station.				Corresponding Observations 8 feet above the Sea.				Height Deduced.
	Barometer.	Temperature. Air.	Temperature. Merc.	Point of Deposition.	Barometer.	Temperature. Air.	Temperature. Merc.	Point of Deposition.	
	Inches.	°	°	°	Inches.	°	°	°	Feet.
Jardin di Serra, floor of the upper story of Mr. Veitch's house . .	27.681	49	49	41.5	30.603	65	65	54	2782.6
Basin of the Spring	26.012	46	46	34	30.543	65.5	65.5	53	4453.9
Ridge at the foot of the Pico das Torrinhas	25.948	42	42	36	30.423	64	64	56	4379.7
Summit of the Pico Ruivo. The observations were made eleven feet below the summit, but the computed height is that of the summit itself	24.938	36	36.5	36	30.423	61 5	61.5	58	5438.1

" The results have been deduced in the manner explained in Mr. Daniell's paper, 'On the Corrections to be applied in Barometric Mensuration, published in No. XXV. of the *Quarterly Journal of the Royal Institution;* the barometric differences have been augmented by $\frac{1}{68}$, as 68 inches of mercury in the tube are equivalent to one inch in the cistern; and $\frac{1}{500}$ of the approximate result has been added, as a correction due to the variation in density of the atmosphere, in the latitude of Madeira.

" *Between Madeira and Sierra Leone.*—We entered the trades on the 20th of January, in lat. 24$\frac{1}{3}$° N., and long. 19° W.; between which, and the Cape Verd Islands, the point of deposition varied from 60.5° at eight A.M.; air 69°, and surface water 68.25°, to 62°; air 70.6° to 71.2° at noon, and six P.M., on each of the 21st, 22nd, 23rd, and 24th, days of January. When amongst the islands, between San Vincente and St. Jago, the breeze being rather fresher than customary, and the weather as usual, clear without haze, the point of deposition was as low as 54.8°, air 71.2°, surface water 71.2°.

" In the due-east passage from the Cape Verd Islands to Cape Verd on the Continent, (400 miles,) the point of deposition (as usual at eight, or rather half-past seven, noon and sun-set,) varied from 63° to 64°; air from 70.2° to 71.2°; until at eight A.M. on the 31st, approaching Cape Verd, then twenty-six miles distant, and not in sight, it fell to 61.5°; air 70°; surface water having also fallen from 71° to 69.6°. At noon we were between three and four miles from Goree; deposition 57.5°, air 70°, water 64°. At sun-set, again proceeding, more distant from the coast, to St. Mary's, 60.5°, air 69°, water 65°; and on the following morning, at sunrise, anchored in the Gambia River, twenty miles from St. Mary's; point of deposition 50°, (and at noon 51.5°), air 69°, water of the surface 67.5°; and at thirty-four feet depth, 66.5°, being two feet above a hard, sandy bottom. On the 2nd of February we anchored off Bathurst, higher up in the Gambia River; point of depression 48.5° to 49.5°, air 69°, on the 2nd, 3rd, and 4th; but on the 5th, in the forenoon, something approaching to a Harmattan took place, the wind fresh from the N.N.E. to N.E.; the weather hazy towards the horizon; the point of deposition 37.5°, air 66.5°, and water the same. We sailed in the afternoon, when the wind shifted to the N.W., and the point of deposition rose to 60.5°, air 70°, at six P.M.

"Between the Gambia and the River Sierra Leone, we were usually off, and in sight of, the coast; occasionally waiting off the mouths of the larger rivers. Whilst at anchor off Cape Roxo, (four miles distant,) we had the air 68.5° to 67°; point of deposition varying from 56.5° to 60° in the course of the day, (7th and 8th of February;) surface water 68°. On the 9th of February we passed outside the Bissao shoals to the mouth of the River Grande; air 71° to 70.2°, hygrometer 64°, surface water 66.5; and when anchored off the mouth of the Rio Grande, the air was warmer, (on the 10th and 11th of February,) 73° to 74.5°; being warmed by the river water, the surface of which was 75° off the mouth; and more moist, the point of deposition being 64° at eight A.M., 66° at noon, 68° at six P.M., 69° at midnight.

" *Sierra Leone.*—From the latter end of February to the beginning of April, at half an hour after sun-rise, the point of deposition was usually found from 68° to 71.5°, air 75° to 78°; at which time, and until the sea-breeze sets in, a calm prevails, and causes it to be the most oppressive period in the twenty-four hours. About ten or eleven the sea-breeze commences usually in the N.W., freshening, and becoming more westerly as the day advances. With this wind the point of deposition generally advances about 2° by noon, and the air to the average of 81°: in the afternoon, little or no regular change is perceptible in the point of deposition—not equal to 1°; whilst the air advances to 83° and 84°. Towards the evening the sea-breeze dies away, and the land wind gradually springs up; and at ten P.M. I generally found the point of deposition 5° or 6° less than at half an hour, or an hour, after sun-rise the following morning. The *average maximum* of the hygrometer, *daily*, was 72.7°; the particular observations varying from 70° to 75°; the minimum may be inferred to have been below 73.2°, which is the average minimum of a register thermometer, suspended a foot above the parapet of the bastion of Fort Thornton, in a fair exposure; and which, consequently, was not so low, as if it had been placed on vegetation on the ground. This average is a mean of twenty nights; the particular observations varying from 72.5° to 74°. On two of these nights the thermometer was placed flat on the ground of the parapet; on two others, suspended by a cord, extended from parapet to parapet, across the bastion; in two others, placed in a copper-inverted cone, highly polished on the outside, and blackened within; and in all the other nights, strung across, between the pillars of my transit, (the instrument itself being removed,) on the parapet. The register thermometer was of Six's construction; the cylindrical bulb blackened, and wrapped in black wool; the wooden back hollowed out in the centre. In the various situations it appeared to be alike affected It being the dry season, there was not a tuft of vegetation (excepting trees) within any moderate distance of the fort. That a greater difference does not take place between a thermometer thus exposed, and one under shelter, measuring, as nearly as may be, the true temperature of the atmosphere, may, perhaps, be caused by the peculiar situation of Free-town, immediately inland of which a ridge of mountains rises somewhat abruptly; and the cold air from hence is brought down immediately to Free-town by the land wind, which prevails at night. Free-town has also the sea on the west, and a river of several miles' width on the north; these circumstances combined, seem to explain the reason, why the usual effects of radiation should be lessened, whilst the air introduced is colder than if the whole country were on the same level.

Register of the Height of the Thermometer at Sierra Leone, 1822.

Hours.		Feb 26	27	28	Mar. 1	2	3	5	6	7	8	9	10	11	12	22	Mean temperature of each four hours.	
10 A.M. to 2 P.M.	Max.	85	86.5	81	81	81.8	85	82	83	82	82	81.5	82.8	82	83	84.2	82.85	80.77
	Min.	78	77.5	77	76	78.3	79	79	80	79	78	80	78.5	79	80.5	80.5	78.69	
2 P.M. to 6 P.M.	Max.	86	87	80	81.5	85	85	83	83	82	85.5	84	84.5	82.5	84	86.5	83.97	81.27
	Min.	79	77	77	78	78	79	80	78	77 75	81.5	78	78.5	78	79	80	78.58	
6 P.M. to 10 P.M.	Max.	78.5	79	79	80	80	79	80	79.5	81.5	81	78.8	79	78.5	80.5	80	79.02	78.36
	Min.	74.5	77	74	76	76.4	78	79	78.5	77	80.5	76.5	76	76.5	78.5	78	77.09	
10 P.M. to 2 A.M.	Max.	78.25	77.5	77	80	80	78.75	80.25	79.5	80	81	77.6	78	78.8	81	81	79.24	77.71
	Min.	74	75.75	72.75	77	75.5	77.5	78.25	76 75	75.75	78	75.5	76	76	77	77	76.18	
2 A.M. to 6 A.M.	Max.	78	76	75	80.5	80	79.5	80.2	79.5	78.5	81	81	77.5	77	78.5	82	78.95	76.98
	Min.	73.5	74 5	71.5	76	74.5	76	77.25	75	74.5	75.5	75.5	74.5	75.5	75.5	76	75.02	
6 A.M. to 10 A.M.	Max.	78.5	78.5	78.5	79	80	80	78.2	81	79.5	79	81	79	79.8	81.5	82.4	79.73	77.97
	Min.	75.5	76.25	72.5	75.5	77.5	77	75.5	77	75	77	76	76	77	78.5	76.8	76.2	
True mean daily temp. . .		78.23	78.54	76.3	78.38	78.92	79.46	79.39	79.23	78.54	80	78.79	78.36	78.38	79.79	80.37	78.85	
Mean of the extremes . .		79.75	80.7	76.26	78.5	79.75	80.5	79.25	79	78.25	80.5	79 75	79.5	79	79.75	81.25	79.45	

A (Six's) Registering Thermometer was suspended in a copper cylinder, highly polished on the outside, and pierced with holes in the top and bottom, to admit a free current of air; the cylinder itself was suspended five feet from the ground, in a thorough draft under a roof on the parapet of the bastion of the fort: a mean of the register of the thermometer at 8¼ A.M. is *under* the true mean temperature, and at 9 A.M. as much above it.

" *Details of a Barometrical Measurement of the Sugar-loaf Mountain.*—The Sugar-loaf, so called from its shape, is the highest point of the mountain district of the colony, included as yet within the limit to which cultivation has extended. This district is the site of the twelve most interesting settlements of liberated Africans, from the principal of which, Regent-town, it is distant about three miles, being altogether about eight or nine from Free-town, the seat of government: a road has been opened by the inhabitants of Regent-town, by which the summit is accessible, and has been sufficiently cleared of its forest-trees to admit the view around. In the continuation of the Sierra towards the south, at about twenty miles' distance, the land appears to attain a greater general elevation than in the neighbourhood of the Sugar-loaf, and there are several points, especially, which are probably much higher: to these there is as yet no road, but from the very rapid advance which the colony is making in population and in settlement, it cannot be doubted that these points must very shortly be necessarily included in the Colonial Survey.

" Dr. Nicol, deputy-inspector of army hospitals, was kind enough to allow me the use of a stationary barometer, in excellent order, made by Cary, and the property, I believe, of the College of Physicians: it is the same instrument which has since accompanied Captain Laing in his very interesting excursion to the Soolima country, in which the Niger takes its rise, and which has enabled him to ascertain satisfactorily the elevation at which that river originates its yet unknown course. The accordance of the portable barometer with the stationary was examined before and after the observations for the measurement; the latter was placed in the room in Fort Thornton, in which my pendulum experiments were made, and its height, consequently, above half tide, carefully ascertained by levelling, was known, with tolerable precision, to be 190 feet. The variations in the density and temperature of the atmosphere, and in the point of deposition of moisture, as indicated by the hygrometer, were observed at this spot by Captain Laing, at stated periods, with a chronometer, on the 28th of March, so as to be simultaneous with such as should be made at elevations.

I shall confine myself to stating the data necessary for the calculation of the heights of the clergyman's house at Regent-town, and of the summit of the Sugar-loaf. At the first of these stations, the barometer, having been suspended above an hour, five feet below the gallery which surrounds the clergyman's house, shewed at seven, A.M., on the twenty-eighth March, 29.017 in., th. 74.5°, and the point of deposition 57°; the corresponding observations at Fort Thornton were 29.820 in., th. 79.5°, and the point of deposition 66°. At eleven, A.M., on the same day, the barometer being suspended in the shade, at the summit of the Sugar-loaf, the cistern 1½ foot below the highest point, was suffered to remain until twelve o'clock, that the mercury might acquire the temperature shewn by the attached thermometer; when the observations registered were 27.560 in., th. 82.2°, and the dew point 70°,—those at Fort Thornton being 29.795 in. th. 84°, and the dew point 70°, also.

" The mercury being reduced to the same temperature at the upper and lower stations, and $\frac{1}{68}$ of the differences in the heights of the column being added, on account of the respective diameters of the tube and cistern of the barometer; the true differences are, between Fort Thornton and Regent-town .8 in., and between Fort Thornton and the Sugar-loaf 2.263 in., at the temperatures of the air, and under the pressure of the amount of atmospheric vapour specified above. The approximate heights due to these differences being corrected for the latter circumstances, it results that the floor of the gallery of the clergyman's house at Regent-town is 983.6 feet, and the summit of the Sugar-loaf, 2521.6 feet above the sea.

" I have taken the liberty to add (though without permission) an extract of a letter which I have received, since my return to England, from Thomas Stuart Buckle, esq., engineer and surveyor of the colony, stating the result of a comparative geometrical measurement. " I was much gratified to find, on computing the altitude of the Sugar-loaf, from the trigonometrical observations that I had taken, that the result differs from your barometrical measurement only a few feet: I make its height 2493 feet; the height of Leicester Mountain I computed to be 1954, and it was sufficiently satisfactory, on taking into account the distance of the Sugar-loaf from Leicester Mountain, and the excess of its height above that of Leicester Mountain, that the result of the latter was 537 feet, which, added to 1954, amounts to 2491, differing from the former calculation only two feet."

"I shall here add the barometric measurements of well-known places in the islands of Ascension, Trinidad, and Jamaica; but I am not aware of any previous results with which to compare them.

"*Height of the Mountain-house at Ascension.*—July 9th, 1822, at 9ʰ 30ᵐ A.M., a barometer, seventeen feet above the sea, in a room in the Barrack-square at Ascension, stood at 30.165 in., the temperature of the air and mercury being 83°, and of the point of deposition 68°; whilst, at the same time, another barometer, three feet above the floor of the Mountain-house, stood at 27 950 in., the air and mercury 70.3, and the point of deposition 66.5. From these data, the floor of the Mountain-house would appear 2221.8 feet above the sea.

"The upper barometer was then taken to the summit of the island, but the registry at that height has been mislaid; it was 27.3 and some hundreds, being less than 700 feet above the Mountain-house; consequently, the highest part of Ascension is under 3000 feet: on returning from the summit, the barometer was replaced three feet above the floor of the house, and allowed to remain until the mercury should have acquired the temperature of the air, when, at 1ʰ 30ᵐ PM., its height was 27.937 in., air and mercury 72°, point of deposition 68°, and in the lower barometer 30.137 in., air and mercury 84.5, point of deposition 71°, whence the height of the floor of the Mountain-house results 2219 feet above the sea, being three feet less than the first measurement

The mean, consequently, or 2220.5 feet, is considered the correct elevation.

"*Height of the Block-house at Fort George, Trinidad.*—October 9th, 1822, at 8h 30m A.M., a barometer, four and a half feet above the foundation of the Block-house, stood at 29.000 in., the air and mercury being 76.5, and the point of deposition 76.5 also, with slight rain. The corresponding height of the barometer, at the same time, in the Protestant church in Port Spain, twenty feet above the sea, was 30.058 in., air and mercury 82°, and the point of deposition 77° Whence the foundation of the Block-house would appear 1067 feet above the sea.

"*Height of Mr. Robert Chisholm's House, in the Port-Royal Mountains, Jamaica.*—October 31st, at 4h 30m P.M., a barometer, suspended against the wall of Mr. Chisholm's house, two feet above the ground, stood at 25.967 in., the air and mercury being 68.5, and the point of deposition 68.5 also; and on the 2d of November, at six A.M., at 25.963 in., the air and mercury 65°, and the point of deposition 60° The corresponding observations at Port Royal, at the same hours, eight feet above the sea, were—

Oct. 31,—	Bar. 30.007;	Air, 82.5;	Merc., 84.5;	Dew point, 77
Nov. 2,	30,023	78.	78.	72

Whence the height of the ground on which Mr. Chisholm's house stands, results respectively, 4087.9 feet, and 4072.7 feet, the mean being 4080.3 feet above the sea.

"All the observations at heights were made with

the same portable barometer; $\frac{1}{68}$, therefore, is added throughout to the barometric differences on account of the ratio of the diameters of the tube and cistern. The height of the column of mercury, in the upper and lower barometer, under equal pressure, was in all cases carefully examined, and the difference, if any, allowed as an index error to the lower barometer. I have great pleasure in remarking, that I found much less difficulty than I had anticipated, in getting corresponding observations made with the hygrometer, on the correctness of which I could sufficiently depend: the ingenuity in the principle of this instrument, and the simplicity of its application, together with the decisive nature of the results which it gives, independent of the labour, and, at best, the uncertainty of formulaic deduction, form its great advantage over the methods by evaporation, or the indications of hygroscopic substances: these particulars excite an interest in its trial, in persons to whom it was previously unknown, which is probably the reason that the distrust, which is almost always, in the first instance, expressed of precision in the observation itself, is found to give way in practice so much sooner than might be supposed. It may be useful also to travellers in warm climates, to add a remark from my own experience, that in ascending elevations, or in journeying inland over rough roads, the ether carries perfectly well in a bottle in the waistcoat-pocket, with a common cork capped with leather; and that the expenditure of ether altogether will probably fall much short of the estimate, as, with ordinary care, very little will be wasted."

Indications of Daniell's Hygrometer, in the Gulf of Guinea, between the 1*st and the* 15*th of May.*

May 1822.			Air.	Point of Deposition.	Proportion of Moisture in the Air to the Quantity which would be required for its Saturation at the existing Temperature.
			°	°	
1	9½	A.M.	78.5	74.5	87.5 parts in 100.
	3	P.M.	81	74.7	80.6
	6	P.M.	81	76	84.6
2	6½	A.M.	79.5	76	89.3
	2½	P.M.	82	75	79.5
	6	P.M.	83	75	77.4
3	6	A.M.	79	76	90.7
	3	P.M.	83.5	77	81.3
	6	P.M.	82.5	77	83.9
4	6	A.M.	81.5	77	86.3
5	7	A.M.	81.5	76	83.4
	2	P.M.	83.5	77	81.3
6	7	A.M.	81.5	76	83.4
	2	P.M.	83.5	77	81.3
8	10	A.M.	80	76	87.6
9	3	P.M.	86	79	80.1
11	6	P.M.	83	80	90.9
12		P.M.	79	Rain	100
13	9	A.M.	80	76	87.6
15	7½	A.M.	82.5	78	86.6

"At St. Thomas's, between the 26th of May and the 12th of June, the range of the hygrometer, observed generally three times a day, in a house by the sea-side, with an extensive swamp behind it, was limited to 71° and 74.5°; the temperature of the air ranging from 74 and 75 at night, to 85 and 86 in the day: (the nights being cooled by the air descending from the high land of the interior;) the average moisture, relatively to that of the Gulf of Guinea, generally, and away from the land, may be considered at a mean between the extremes corresponding to the observation above mentioned; *i. e.*, 79.6, to 85.1.

"On passage between the River Gaborn, on the coast of Africa, under the equator, and the Island of Ascension, in 7° south latitude, and midway between Africa and America, I obtained the following results:—

June.	Time.	Air.	Surface Water.	Point of Deposition.	REMARKS.
			°	°	
17	8 A.M.	74.5	73	69	Roll's Isl. E. N. E.
18	8 A.M.	74	74.2	69	Wind during the passage fresh from S. to S. S. W. The ship passed from the equinoctial current into its north-westerly feeder on the morning of the 22nd, when the surface water resumed the ordinary ocean temperature.
19	8 A.M.	73.5	72.5	68.5	
20	8 A.M.	73.5	72.8	67.5	
21	8 A.M.	74	77.5	66.5	
22	8 A.M.	77	77.5	69	
	4½ P.M.	75.5	77.3	68.5	
23	8 A.M.	75.5	78.2	67.5	
24	8 A.M.	76	78.2	66.5	
	4 A.M.	77	78.2	69.5	
25	7 A.M.	76.5	76.5	Close in with Ascension.	

"During ten days' residence at Ascension, the thermometer, sheltered, varied between 75° and 84° during the 24 hours. The point of deposition from 65°, before sunrise, to 70.5 once, and generally to 69 and 70 in the afternoon. The details of a barometrical measurement of the height of Ascension have been given above

"*Bahia.*—Place of observation, a plateau unsheltered, between the consul's house, and the edge of a cliff 213 feet above the sea, facing the harbour of St. Salvador.

"The following Table contains the observations for the mean temperature of the air, made with a transparent register thermometer, suspended in a polished copper cylinder, with holes at the top and bottom, in a fair exposure to the wind, but sheltered from the heavens, eight feet above the ground:—

Days.	Time.	Temperature of the Air.		Means.	REMARKS.
July 24 and 25	Sun-set to sun-rise	76°.	69°.5	75.75	
	Sun-rise to sun-set	82.	75.5		
— 25 and 26	Sun-set and midnight . .	75.5	71.	74.	It was almost constantly calm and clear, until the 28th, when southerly winds set in with frequent rain—the air became drier, and the temperature less.
	Midnight and sun-rise . .	71.	68.		
	Sun-rise and noon	80.5	67.5		
	Noon and sun-set	82.	76.		
— 26 and 27	Sun-set to midnight . . .	76.	72.	75.5	
	Midnight to sun-rise . . .	72.	73.5		
	Sun-rise to noon	83.	73.		
	Noon to sun-set	82.	73.		
— 27 and 28	Sun-set to midnight . . .	73.	70.	72.5	
	Midnight to sun-rise . . .	73.	68.5		
	Sun-rise to noon	78.	69.		
	Noon to sun-set	77.	71.		
— 28 and 29	Sun-set to midnight . . .	72.	67.	71.5	
	Midnight to sun-rise . . .	70.5	64.		
	Sun-rise to noon	78.	71.		
	Noon to sun-set	77.	73.		Southerly wind, with frequent rain.
— 29 and 30	Sun-set to midnight	73.	71.5	73.	
	Midnight to sun-rise . . .	72.	67.		
	Sun-rise to noon	80.	69.5		
	Noon to sun-set	80.	73.		
— 30 and 31	Sun-set to midnight . . .	73.5	68.5		
	Midnight to sun-rise . . .	72.	65.		

"From the twenty-third to the twenty-seventh, the point of deposition varied in the day, *i. e.*, from six A.M., to six P.M., between 66°, and 70.5°, being on each day either 70° or 70.5° in the afternoon.—After the southerly winds set in, on the twenty-eighth, with occasional rain, the point of deposition varied from 61° to 64°, at the same hours, and continued so until the end of the month, except, of course, when the rain was actually falling.

"On the passage along the Brazil coast, between Bahia and Pernambuco, generally in sight of land, the temperature varied between 74.4° and 78.4°, the point of deposition between 66.5° and 70.3°, and the surface water between 77.1° and 78°, all being observed at three periods of the day, before eight, after noon, and before sun-set.

"Between Pernambuco and Maranham, also along the Brazil coast, temperature 76.5° to 78.5°, surface 77.8° to 78.2°, and the point of deposition from 70° to 74°.

"At Maranham we inhabited a house in the middle of the city, without a garden or court; consequently, we observed only the hygrometer and a thermometer in our upper room in the house, open in every direction, and surrounded by a gallery and viranda; the thermometer, at midnight, was generally 76° or 77° At six A.M., 80°, at noon 84°, and at sun-set 82°. The point of deposition very regularly 71° at midnight, and rising to 73° in the day, being only once observed so low as 69°.

"Between *Maranham* and *Trinidad*, the air, during the day, from 79° to 83°, the point of deposition from 73° to 75.5°—the surface water from 83° to 84°.

" *Port Spain, Trinidad.*—A transparent register thermometer was suspended in a copper cylinder, pierced with holes at the top and bottom, in the centre of the belfry, in the tower of the new Protestant church, open to the wind on the four sides, by windows fitted with Venetian shutters. The pavement of the church was eighteen feet above the level of the sea: the *maximum* and *minimum* of the thermometer were registered each day at noon.

Date.	Air.		Mean.	
	°	°	°	
Sept. 23 and 24	74.	88.	81.	
24 and 25	74.	86.	80.	
25 and 26	73.5	86.	79.75	
26 and 27	74 75	86.25	80.5	
27 and 28	74.	84.5	79.5	
28 and 29	73.	85	79.	
29 and 30	75.	87.5	81.25	
30 and 1 Oct.	74.	85.5	79.97	
1 and 2	72.	88.	80.	
			80.11	Lat. 10° N.

" The sun had passed the corresponding declination, and had arrived at the equator.

" Between *Trinidad* and *Jamaica*, in the run across the Caribbean sea, the following were the observations:—

Date.	Time.	Air.	Point of Deposition.	OBSERVATIONS.
Oct. 11	8 A.M.	83.2	77.5	Clouded, strong breeze.
—	2½ P.M.	83.	78.5	Clouded, less wind.
12	8 A.M.	82.	76.5	Clear sunshine, fresh breeze.
13	8 A.M.	83.	77.5	Sunshine, with clouds.
14	8 A.M.	82.	78.	Sunshine, with clouds.
15	8 A.M.	82.	rain	Heavily clouded.
16	8 A.M.	83.4	77.5	Sunshine and fine.
17	8 A.M.	82.	78.	Fine.
—	Noon.	82.	77	

" *Port Royal, Jamaica.*

" *For the Mean Temperature.*—A register thermometer was suspended in a copper cylinder, pierced with holes in the top and bottom, in a free current of air, in Fort Charles, sheltered by a platform overhead, and exposed to the draft of the sea breeze through an embrazure, in which the thermometer was placed. It was 200 feet distant from the windward shore of Port Royal, and the height above the sea, eight feet.

Date.	Extremes Registered.		Mean.	OBSERVATIONS.
Oct. 23 to 24	83.5	77	80.25	Weather continually clouded, and esteemed unusually cold.
24 to 25	84.5	78.	81.25	
25 to 26	86.	76.	81.	
26 to 27	87	76.	81.5	
27 to 28	86.	76.	81.	
28 to 29	86.5	76.	81.25	
29 to 30	87 5	76.5	81.5	
Nov. 4 to 5	86.	76.	81.	
			81.17	

"*Hygometry.*—Between the twenty-first October, and the thirtieth, the maximum of the point of deposition varied on different days, between 76° and 78°."

The particulars of a measurement of the mountains of Jamaica have been already given.

METEOROLOGICAL OBSERVATIONS

IN BRAZIL, AND ON THE EQUATOR.

BY ALEXANDER CALDCLEUGH, ESQ.

I AM indebted to the obliging attention of Mr. Caldcleugh for the following interesting particulars of the climate of Brazil, and observations in the northern and southern trade winds:—

"The summer at Rio de Janeiro begins about the months of October or November, and lasts until March or April. This is the wet season, but the rains by no means descend from morning till night, as in some other tropical countries, but commence, generally, every afternoon, about four or five o'clock with a thunderstorm. The heaviness of the rain can only be conceived by those who have been in these latitudes. This fall naturally arrests the sea-breeze, and the succeeding night is dark and cloudy. Formerly these diurnal rains came on with such regularity, that it was usual, in forming parties of pleasure, to arrange whether they should take place before or after the storm. During this period of the year there is seldom, if ever, a deposition of dew.

"From April until September very little rain falls: vegetation almost stops, and to the eye of every one who has not just arrived from Europe, a wintery appearance is discernible. The land and sea breezes do not succeed each other with the same regularity, and are, besides, more frequently disturbed by violent gusts from the S.W., imagined to be the tails of those destructive winds, the Pamperos of the River Plate. The nights are beautifully clear; Venus casts a shadow, and the southern constellations are seen in all their beauty. The dews, as might be expected, are at this season very copious. The annual mean height of the barometer in Rio de Janeiro is about 30.275, and of the thermometer a fraction above 73° Fahrenheit.

The particular observations upon the climate which I was enabled to make during my residence in the country, are contained in the following Journal:—

Mr. Caldcleugh's *Meteorological Journal, commencing at Rio Janeiro,* 1*st August,* 1821, *and continued on the Route to Villa Rica.*

Day of the Month.	Locality.	Face of the Country.	Barometer noon.	Thermometer.	
				Attd.	Detd.
Aug. 1	Rio de Janeiro	The Country about Rio abounds in cone-shaped hills, clothed to their summits with wood.	The state of the Barometer was not recorded during this period.		71
2	,,				70½
3	,,				71
4	,,				69
5	,,				70
6	,,				72
7	,,				73
8	,,				73
9	,,				72
10	,,				72½
11	,,				72
12	,,				72½
13	,,				73
14	,,				73½
15	,,				73
16	,,				74
17	,,				74
18	,,				73
19	,,				75
20	,,				74
21	,,				74½
22	,,				74
23	,,				77
24	,,				73
25	,,				74
26	,,				75
27	,,				75
28	,,				79

Dew Point.	Difference.	Weight Vapour in Grains in Cubic foot.	REMARKS. Observations at 12 o'Clock, when not otherwise expressed.
60	11	5.560	Fine.
59	11½	5.467	Fine.
62	9	6.029	Fine, a little cloudy.
58	11	5.298	Fine.
56	14	4.939	Fine, regular breezes.
55	17	4.756	Fine.
57	16	5.080	Fine.
56	17	4.909	Fine.
57	15	5.089	Fine.
55	17½	4.756	Fine.
55	17	4.756	Fine.
56	16½	4.918	Fine.
58	15	5.251	Fine.
57	16½	5.080	Fine.
59	14	5.434	Fine.
59	15	5.425	Fine.
58	16	5.242	Fine.
66	13	6.625	Fine.
62	13	5.982	Fine—Late sea breeze.
63	11	6.184	Fine.
60	14½	5.616	Fine.
57	17	5.070	Fine.
69	8	7.430	Late sea breeze—Rain at night.
66	7	6.824	Fine.
64	10	6.387	Fine.
65	10	6.493	Fine.
67	8	6.909	Fine.
74	5	8.715	Warm, heavy rain at night.

Day of the Month.	Locality.	Face of the Country.	Barometer noon.	Thermometer. Attd.	Thermometer. Detd.
Aug. 29	Rio de Janeiro				74
30	Left Rio de Janeiro		29.753	75	76
31	Porto d'Estrella	Swampy.	29.801	75	75
Sept. 1	Mandioca		29.654	71	70
„	Pass of the Organ Mountains .	Extremely mountainous and thickly wooded.	27.152	67	68
2	Padre Corrêa		27.752	66	65
3	Pampulha		28.152	68½	67½
4	Three leagues in advance . . .		29.250	68	66
„	Over the Valley of the Paraiba		28.202	80½	80
„	Bank of the Paraibuna		29.902	78	78½
5	„		29.150	78	78
6	Mathias Barbosa		28.602	68	67
„	Morro de Mideiras		27.650	74	72
7	Alcaide Môr		27.701	65	65½
„	Chapeo d'Uvas		27.600	74	75½
8	„		27.751	64	65
9	Montiqueira		27.351	59	58
„	Height of the Serra		26.601	64½	64
10	Barbacena		26.552	50	52
11	Four leagues in advance . . .	Open plain with little timber but the pine.	. . .	. .	58
12	Queluz Villa		27.003	60½	59½
13	Congonha do Campo		27.302	59½	59½
14	Capao d'Olanda Topaze Mine	Very mountainous country with little wood.	26.250	65	66
15	Villa Rica		26.370	64	63
16	„		26.400	65	65½
17	„		26.420	74	75
18	Mariana		27.901	68	67
„	Itacolumi of Marianna		26.254	70½	73½

Dew Point.	Differ-ence.	Weight Vapour in Grains in Cubic foot.	REMARKS. Observations at 12 o'Clock, when not otherwise expressed.
68	6	7.233	Rainy in squalls.
65	11	6.482	Fine, but sultry and cloudy.
62	13	5.982	Gloomy, heavy rain at night.
60	10	5.660	Fine, thunderstorm.
61	7	5.877	Slight rain, 1-80th.
57	8	5.156	Gloomy, 7 A.M.
58	9	5.303	Gloomy morning.
54	12	4.664	Fine, most oppressive sun.
75	5	9.013	Noon observation.
72	6½	8.184	Evening—Temperature of the river 71°.
68	10	7.183	Morning cool and gloomy.
59	8	5.496	Slight rain, which made the track slippery.
67	5	6.945	Clear about one o'clock.
60	5½	5.710	Dull and cold morning.
68	7½	7.221	Fine evening.
61	4	5.908	Cloudy, afterwards heavy rain.
52	6	4.420	Severely cold morning, mist and rain.
59	5	5.525	About eleven A.M. foggy.
50	2	4.179	Morning severely cold.
. .	. .	. . .	Felt the cold severely.—From a variety of causes prevented examining the Barometer or Hygrometer.
50	9½	4.120	Thick fog, afterwards oppressive sun.
49	10	3.988	Clear morning.
52	12	4.352	Bright morning.
54	9	4.689	Fine.
57	8	5.156	Fine—Observation made in the morning.
68	7	7.221	Fine, idem at noon.
62	5	6.072	Fine.
67	5½	6.933	At one P.M.

Day of the Month.	Locality.	Face of the Country.	Barometer noon.	Thermometer.	
				Attd.	Detd.
Sept. 18	Maynerte		27.950	79	76
19	,,		28.208	68	69½
20	Mangeleguas		27.802	75	74
21	Bandeira		27.601	68	57
22	Villa Rica		26.394	69	69½
23	,,		26.380	73	73½
24	,,		26.412	72	72
25	,,	Very mountainous country with little wood.	26.392	71	71½
26	,,		26.361	69	69½
27	,,		26.376	70	70
28	,,		26.381	70	70½
29	,,		26.400	73	74
30	,,		26.390	73	73
Oct. 1	,,		26.416	72	71½
2	,,		26.382	70	69½
3	Coxo de Agua				78
4	Congonha de Sabará				83
5	Sabará	Abounding in wood.			87
6	Caete		The barometer was left at Villa Rica during this excursion.		78
7	St. Joao				71
8	Inficionado				70
9	Vamos vamos				71
10	Villa Rica		26.378	67	67
11	,,		26.408	66	66½
12	,,		26.400	67	67
13	,,		26.398	68	68½
14	,,		26.399	68½	68
15	,,		26.406	70	70½

Dew Point.	Differ-ence.	Weight Vapour in Grains in Cubic foot.	REMARKS. Observations at 12 o'Clock, when not otherwise expressed.
68	8	7.208	In the evening.
65	4	6.567	Thick fog.
70	4	7.715	Very thick fog.
63	4	6.265	Cloudy morning—Torrents of rain in the afternoon.
67	2	6.988	Gloomy. Noon observation.
70	3	7.728	Fine.
69	3	7.495	Very foggy.
68	3	7.277	Clearer.
65	4	6.567	Cloudy.
65	5	6.556	Sun obscured.
66	4½	6.863	Cloudy, light rain.
68	6	7.233	Morning foggy, afterwards clear.
69	4	7.482	Fine, a little foggy.
68	3½	7.277	Cloudy and thick mist.
67	2½	6.988	Thick mist—Dreadful storm.
74	4	8.746	At two P. M. great thunderstorm.
80	3	10.536	At three P. M. idem.
85	2	12.210	Very warm—Violent storm at night.
74	4	8.746	Storm of thunder and rain at four P. M.
65	6	6.544	Fine.
65	5	6.556	Foggy.
64	7	6.427	Foggy.
64	3	6.471	Thick mist.
62	4	6.082	Idem.
63	4	6.265	Idem.
65	3	6.587	Idem.
62	6	6.061	Idem.
65	5½	6.556	Idem.

"On the 15th October I began to retrace my steps to Rio de Janeiro. Having left the barometer at Villa Rica, I made no kind of observations on the weather. The rains having commenced, the roads were in some places in almost an impassable state, and I scarcely think the barometer could have escaped, from the many falls my mule experienced.

"I had imagined that the great humidity at Rio proceeded from the saline particles blown over by the sea breeze; but, on examining the foregoing register, it will be remarked, that there was more vapour in the air, on the 19th and 23th of August, before the sea breeze had commenced, than on the preceding days. I have no doubt, therefore, that when the land breeze prevails all day, which sometimes, though fortunately rarely, happens, the most vapour is contained in the air; and it seems to me this must be the case.

"Beyond the Serra de Montigueira, the track leaves the mountainous and thickly-wooded country, and crosses a high table land, where the pine is the only tree that seems to flourish. The height of this, from the average of the barometrical observations above the sea, may be about 3720 English feet. Baron Humboldt gives the lower limit of the Mexican Pine (19 N. Lat.) at 1150 metres = 3769 English feet. I have seen this species growing in still lower situations in Brazil, but certainly with not so much luxuriance.

"The means of the observations made at Villa Rica, are as follows:—

"Bar. 26.393—attached and detached thermometer 69½°—dew point 65°, and grains of vapour in cubic foot 6.577. Its consequent height above the sea 3969 feet.

"The mean quantity of vapour, observed by Mr. Daniell, for the two years ending with the summer of 1821, was, grains 3.652, not much more than half the mean at Villa Rica. The prevailing winds there were south and south-east.

"I remarked invariably the barometer to stand lower, and the quantity of vapour more considerable, in the evening than the following morning. When overlooking some of the thick woods, it was curious to see the warm vapour ascending like smoke from particular spots, where the foliage did not form a mechanical obstruction.

"On the excursion made from Villa Rica to Sabarà, it will be seen that violent thunder-storms were experienced almost daily: nothing causes so much attention to be paid to the weather as being exposed to its changes; and I could not help noticing the way these storms commenced. The sky was perfectly clear until about two or three o'clock, when some light white clouds were seen approximating the sun with great rapidity. Sometimes they all passed, but if one lingered, as if within its influence, thunder was heard, and in a few minutes no remains of a blue sky were visible. The storm commenced directly, and the change that took place in the temperature often caused a kind of whirlwind.

"As after all, perhaps, we must search for the cause of that singular excrescence, the goitre, or wen, in the state of the air or vicissitudes of climate, it may not be irrelevant to mention that I met by far the greater number of persons afflicted with this complaint near Sabarà.

"From the degree of cold in the province of the mines, the hue of the negroes is much deeper than in Rio de Janeiro. The *Mineiros* who come down, complain much of the heat, and have their health affected. This may proceed, however, from other causes, such as excess in fruits, which are unknown in their province, and a mode of life entirely different. I am inclined to think, on the whole, that foreigners would consider the coast more healthy than the interior.

"Having reached Rio de Janeiro, I embarked on board His Majesty's ship Owen Glendower, for England, on the 22d November, and as soon as the usual sickness had abated, recommenced my observations with the hygrometer. The Hon. Captain Spencer, whose only aim seemed that of rendering all under his command and on board his ship perfectly happy, and in which it is almost superfluous to say he was most successful, dedicated a portion of his time to, and took considerable interest in, these experiments. Many of them recorded here were conducted by him, and he indeed suggested an improvement in the instrument, (that of colouring the bulb,) which on our arrival in England we found had been already contrived.

"The hygrometer was accidentally broken on the 27th December, and, being provided with only one, my observations ceased on that day. When I commenced using the instrument, I was almost afraid to touch it, from its apparent delicacy, but was soon convinced, from the many rude shocks it underwent, that it was stronger than I had imagined; more than common carelessness, indeed, is required to break it. I may be permitted to add, that I think no traveller will find any inconvenience from carrying this hygrometer, or its accompaniment, a small stock of ether; the latter I usually placed among my linen.

"Although the observations, of necessity, ended here, I had the gratification of thinking they were continued through the south-east trade, south of the equator, and until we were on the northern limit of the north-east trade, which it is well known prevails on the northern side. On examining the register it will appear, that on the days when the trades were fresher, there was a slight diminution of vapour; and that, as we approached near the equator, we approximated the point of saturation, the precise position of which, probably, varies according to the longitude and season, as is the case with the trades themselves.

"In these winds there is something so exhilarating, that one with difficulty believes so much vapour exists as the hygrometer indicates. Baron Humboldt, who did not proceed farther south than ten degrees north latitude in crossing the Atlantic, marks eighty-six degrees of Saussure's instrument in that altitude.

"A set of experiments conducted on board some

of the foreign ships that endeavour to pass the line in improper longitudes, and consequently experience calms of many days' duration with much rain, would prove particularly interesting."

METEOROLOGICAL OBSERVATIONS, made on board His Majesty's Ship OWEN GLENDOWER, Captain the Hon. R. C. SPENCER, during a voyage from Rio de Janeiro to Spithead, commencing 1st December.

Time of the day.	Latitude S.	Longitude W.	Winds.	Remarks, State of the Weather, &c.	Barometer.	Ther.	Dew Point.	Difference.	Grains of Vapour in Cubic Foot.	Time of the day when Hygrometer was used.
8 A. M. noon 8 P. M.	24°.25	25°.24	N N N E	*Saturday* 1st *December*.—Fine pleasant weather with steady breeze; towards sunset there were many clouds, chiefly nimbi, cumuli, and strati.	33.09 30.10 30.08	76 76	73 72	3 4	8.488 8.212	1 P. M. 6 P. M.
8 A. M. noon 8 P. M.	24.17	23.12	N ½ E N W by N	*Sunday* 2d *December*.—Rather fresh breeze and fine weather. Evening a few scattered cumuli, mostly in the south-east quarters.	30.05 30.06 30.08	76 76	70 70	6 6	7.688 7.688	10 A. M. 3.3 P. M.
8 A. M. noon 8 P. M.	23.25	21.38	N N by W Calm	*Monday* 3d *December*.—Light breezes, wind dying away fast. Evening calm and fine; a few dispersed clouds.	30.09 30.10 30.13	74 77	69 72	5 5	7.469 8.198	9 P. M. 4 P. M.
8 A. M. noon 8 P. M.	22.31	21.46	E N E by E N E by N E N E	*Tuesday* 4th *December*.—Light breezes and fine weather; sky perfectly clear until the evening, when the wind died away clouds gathered thick all round.	30.11 30.11 30.09	76½ 76½	71 70	5½ 6½	7.951 7.689	9 A. M. 3 P. M.
8 A. M. noon 8 P. M.	20.57	22.24	N N E E N E E by N N E	*Wednesday* 5th *December*.—Cloudy dull weather early, but afterwards beautifully clear, with a light breeze; not a cloud to be seen in the evening; breeze rather fresher.	30.09 30.09 30.07	76 76	70 72	6 4	7.640 8.213	9 A. M. 3 P. M.
8 A. M. noon 8 P. M.	18.52	22.49	East E by S got the trade about noon	*Thursday* 6th *December*.—In the morning a squall of wind and rain for a quarter of an hour; weather finer than yesterday. In the evening fresh breeze and steady.	30.10 30.11 30.11	77 79	71 70	6 9	7.937 7.676	6 A. M. 3 A. M.
8 A. M. noon 8 P. M.	15.37	22.48	East E S E E by S E S E	*Friday* 7th *December*.—Fresh trade wind and cloudy; fine at noon and clear; evening a few clouds scattered, chiefly cumuli.	30.12 30.10 30.08	76 76	72 70	4 6	8.213 7.689	9 P. M. 3 P. M.
8 A. M. noon 8 P. M.	11.54	22.33	E S E E by S S E by E	*Saturday* 8th *December*.—Trades varying in point of strength all day; weather fine all the forenoon; in the evening fresh trade and cloudy weather.	30.07 30.07 30.01	76 77	70 71	6 6	7.689 7.937	8½ A. M. 3 P. M.
8 A. M. noon 8 P. M.	8.6	22.6	E S E E by S S E S E by S	*Sunday* 9th *December*.—Fine weather; fresh trade wind; in the afternoon fresh breezes and cloudy; wind drawing aft; evening fine weather; many flying fish.	30.02 30.01 29.97	78 77	72 71	6 6	8.185 7.937	8¼ A. M. 3 P. M.
8 A. M. noon 8 P. M.	4.55	22.	S E by S E S E S E by N S E by S	*Monday* 10th *December*. Trade tolerably fresh, but fell off in the evening; cloudy, small halo round the moon.	29.96 29.96 29.96	79 79½	74 73	5 6½	8.732 8.438	9 A. M. 3 P. M.
8 A. M. noon 8 P. M.	2.11	22.3	S E S S E S E by E	*Tuesday* 11th *December*. Light winds; breeze lighter in the evening; a few straggling clouds.	29.96 29.97 29.98	80 79	75 74	5 5	9.013 8.732	8¾ A. M. 3 P. M.
8 A. M. noon 8 P. M.	0.8 N	21.53	S E by E S E by S E by S S E	*Wednesday* 12th *December*. Light trades and fine weather; in the evening wind fresher; a few scattered clouds.	30. 29.98 29.95	79½ 79½	76 76	3½ 3½	9.334 9.334	8.50 A. M. 3 P. M.
8 A. M. noon 8 P. M.	2.14	22.	S S E E by S S E	*Thursday* 13th *December*. The wind light all day; weather clear; towards evening fresher and some clouds appeared; nimbi, cumuli, and strati.	29.92 29.92 29.92	80 82	78 79	2 3	9.958 10.263	8¾ A. M. 3½ P. M.

[*To face Page* 348.

The material originally positioned here is too large for reproduction in this reissue. A PDF can be downloaded from the web address given on page iv of this book, by clicking on 'Resources Available'.

METEOROLOGICAL OBSERVATIONS, &c., continued.

Time of the day.	Latitude N.	Longitude W.	Winds.	Remarks, State of the Weather, &c.	Barometer.	Ther.	Dew Point.	Difference.	Grains of Vapour in Cubic Foot.	Time of the day when Hygrometer was used.
8 A. M. noon 8 P. M.	4°.19	28°.8	S S E N E by N E by S variab.& calm	*Friday* 14th *December*. Rain early in the morning; trade left us at five A.M.; the weather continued cloudy and dull all day; wind light and unsettled; much lightning after dark.	29.92 29.92 29.92	78 80	75 78	3 2	9.044 9.958	8.50 A. M. 3.30 P. M.
8 A. M. noon 8 P. M.	5.18	22.14	Calm S S W E by S N E by E	*Saturday* 15th *December*. This day set in with a heavy squall of rain, which went astern; lightning; after a few intervals of calm we got the north-east trade; a few drops of rain in the evening.	29.90 29.88 29.86	81½ 80½	76½ 78	5 2½	9.405 9.949	8.50 A. M. 3.30 P. M.
8 A. M. noon 8 P. M.	6.56	23.51	E N E N E by E N E by E N N E	*Sunday* 16th *December*. Dull and cloudy; dark black clouds; fresh trade; in the evening fresher; weather cloudy.	29.93 29.90 29.93	82 80	78 77	4 3	9.924 9.638	8.40 A. M. 1.45 P. M.
8 A. M. noon 8 P. M.	9.13	24.58	E by N E N E N E by E E N E	*Monday* 17th *December*. Moderate trade winds with thick cloudy weather, very hot; in the evening many nimbi and cumulo-strati.	30.01 29.97 30.	80 80	77 75	3 5	9.638 9.013	8.50 A. M. 4 P. M.
8 A. M. noon 8 P. M.	11.41	26.55	N E by E N E N E by E	*Tuesday* 18th *December*. Still cloudy; dark bank of clouds; some strati and cumuli; fresh trades, a nimbus or two.	30.03 30.01 30.04	77 79	73 73	4 6	8.474 8.445	8.55 A. M. 3.40 P. M.
8 A. M. noon 8 P. M.	14.18	29.1	N E N E by E E N E E by N	*Wednesday* 19th *December*. Strong trades and fine weather with a few scattered clouds; in the evening a dark thick bank of clouds; some cumuli.	30.07 30.03 30.04	76 76	70 72	6 4	7.689 8.213	8.55 A. M. 3.55 P. M.
8 A. M. noon 8 P. M.	17.11	30.12	E by N E N E E	*Thursday* 20th *December*. Weather similar to that of yesterday until the afternoon, when the wind lulled a short time, and a small quantity of rain fell; breeze freshened and afterwards cloudy.	30.11 30.10 30.11	76 77	73 74	3 3	8.482 8.762	8.45 A. M. 4 P. M.
8 A. M. noon 8 P. M.	20.8	31.3	East E by N E by N E N E	*Friday* 21st *December*. Strong gales and squally with thick clouds; in the afternoon weather fine; cumuli over the horizon; cumulo-strati above.	30.17 30.16 30.20	75 75	72 72	3 3	8.227 8.227	8.50 A. M. 3.55 P. M.
8 A. M. noon 8 P. M.	22.25	32.15	E by N N E by E N E N E by N	*Saturday* 22d *December*. Moderate breezes and fine weather; two or three nimbi at different times, but no rain; cumuli.	30.21 30.21 30.24	74 74	71 68	3 6	7.978 7.234	8.50 A. M. 4 P. M.
8 A. M. noon 8 P. M.	24.30	34.18	N E E N E E by N N E	*Sunday* 23d *December*. Light winds and very fine weather, as the day advanced wind became lighter; some clouds; one large nimbus, the rest cumuli; sun set clouded.	30.28 30.29 30.31	70½ 74	67 68	5½ 6	6.970 7.233	8.45 A. M. 3.30 P. M.
8 A. M. noon 8 P. M.	26.1	35.43	N E E N E N E by N N by W N E	*Monday* 24th *December*. Cloudy with showers of rain and a light variable breeze in the morning; breeze stronger in the afternoon, the weather clearer; light again and cloudy; cumuli and cumulo-strati chiefly.	30.34 30.34 30.34	72 73	70 66	2 7	7.742 6.822	8¾ A. M. Gulf weed 3½ P. M.
8 A. M. noon 8 P. M.	26.36	36.41	N N N E N E by N N E	*Tuesday* 25th *December*. Thick cloudy weather and variable wind in the forenoon; the breeze fresher and less cloudy; in the evening high cumuli and black bank clouds.	30.34 30.34 30.35	70½ 72	66 67	4½ 5	6.858 6.946	8¾ A. M. Gulf weed 3¾ P M.
8 A. M. noon 8 P. M.	27.50	38.26	N N E E N E N E E N E E	*Wednesday* 26th *December*. Weather dull and wind light, but squally; evening fine, but breeze very light.	30.38 30.35 30.31	70½ 69	67 65	3½ 4	6.970 6.568	8.50 A. M. Gulf weed 3.50 P. M.

The material originally positioned here is too large for reproduction in this reissue. A PDF can be downloaded from the web address given on page iv of this book, by clicking on 'Resources Available'.

REMARKS

UPON THE BAROMETER AND THERMOMETER,

AND THE

MODE OF USING METEOROLOGICAL INSTRUMENTS IN GENERAL.

BY offering the following remarks upon meteorological instruments, I would not wish it to be supposed that I claim, for the observations which I have hitherto recorded, a greater degree of precision than attention to the usual precautions has been sufficient to confer: but in the course of my experiments, the necessity of much greater care and method has become strongly impressed upon my mind, and I think that it may not be wholly without its use, to indicate such measures as the result of my experience suggest, as likely to ensure that degree of perfection, of which the science of meteorology is doubtless susceptible. I have little of novelty to offer upon the subject; but if, by repeating well-known observations, I can contribute to excite that attention to them which is absolutely necessary to success; if the numerous observers of atmospheric phenomena may possibly be thus engaged to that strict co-operation, which alone can prevent their daily labours from proving abortive, a great and important object will be attained.

Much of my attention has lately been given to the manufacture of barometers. The committee of the Royal Society, appointed to take into consideration the state of the meteorological instruments, did me the honour to request that I would attend to the construction of a new barometer for their apartments; and as, in the course of the close attention which I paid to the most minute details, I had occasion to make many practical remarks, I cannot, I think, do better than here introduce the account of the process which I had prepared for the society.

In the course of the experiments I was led to a new method of filling the tube, which I flatter myself may prove generally useful, and tend, by the facilities which it affords, to the perfection of the instrument.

Previous to commencing the operation, some ex-experiments were undertaken to ascertain the practicability and effect of introducing the metal, after the air had been abstracted, as nearly as possible, by means of an air-pump, and the mercury and interior surface had been exposed to the desiccating influence of a large surface of sulphuric acid. For this purpose a barometer tube was fitted with a stop-cock, which was screwed into the under surface of a pump-plate; on the upper surface stood a glass dish, perforated in the centre, and containing the acid. In this was placed a stand with glass legs, which received a funnel, the stem of which being drawn out into a capillary tube, passed down into the mouth of a small paper cone, resting upon the tube. The aperture at the upper part of the stem was closed by an iron plug, ground to fit, between which and the capillary opening was placed some cotton. The glass funnel was filled with clean mercury, carefully boiled, and many times filtered, and the whole was covered with a glass receiver. Through a collar of leather, in the upper part of the receiver, passed an iron rod, which moved freely up and down, and fitted into a screw in the plug before mentioned, by which means it could be drawn up, and re-placed, at pleasure. The apparatus being thus arranged, the pump was worked, and the air exhausted from the receiver and tube. Air was at first given off from the surface of the acid in abundance, and a few bubbles passed up from between the mercury and the glass; but none appeared upon the surface of the mercury. When the rarefaction had been carried as far as possible, the siphon-gauge stood at about half an inch. The iron plug was carefully withdrawn, and the mercury began to trickle very gradually into the tube. In its fall it was broken into small globules, many of which adhered to the sides of the glass; and, notwithstanding the utmost precaution and frequent repetitions of the experiment, the column of mercury, as it rose, contained very minute cavities, which decreased in size as the weight increased; and when the pressure of the atmosphere was restored, were only discernible upon very close examination. When the air was again extracted, they returned to their former size, and again diminished upon its restoration. The difficulty of getting rid of these cavities appears to me to arise chiefly from their form: for the mercury, assuming the shape and properties of a dome round the bubble, resists a degree of pressure which would otherwise cause it to run together.

To avoid this mechanical action of the fall of the mercury, the apparatus was varied as follows: A small tube was passed down to the bottom of the barometer-tube, and was fastened at the top by a piece of cork, to prevent its coming in contact with the sides. The lower aperture had been lessened, and the small paper funnel was inserted into the upper end. The exhaustion having been made as before, the mercury was allowed to trickle down the interior of the inner small tube, from the bottom of which it issued slowly; and gradually rose in the larger, in perfect and uninterrupted contact with the glass. When the tube was full; the air was let into the receiver, and the tube detached from the plate. To prevent the possibility of the disengagement of any particles of air, which might be entangled in the mercury of the small tube, its orifice was hermetically sealed by a lamp, and the tube itself full of mercury, carefully withdrawn from the large one. The closest examination, with a magnifying glass, of the barometer-tube so filled, failed to detect the minutest air-bubble, and the surface everywhere was as resplendent as that of the most perfect mirror. The application of heat produced no alteration in this appearance, nor any traces of either air or moisture. The small tube, upon inspection, was found to contain very minute, and scarcely visible, specks, like those of the tubes filled in the first method; but these were, of course, diminished in quantity, in proportion to the diminution of the tube in which they were formed.

The success of this experiment was so great, that, in any common case, it would scarcely have been thought necessary to subject the barometer to the troublesome and hazardous process of boiling the mercury; but upon this occasion it was resolved, that no possible precaution should be omitted.

The tube, which was selected for the society's barometer is 33¾ inches long, its exterior diameter 0.86 inch, and the diameter of its bore 0.530 inch.

These measures were taken at the upper extremity, and it is very regular for 14 inches, but enlarges a little from that point downwards. It is ground flat at the lower end.

Many tubes were destroyed, after all the trouble bestowed upon their mensuration and filling, by the after-process of boiling, which, in tubes of such large capacity, was found to be very troublesome and hazardous, and required the glass to be of a red-heat. The above dimensions are those of the barometer now complete.

The cistern is turned in well-seasoned mahogany, and there is a small cavity in its bottom to receive the end of the tube, which rests upon it: a groove communicates with the cavity, to ensure the free passage of the mercury. By means of the float in front, the level may be very accurately taken. Fifty inches, measured in the upper part of the tube before it was sealed, in four equal proportions, raised the float exactly half an inch; the correction, therefore, for the capacity of the cistern, is $\frac{1}{100}$th.

The cistern being accurately levelled, and the tube and thermometer both in their places, the quantity of mercury was adjusted to the upper edge of the black line on the stem of the float: a card gauge of nearly the diameter of the cistern was then fitted to slide upon the tube, which was fixed perpendicularly in its place. The lower edge of the gauge was made to coincide with the surface of the mercury on both sides, and at its contact with the glass, two distinct marks were scratched upon the tube. From these marks, twenty-nine inches were measured off by a brass dividing engine, which was formerly the property of the late Mr. Cavendish, and at that distance, another distinct mark was made.—The utmost care was taken to read off these distances by means of lenses, and the temperature of the scale, the glass, and the mercury, was 54°.

It being deemed too hazardous an experiment to attempt to boil, at once, so large a body of mercury as would be contained in a tube of this capacity, it was resolved to perform the operation in two portions, and under the diminished pressure produced by an air-pump. Accordingly, seventeen inches of mercury were introduced into the tube with all the precautions above described. During the exhaustion, no air was disengaged from the mercury, care having been taken to fill the funnel without agitation. The contact with the glass, while the siphon gauge of the pump stood at .4 in., was perfect, and the appearance of the tube when detached as bright and compact as could be wished. The air was again exhausted, and by means of a stop-cock, the vacuum preserved. The tube was then gradually heated before a fire, and the boiling, afterwards, cautiously begun over a large spirit lamp. The upper part of the column was first strongly heated, and when it had arrived at the point of ebullition, the boiling was slowly continued downwards. When it had reached the bottom, it was again as gradually conducted to the top. The bubbles of vapour freely passed, with the assistance of a slight degree of agitation, from one end of the column to the other; and very bright flashes of green light accompanied their extrication. One minute globule of air alone was detected during the heating, notwithstanding the diminished pressure, and this was readily extricated; and there was not the slightest condensation of moisture visible in the cold portion of the tube.

The cooling was conducted with all the precautions used in heating; and to allow the mercury to resume the temperature of the air, the completion of the process was deferred till the next day. After fifteen hours' repose, upon opening the stop-cock, the exhaustion was found to have been perfectly maintained: the apparatus was again arranged, the small tube being made just to touch the surface of the mercury in the larger. A quantity of mercury was then introduced as before, which was found after the interior tube had been withdrawn, to amount to twelve inches and a half. The appearance was as perfect as in the first operation, except that two or three very minute specks appeared at the junction of the two portions. These scarcely-visible air-bubbles had probably been introduced in extracting a very small particle of cement, which had fallen down; the rod with which this was done having been passed about $\frac{1}{10}$th of an inch below the surface of the mercury. They disappeared under atmospheric pressure. The whole length of the mercurial column was now twenty-nine inches and a half, leaving only three inches and a half of the tube unoccupied, which was deemed but barely sufficient to prevent the communication of the heat melting the cement and destroying the exhaustion. When the air was abstracted, the junction was again just discernible. The boiling was begun as before from the top, and carried downwards to about two inches below the union of the two portions. The little air-bubbles were visibly expanded and easily passed up: the boiling was continued for a considerable time, and large bubbles of mercurial vapour, accompanied with bright green light, freely traversed the whole column. The tube was then suffered gradually to cool. Its appearance was compact and bright: a very slight haziness or discoloration was observable at the junction, but not the slightest indication of air even under exhaustion. No precipitation of moisture was perceptible in the cool portion of the tube.

It was the original intention to have completed the filling of the barometer, by boiling the remaining three inches and a half; but, upon consideration of all the circumstances, and, especially of the necessity there would be of performing this under atmospheric pressure, it was concluded not again to expose the tube to so much risk. The column already boiled comes within the range of the atmospheric oscillations, and the utmost care was taken in filling the remainder, as before, *in vacuo*. The last portions of mercury were introduced hot, and the whole was left for forty-eight hours, to take the temperature of the air. The tube was then carefully inverted in the cistern; but the mercury, notwithstanding its great body, did not descend till after it had received two or three smart concussions. This, I believe, to be the most certain proof of the complete displacement of every particle of air. The adjustments, of the scale, with its nonius to the upper mark upon the tube, and of the quantity of mercury in the cistern to the line upon the float, were now easily made, and the instrument was fixed in its proper situation. Whenever the mercury vibrates in the tube, a beautiful green light flashes through the vacuum, and the crackling sound of electric excitation is heard when the finger is presented to it. Electric attraction and repulsion are also exhibited by presenting a piece of gold leaf to its influence.

Every thing has been studied in this instrument to render accuracy attainable, with as little trouble as possible to the observer. The diameter of the tube renders the correction for capillary action, almost unnecessary—the correction for the capacity of the cistern has been contrived to be $\frac{1}{100}$th of the result above or below the neutral point, 30.576 – and a scale is engraved upon the front, of the correction to be applied for the expansion of mercury and mean dilatation of glass; by which the observation may be at once reduced to the standard temperature of 32° A small thermometer in front of the instrument dips into the mercury of the cistern. The specific gravity of the mercury employed was carefully ascertained at the Royal Institution, by Mr. Faraday. The temperature of the metal and the water were both 40°, and 1000 grains of the former displaced 73.4 grains of the latter: hence $\frac{1000}{73.4} =$ 13.624.

One of my chief objects during these experiments has been to ascertain the agreement of different barometers made with equal care and independently graduated, after all the necessary corrections have been made for accidental differences. I, therefore, attended particularly to the construction of a mountain barometer for my own use, which was filled *in vacuo*, and afterwards boiled. After the process, the tube was perfect in appearance, the mercury adhered when reversed, and the electric light was very visible. The graduation was made with every care from the surface. The interior diameter of the tube is 0.15 in., and the correction for the capacity of the cistern $\frac{1}{41}$ – the neutral point 30.180.—I shall here give the details of three separate comparisons of these two instruments.

Royal Society's Barometer.		Mountain Barometer.	
30.576	Temp. of Mer. 50°.	30.526	Temp. of Mer. 50°.
–.047	Correction for Temp.	–.047	Correction for Temp.
30.529		30.479	
+.006	Capillary Action.	+.088	Capillary Action.
30.535		30.567	
		+.009	Capacity of Cistern.
30.535		30.576	
29.872	Temp. of Mer. 64°.	29.849	Temp. of Mer. 70°.
–.007	Capacity of Cistern.	–.008	Capacity of Cistern.
29.865		29.841	
–.082	Temp. of Mer.	–.098	Temp. of Mer.
29.783		29.743	
+.006	Capillary Action.	+.088	Capillary Action.
29.789		29.831	

Royal Society's Barometer.		Mountain Barometer.	
29.756	Temp. of Mer. 63°.	29.742	Temp. of Mer. 72°.
−.008	Capacity of Cistern.	−.010	Capacity of Cistern.
29.748		29.732	
−.078	Temp. of Mer.	−.102	Temp. of Mer.
29.670		29.630	
+.006	Capillary Action.	+.088	Capillary Action.
29.676		29.718	

The results of these comparisons disappointed me at first, as I had been induced to expect a much nearer accordance, after all the pains that had been taken. Upon reflection, however, I am inclined to think that the apparent discordance is in favour of the instruments, and that the difference points to an error in one of the corrections which has been overlooked. In the first place, it will be remarked that the difference .040 in. is constant, and its cause, therefore, is probably to be sought in the only constant correction, namely, that for capillary action. The quantities allowed have been taken from Dr. Young's Table * of the depression of mercury in barometer tubes, which was calculated from experiments. But these experiments were made with tubes in which the mercury had not been previously boiled, and a little consideration will be sufficient to shew that the results must have been very much influenced by this circumstance.

The phenomena of capillary depression depend upon the balance of the attraction of the particles of the fluid for each other, and for the

* Young's *Nat. Phil.*, vol. ii., p. 669.

solid of which the tube is composed. The attraction of mercury for glass is well known to increase as the contact becomes more perfect; and, indeed, all the phenomena attending the boiling of a barometer-tube prove that this is the case. The depression in a tube, from which the air has been thoroughly expelled, must therefore necessarily be less than in one which has been filled without this precaution. Professor Casbois, of Metz, long ago remarked, that the depression of mercury in tubes of glass depended upon the imperfection of the contact; and M. de Luc, speaking of the same fact, observes—" MM. Cassini de Thury et Le Monnier employèrent des tubes de différens diamètres, et cependant ils ne trouvèrent les différences dont parle M. de Plantade que dans les tubes que n'avoient pas été chargés au feu."—*De Luc, Recherches sur l'Atmo.*, tom. i. p. 95.

The comparison above made would seem to indicate, that the depression is decreased one half by boiling, and by diminishing the correction accordingly, the two instruments exactly agree. By a comparison of several others, this estimate is greatly confirmed; and I have lately had an opportunity of putting it to a decisive test. Captain Sabine, before his departure for the North Seas, requested me to assist at an examination of his barometers: two were of the mountain construction, with iron cisterns, by Newman; and one was a marine barometer, by Jones. They had all been independently graduated from the surface of the mercury, and boiled. The interior diameters of the two first were .15 inch, and the correction for the capacities of the cisterns $\frac{1}{54}$th. The diameter of the last .31 inch, and the correction for the capacity of the cistern $\frac{1}{11}$th. The neutral points of the three were the same, *viz.*, 30.400. The following is the comparison of the three instruments with the one which had been already compared with that of the Royal Society. The latter I shall call 1; the two other mountain barometers, 2 and 3, and the marine barometer, 4. The heights are the means of four observations, taken independently by different observers, who never differed more than .005 inch.

No. 1.	No. 2.	No. 3.	No. 4.	
30.1835	30.1730	30.1845	30.1937	Temperature of Mercury alike.
.0000	−.0042	−.0039	−.0187	Capacity.
30.1835	30.1688	30.1806	30.1750	
+.0880	+ 0880	+.0880	+.0280	Capillary Action, by Dr. Young.
30.2715	30.2568	30.2686	30.2030	

SECOND COMPARISON.

30.1130	30.1045	30.1087	30.1202	
−.0016	−.0054	−.0052	−.0254	Capacity.
30.1114	30.0991	30.1035	30.0948	
+.0880	+.0880	+.0880	+.0280	Capillary action.
30.1994	30.1871	30.1915	30.1228	

It will be observed, that the three mountain barometers agree very closely together, the greatest difference from their average height being .008 inch, while the difference of the marine barometer, from the same height, is, in the first comparison, 062 inch, and in the second .070 inch. If, however, we substitute half the correction for the capillary action, as in the case of the Royal Society's barometer, the difference is decreased one-half.

No. 1.	No. 2.	No. 3.	No. 4.	
30.1835	30.1688	30 1806	30.1750	
+.0440	+.0440	+.0440	+.0140	Half the Correction for Capillary Action.
30.2275	30.2128	30.2246	30.1890	
30.1114	30.0991	30.1035	30.0948	
+.0440	+.0440	+.0440	+.0140	Half the Cor.
30.1554	30.1431	30.1475	30.1088	

The remaining discrepancy I have some reason for believing, is in the measurement of the neutral point of the marine barometer.

Thus we see that the application of half the correction for capillary depression, derived from experiments upon unboiled tubes, is most applicable to boiled barometers; and from its application to tubes of the greatly differing-diameters, .53 inch, .31 inch, .15 inch, we may pretty safely conclude, that the proposition is universal.

The following Table gives the results of the experiments of Lord Charles Cavendish upon capillary depression, the correct calculation of the same by Dr. Young, and the probable amount in boiled tubes. :—

TABLE I. *Correction to be applied to Barometers for Capillary Action.*

Diameter of Tube.	Cavendish.	Young.	Amount in boiled tubes.
Inch.	Inch.	Inch.	
.60	.005	.0045	.002
.50	.007	.0074	.003
.45		.0100	.005
.40	.015	.0139	.007
.35	.025	.0196	.010
.30	.036	.0280	.014
.25	.050	.0404	.020
.20	.067	.0589	.029
15	.092	.0880	.044
.10	.140	.1424	070

During my experiments upon the filling and boiling of the barometer-tubes, my attention was particularly directed to the assertion of Sir Humphry Davy, (*Phil. Trans.*, 1822, p. 74), that "there is great reason to believe that air exists in mercury, in the same invisible state as in water, that is, distributed through its pores;" and to the disheartening fact, (if proved), that absorption of air "may explain the difference of the heights of the mercury in different barometers; and seems to indicate the propriety of re-boiling the mercury in these instruments, after a certain lapse of time." It is with much diffidence that I am compelled to differ from the high authority of the President upon this interesting point: but there is one observation which I made, which, I think, nearly disproves the supposition. All fluids, which are known to absorb air into their pores, invariable emit it when the pressure of the atmosphere is removed: but, upon an attentive examination of large bodies of mercury, variously heated in the vacuum of an air-pump, I never saw a bubble of air given off from the surface of the metal. Air will rise from the contact of the mercury with the glass in which it is contained, in exact inverse proportion to the care with which it has been filled, but it never rises from the surface of the mercury alone. The difficulty of properly filling a barometer-tube, I attribute to the attraction between the glass and the air—not to that between the mercury and air; and I believe that air will insinuate itself a little way between the glass and the metal at the exposed end of a boiled tube, but that this cannot happen if the end be plunged in mercury; and, consequently, that no deterioration of barometers is to be apprehended from this cause. Such a deterioration, indeed, if it had existed, must, long ago, have been detected from the instruments themselves; for, although the register of the Royal Society is not in such a state as to enable any one to reason upon its conclusions, that of the Royal Observatory of Paris, and some others, must have disclosed the fact.

With respect to the method of filling a barometer-tube *in vacuo*, recommended above, I have little doubt that it is as accurate as the method of boiling, if performed with proper care; and it is infinitely less troublesome and hazardous. The electric light is as strong in the tube, and its appearance is in every way as perfect. There is, however, one precaution which it is proper to take, *viz.*, to boil about an inch of mercury in the lower end of the tube, as this will prevent that concussion of the metal in its fall, which, breaking it into globules, is apt to entangle any of the residual air. At all events it will be a great improvement upon the common method, which merely consists in passing a large bubble of air up and down the tube, to collect together the smaller particles which adhere to the glass.

Indeed it is high time that more attention should be paid to the construction of meteorological instruments in general. The generality of observers are little aware of the serious inaccuracies to which they are liable. In the shops of the best manufacturers and opticians I have observed that no two barometers agree; and the difference between the extremes will often amount to a quarter of an inch: and this, with all the deceptive appearance of accuracy, which a nonius, to read off to the five hundredth part of an inch, can give.

The common instruments are mere play things, and are, by no means, applicable to observations in the present state of natural philosophy. The height of the mercury is never actually measured in them, but they are graduated one from another, and their errors are thus unavoidably perpetuated. Few of them have any adjustment for the change of level in the mercury of the cistern, and in still fewer is the adjustment perfect: no neutral point is marked upon them, nor is the diameter of the bore of the tube ascertained: and in some the capacity of the cisterns is perpetually changing from the stretching of a leathern bag, or from its hygrometric properties. Nor would I quarrel with the manufacture of such play-things; they are calculated to afford much amusement and instruction; but all I contend for is, that a person, who is disposed to devote his time, his fortune, and oftentimes is health, to the enlargement of the bounds of science, should not be liable to the disappointment of finding that he has wasted all, from the imperfection of those instruments, upon the goodness of which he conceived that he had good grounds to rely. The questions, now of interest to the science of meteorology, require the measurement of the five hundredth part of an inch in the mercurial column; and, notwithstanding the number of meteorological journals, which monthly and weekly contribute their expletive powers to the numerous magazines, journals, and gazettes, there are few places, indeed, of which it can be said that the mean height of the barometer for the year has been ascertained to the tenth part of an inch. The answer of the manufacturer to these observations is, that he cannot afford the time to perfect such instruments. Nor can he, at the price which is commonly given; for few people are aware of the requisite labour and anxiety. But who would grudge the extra-remuneration for such pains? Not the man who is competent to avail himself of its application. Let the manufacture of play-things continue, but let there be also another class of instruments which may rival in accuracy those of the astronomer.

It will, no doubt, be a part of the plan of the Committee of the Royal Society, to establish a standard barometer, and to afford every facility of comparison with it; so that any person, for scientific purposes, may have an opportunity of verifying an instrument: and it is to be hoped that they may proceed one step further, and take measures for ascertaining the agreement of the instruments at all the principal observatories, not only in this country, but in other parts of the world.

Nor is it in the construction of barometers only that the meteorologist has to complain of that want of accuracy which is so essential to the progress of his science: the same carelessness attends the manufacture of the thermometer. Few people are aware that they are all, even those which bear the first makers' names, made by the Italian artists, who graduate them one from another, and never think of verifying the freezing and boiling points. The bulbs are all blown with the mouth, and very little attention is paid to the regularity of the tube. The register thermometers particularly, are shamefully deficient. Those of Six's construction are often filled with some saline solution instead of alcohol; and in the best, the spirit is not exposed long enough *in vacuo*, to disengage the air with which it is mixed. The consequence is, that it is liable to become liberated, and, of course, interferes with the results. The original directions of the inventor have also been departed from, as to the proportions of the different parts, and as to the construction of the *indices*.

Those upon Rutherford's plan are universally sealed with air in their upper parts, which acts as a spring against the expansion of the column. The iron index of one is liable thereby to become oxidated, and adheres to the glass, when the mercury passes it, and it becomes entangled; while the spirit of the other being unavoidably mixed with air, when the pressure is decreased by cold it is disengaged. The air may be again dissolved by increasing the pressure before a fire, and passing the bubble backwards and forward, and, in a state of solution, it does not appear to interfere with the equability of the expansion. This, however, is not certain; and, at all events, it is liable to re-appear, and is very troublesome. These imperfections are, by no means, necessary consequences of the construction of the instruments, although the makers are very willing that they should be so considered; but it requires great care and attention to guard against them. The general mounting of the meteorological thermometers is exceptionable in every way; buried as they are in a thick mass of wood, and covered with a clumsy guard of brass, they can but very slowly follow the impressions of atmospheric temperature.

The establishment of a perfect standard thermometer, which shall be accessible to all who may wish to consult it, will also, doubtless, be another object of the Committee of the Royal Society.

With respect to the change in the freezing point, which takes place in time in the best thermometers, I have lately had an unexceptionable opportunity of confirming the assertions of the French and Italian philosophers. Mr. Jones has obligingly put into my hands two thermometers of the late Mr. Cavendish, which have evidently been constructed with much care. The mercury in the balls of both flows freely into the tubes when reversed; and when suffered to fall sharply, strikes the ends with a metallic sound. The same *click* may be heard in the bulbs when it is permitted to fall back, and the cavity closes without the slightest speck. These indications of a well-boiled tube are rarely to be met with in the common thermometers of the present day. They are mounted upon common deal sticks, and the graduation, which is only continued for a few degrees about the freezing point, is engraved upon a small slip of brass. The degrees are very large, and they are distinctly divided into tenths. Each degree of No. 1, occupies a space of .208 inch, and of No. 2, .130 inch. The scratch upon the glass for the freezing point is very visible in both. It is difficult to say for what purpose they were originally made, but evidently for some experiments upon the freezing point of water; and if they had been expressly constructed to verify the present point, they could not have been better contrived for the purpose. The bulbs of both were plunged into pounded ice, in which they were left for half an hour, and the height of the mercury was carefully taken by two observers with the aid of magnifying glasses. The result of the examination was, that in No. 1. the freezing point upon the scale was 0.4 degree too low, and in No. 2, 0.35 degree. There can be little doubt, I think, that the right cause of the phenomenon has been assigned, *viz.*, the change of form and capacity which the glass undergoes from the pressure of the atmosphere upon the *vacuum* of the tube.

But attention to the perfection of instruments will be all in vain, without a proper degree of care and system in making and recording the observations. Observers would render a much greater service to science by devoting less of their time to the actual inspection of their instruments, and more to applying the proper corrections. If the meteorologist plead want of leisure, instead of daily observations, let him record the atmospheric changes of every second or third day, but let what he does record be correct. The proper hours of the day for observation are indicated by the barometer: the maximum height of the mercurial column is about 9 A.M., the mean at 12, and the minimum at 3 P.M. If a person have time to make three observations in the day, these are the hours which he should select: if circumstances only allow of his observing twice, 9 A.M. and 9 P.M. are the proper intervals: if only once, noon is the time. These fortunately happen to be, probably, the most universally convenient hours that could have been selected. In national observatories, it would not be too much to expect, that observations at 3 A.M. should be added to the preceding. Even those who merely consult the barometer as a weather-glass, would find it an advantage to attend to these hours; for I have remarked, that much the safest prognostications from this instrument may be derived from observing when the mercury is inclined to move contrary to its periodical course. If the column rise between 9 A.M. and 3 P.M., it indicates fine weather; if it fall from 3 to 9, rain may be expected.

But the meteorologist, who wishes to confer a real

benefit upon science by his labours, has a much more tedious duty to perform than this. After taking the height of the barometer at the prescribed times with all possible caution, he will take care to make the proper corrections of the observations. If his instrument be not furnished with a contrivance for adjusting the level of the mercury, he will correct it according to the relative capacities of the tube and cistern: he will add the proper quantity for capillary depression, according to the diameter of the tube: and he will then reduce the height to what it would have been, if the mercury had been of the standard density at the temperature of 32°.

For the purpose of facilitating this last operation, I shall here subjoin a Table of the proper correction, calculated by Mr. Rice from the experiments of MM. Du Long and Petit, upon the expansion of mercury and mean dilatation of glass:—

TABLE II. *Correction to be applied to Barometers for Expansion of Mercury and Mean Dilatation of Glass.*

Temp.	Inches. 28.	Inches. 28.5	Inches. 29.	Inches. 29.5	Inches. 30.	Inches. 30.5	Inches. 31.	Inches. 31.5
25	+.017	.017	.017	.018	018	018	.019	.019
30	+.005	.005	.005	.005	.005	.005	.005	.005
35	−.007	.007	.007	.008	.008	.008	.008	.008
40	−.019	.020	.020	.020	.021	.021	.021	.022
45	−.031	.032	032	.033	033	.034	.035	.036
50	−.043	.044	.045	046	.046	.047	.048	.049
55	−.055	.056	.057	.058	.059	.060	.061	.062
60	−.067	.068	.069	.071	.072	.074	.075	.076
65	−.079	.081	.082	.083	.085	.086	.088	.089
70	−.091	.093	.094	.096	.098	.100	.101	.103
75	−.103	.105	106	.109	.111	.114	.116	.118

By calculating the monthly means, the observer will give a still greater value to his co-operation.

Attention to these directions, in addition to the benefit which it would confer upon meteorology, would also facilitate the purposes of barometric levelling; in return for which, the detached operations of barometric mensuration should, if possible, be performed with a due regard to the prescribed hours of the meteorologist.

The observation of the barometer almost necessarily implies an inspection of the thermometer, and the height of that instrument should be recorded at the same periods; in addition to which the *maximum* and *minimum*, by register thermometers, should be carefully noted. The proper precautions to be taken in placing the instruments for this purpose, are now so well understood that it is needless to repeat them: they are summed up by saying, that they should be sheltered from every species of radiation. The register of the force of radiation in a reflector (as described in the Essay upon Radiation,) and the power of the sun's rays upon black wool would also be particularly interesting; in addition to which, those who have the opportunity should not neglect the variations of the temperature of the sea and other deep bodies of water.

The periods of the barometric observations are also well adapted to those of the hygrometer; but the mean pressure of the aqueous atmosphere should be calculated from the dew-point at 3 P.M., and the lowest temperature at night of the sheltered thermometer. The prognostications to be derived from this instrument have been already described in the Essay upon the Hygrometer, and to these I shall only now add, that by comparing the dew-point with Table III. of the Essay upon the Climate of London, an accurate estimation may be formed of its accordance with the mean, and of the consequent probability of precipitation, change of wind, &c.

With respect to the rain and evaporation gauges, and the vane, I can add nothing to the full directions given by Mr. Howard; and can only lament with him, that some effectual means have not yet been adopted for measuring the force of the aërial currents as well as their direction. Nor have I, at present, any thing to offer upon the extremely important subject of atmospheric electricity. This interesting department of the science has been almost totally neglected, and it is much to be wished that some competent person would devise the proper means of prosecuting an experimental investigation of the subject.

In concluding these observations, I must not, in justice, omit to state, that in all the practical details with which I have been engaged, I have met with the most ready and able assistance from Mr. Newman. He entered fully into all my views with respect to the improvement of meteorological instruments, and has bestowed much time and attention upon executing the hints which I have suggested. His portable barometers with iron cisterns may be depended upon for the nicest experiments.

I must terminate these remarks, as I began, by an urgent recommendation to meteorologists to use standard instruments, to observe them with care, and to make all necessary corrections for accidental differences; and, above all, to keep their tables upon the same scheme. Much curious information is dependant upon such an extensive plan of comparative observation; and without it the observer does little more than accumulate an overwhelming mass of crude and incorrect materials, already too large for arrangement and correction. The example has been set by the Royal Academy of Sciences of Paris, and no better model can be taken than the Meteorological Journal kept at their observatory.

BAROMETRICAL EXPERIMENTS

UPON HEIGHTS.

THE late Professor Playfair, in his elaborate Essay upon Barometrical Measurements, has suggested * the idea of fixing two barometers, the one at the top, and the other at the bottom of a high tower or hill of moderate elevation; to be observed at the same instant, together with their corresponding thermometers, for the purpose of computing from the variation of the difference of their heights the quantity of moisture dissolved in the air. "The height at which the one barometer," he observes, "should be placed above the other, ought not to be so small that the unavoidable errors of observation (which may amount to five feet) may be considerable in respect of the whole; nor so great as to introduce error from other causes. It ought not, therefore, to be less than 100, nor much greater than 500 feet." He concludes, "Nor can this application of the barometer fail of leading to some useful conclusion; for if, on trial, it shall be found that the operation of humidity in changing the specific gravity of the air is overruled or concealed by the action of more powerful causes, the discovery, even of this fact, will give a value to the observations."

This suggestion is no longer required for the pur-

* Works of John Playfair, Esq., vol. iii., p. 85.

poses of hygrometry, as we have now the means of accurately appreciating the effects of moisture upon the air: but there is no doubt that it might be applied to the discovery of other atmospheric influences. For this purpose it is now particularly fitted, as the before unknown hygrometric correction may be independently applied with certainty; and any other disturbances are disengaged from this source of ambiguity.

I have long wished for an opportunity of making the attempt with all the requisite precautions; but, as it is one which requires patient and very careful co-operation, I have not been able to execute the details satisfactorily. The following experiments, however, though necessarily deficient in precision, may not be without interest, and their results may possibly induce others to undertake an investigation which promises amply to repay a patient pursuit.

I have been extremely anxious to ascertain, in the first place, to what degree of precision it is possible to arrive in barometrical mensuration, in different states of the atmosphere, when the corrections for temperature and vapour have both been made; and I availed myself of a long residence in the neighbourhood of Box-hill, and Leith-hill, in Surrey, to decide the point, as far as these heights would permit me; and also to select a series of stations for further experiments. The great disadvantage that I have had to contend with, consists in the want of contemporaneous observations; in lieu of which I have been obliged to substitute the mean of two observations, at the lower station, at setting out and returning. I shall here give the details of four measurements, of Leith-hill, to shew the degree of uncertainty which attaches to this method of proceeding. The heights of the barometer are corrected for all adventitious circumstances, every precaution was taken in observing them, and the instrument is one upon the accuracy of which I can confidently rely.

The lower station was at the foot of Box-hill, about forty-five feet above the bed of the river Mole; and the upper station, the tower upon Leith-hill, about seven miles distant:—

TABLE I. *Barometrical Measurements of Leith Hill.*

Date. 1822.	Barometer.		Temperature.		Results.		OBSERVATIONS.
	Lower Station.	Upper Station.	Air.	Dew Point.	Height in Feet.	Weight of Col. of Dry Air.	
June 24	29.924	—	71	58	—	—	Very fine, with some heavy clouds and heat drops. Not clear.
return	29.926	—	74	60	—	—	
—	—	29.057	69	59	841	.944	
July 1	29.993	—	66	44	—	—	Very fine and clear.
return	29.995	—	66	47			
—	—	29.128	64	43	823	.933	
August 12	29.748	—	65	64	—	—	Close and damp, with small rain.
return	29.729	—	68	62	—	—	
—	—	28.875	65	60	842	.936	
Dec. 30	29.787	—	29	15	—	—	Very fine and very cold.
return	29.750	—	29	15	—	—	
—	—	28.819	26	15	838	.940	
					836	.938	Means.

The differences from the mean, exhibited by this Table, must be acknowledged to be very small, being only six feet in 836 feet, or .006 in. of mercury in the weight of the intercepted column of air, corrected for vapour and temperature.

The next series of Experiments was made upon a less elevation, but one which offered the advantage of easier access, and a smaller interval between the observations. The lower station was the same as before, and the upper, a clump of trees upon Hedley Heath, which form a very conspicuous landmark for a large extent of surrounding country This height I divided into three stations, one above the other, for the purpose of ascertaining whether the parts of a height, so measured in divisions, would correspond with the direct measurement at two observations; in what part the error, if any, was most likely to occur; and also the effect of difference of position with regard to the surrounding hills. The first station above the point of departure was in a deep ravine, below the military road which leads to the top of Box-hill. It is surrounded, except at a narrow entrance, by very steep hills, those on each side being about 200 feet high. It was selected for the purpose of ascertaining whether a difference in the velocity of the wind passing over such a hollow, would produce any difference in the pressure of the atmospheric column. The second station was on the top of Box-hill, almost perpendicularly above the point from whence I set out; and the third, the trees before described upon the edge of the hill, which is very steep, and forms part of the boundary of a valley which runs at right angles to the one which is overlooked by the second station. I shall not attempt to give the particulars of the observations, which would occupy too much space, but only the calculated results, and such circumstances as may be supposed to have had an influence in their production. Each height was calculated from the mean of two observations, one made in the ascent and the other on the return. They are included in the following Table :—

TABLE II. *Barometrical Measurements of Hedley Heath, at different Stages of the Height.*

Height of Ravine above the first Station.	Difference from Mean.	Height of Box Hill above the Ravine.	Difference from Mean.	Height of Hedley above Box Hill.	Difference from Mean.	Height of Box Hill by direct Observation.	Difference from Mean.	Height of Box Hill by two Stations.	Difference from Mean.	Height of Hedley by direct Observation.	Difference from Mean.	Height of Hedley by three Observations.	Difference from Mean.	Temp. Air.	Temp. Dew Point.	OBSERVATIONS.
165	+ 7.5	261	− 6.8	159	+ 1.2	427	+ 2.4	426	+ 0.7	586	+ 3.9	585	+ 1.9	33	29	Wind little and very cold.
168	+10.5	272	+ 4.2	161	+ 3.2	441	+16.4	440	+16.7	600	+17.9	601	+18.9	69	60	Wind South and very high.
148	− 9.5	268	+ 0.2	147	−10.8	416	− 8.6	416	− 9.3	563	−19 1	563	−20.1	63	60	Wind S.W. and calm.
150	− 7.5	270	+ 2.2	157	− 0.8	421	− 3.6	420	− 5.3	578	− 4.1	577	− 6.1	61	52	Wind W. and little.
156	− 1.5	265	− 1.2	155	− 1.2	416	− 8.6	421	− 4.3	571	−11.1	576	− 7.1	66	52	Ditto ditto
158	− 0.5	271	+ 3.2	168	+10.2	427	+ 2.4	429	+ 3.7	595	+12.9	597	+13.9	81	66	Wind brisk on the Hill, and very hot.
157.5	—	267.8	—	157.8	—	424.6	—	425.3	—	582.1	—	583.1	Means			

Of these observations, the first series was made at a time when the atmosphere was in such a state, as to require the smallest possible correction for temperature and moisture; and it will be observed, that the calculation from them of the height of the highest station, scarcely differs from the mean, the error being less than two feet in 585. The height of the next lower station also corresponds very closely; but in the height of the first station, we have a difference of 7.5 feet in 157. This difference, most probably, attaches to the observation in the ravine: for, if it had been in the first observation, it would have been discoverable in all the results; whereas, omitting the second, all the rest are correct, and the third is deficient exactly the quantity which is in excess in the second.

The last series forms the proper contrast to this, as requiring almost the greatest possible correction for both temperature and moisture: and here we perceive that the first and second heights are very correct, but there is a large error in the third, amounting to ten feet, in 158. The height of Box-hill thus differs from the mean, only 2.4 feet; while that of Hedley-heath differs thirteen feet; the error must, consequently, be included in the last observation.

In the second series of observations we find an error of eighteen feet in the total elevation; but by attending to the analysis, we cannot, as in the two last cases, trace it to any particular station: it is largest at the first, but goes on accumulating at all: during this series, the wind was extremely violent.

The third set of observations exhibits great errors in deficiency in the first and last sections of the elevation, but none in the intermediate. The results of the fourth are all pretty accurate; and those of the fifth present the only instance of any considerable difference between the measurement at one operation, and the measurement in parts.

The result of all the observations taken together prove, that the intermediate station was very considerably less liable to error than either of the extremes, and strongly suggests the following query:—Whether local currents of air, and those deflections of the wind, which are caused by the different directions of different valleys, may not produce various partial adjustments of density, which may have an influence upon barometrical mensuration?

The measurement of a height, in divisions, does not appear, by this analysis, to be liable to any objection; and it possesses this great advantage, when the altitude is very considerable, *viz.*, that we can make a much nearer approximation thereby, to the real specific gravity of the intercepted column of air, than by two observations only. This remark is applicable to the correction for temperature, but much more so to that for moisture; for, as we have seen in the previous investigation, the quantity of vapour does not decrease gradually, like the heat, with the elevation, but continues of nearly equal elasticity to a certain height, and then suddenly decreases considerably. The mean, therefore, of two observations, taken at the bottom of a mountain and at the top, might be very far, indeed, removed from the real state of the aërial column; and the more we multiply observations of the dew-point, the more we diminish the chances of error from this source.

The last set of experiments includes a series of

forty-five observations upon one height, under every possible variation of the atmosphere. The station was that upon Box-hill, which formed the second stage of the previous set. The results I shall divide into different classes, to ascertain the influence of different circumstances, and I shall express them by the length of the column of mercury, which would be the equipoise of the intercepted column of air, supposing it corrected for temperature and moisture. The mean result of all the observations is .4848 inch.

The following Table includes the eleven observations, in which both corrections were at the greatest amount.

TABLE III. *Barometrical Measurement of Box-hill in very hot Weather.*

Length of Column.	Temperature.		Dryness.
	Air.	Dew-point.	
.486	81	66	15
.485	68	59	9
.494	68	60	8
.470	67	61	6
.472	67	64	3
.495	66	56	10
.499	65	55	10
.495	65	57	8
.493	65	60	5
.479	65	59	6
.470	65	52	13
.4852	Mean.		

The average scarcely differs from the general mean.

Seven observations in cold weather, in which the required corrections were very small, and mostly on the contrary side, are included in the following Table:—

TABLE IV. *Barometrical Measurement of Box-hill in very cold Weather.*

Length of Column.	Temperature. Air.	Temperature. Dew-point.	Dryness.
Inch. .492	31	29	2
.494	34	30	4
.487	30	23	7
.486	29	28	1
.490	32	31	1
.474	27	25	2
.467	32	30	2
.4842	Mean.		

The average of these, again, only differs .0006 inch from the same standard. These experiments may, therefore, be regarded as decisive of the adequacy of the corrections for temperature and vapour.

The following eight results were calculated from observations when the moon was nearly upon the meridian:—

TABLE V. *Barometrical Measurement of Box-hill, with the Moon upon the Meridian.*

Length of Column.
482
.483
.469
.474
.479
.486
.480
.470
.4778 **Mean.**

Here we have a small but decided difference, sufficient to strengthen the query already suggested in the first Essay. Does not the position of the moon influence, in some degree, the results of barometrical mensurations? The difference, .007 inch, is in deficiency, and agrees, so far, with the anticipation of the effect.

The position of the sun may, also, be expected to have an influence upon the elastic fluids of the atmosphere,, independent of its heating power; to determine which, the following observations at noon were extracted:—

TABLE VI. *Barometrical Measurement of Box-hill, with the Sun upon the Meridian.*

Length of Column.
.473
.470
.495
.475
.465
.476
470
486
.494
.474
.4778 Mean.

The difference is the same as in the last Table, and points to the same kind of planetary influence. It is sufficient to justify the query—Does not the position of the sun affect the results of barometrical mensurations?

A third disturbing cause we cannot but look for in the operations of the electric fluid.

The following four observations were made when the atmosphere was highly charged, and just before the commencement of violent thunder-storms :—

TABLE VII. *Barometrical Measurement of Box-hill during Thunder-storms.*

Length of Column.
.481
.485
.476
.465
.4767 Mean.

The experiments, it must be acknowledged, are not sufficient to establish the fact; but the mean difference, it will be observed, is more than a sixth of the total result, and strongly calls for further inquiry—Whether the electric state of the atmosphere does not affect the results of barometrical mensurations?

The following Table exhibits the barometrical results in the most opposite states of the wind, *viz.*, when very high and when perfectly calm.

TABLE VIII. *Barometrical Measurement of Box-hill in different States of the Wind.*

Length of Column.	
Wind high.	Calm.
.492	.475
.499	.465
.495	.476
.470	.470
.494	.472
.4900	.4716 Means.

The differences of +.0052 in wind, and −.0132 in calm weather, induce me to conclude my queries by proposing the following question: What is the effect of wind upon barometrical mensurations? If I had had the means of prosecuting these inquiries in the complete manner which the nicety of the subject requires, I would not have suffered them to retain the form of crude speculations: but, under all circumstances, I am not without hopes that this premature publication may be useful. It may possibly illustrate Mr. Playfair's suggestion; it may indicate the objects which it is calculated to illustrate, and it exemplifies the method of proceeding.

METEOROLOGICAL JOURNAL.

NOTE.—All the maxima results have the sign +, and all the minima − prefixed.

1819 September.	Morning. Barometer.	Hygrometer.			Weather.	Afternoon. Barometer.	Hygrometer.			Weather.	Night. Barometer.	Hygrometer.			Weather.
1	29.52	62	42	20	fine	29.59	62	42	+20	overcast	29.62	57	42	15	very fine
2	29.67	63	43	20	overcast	29.67	67	48	19	ditto	——	—	—	—	rain
3	29.59	68	+65	3	showers	29.69	71	61	10	showers	29.75	61	60	1	very fine
○ 4	29.87	66	59	7	very fine	29.85	70	63	7	overcast	29.83	64	64	—	rain
5	29.74	63	62	1	rain	29.75	65	63	2	fine	29.76	56	54	2	very fine
6	29.87	62	46	16	very fine	29.93	66	46	20	very fine	29.96	55	48	7	same
7	29.97	65	63	2	overcast	29.97	71	64	7	lowering	30.00	66	64	2	overcast
8	30.03	70	62	8	fine	30.04	70	60	10	clearing	30.04	62	59	3	very fine
9	30.05	68	59	9	same	30.02	71	58	13	overcast	30.03	63	61	2	overcast
10	30.01	67	59	8	fine	29.96	70	55	15	very fine	29.97	59	56	3	very fine
☾ 11	30.00	63	57	6	overcast	30.02	64	57	7	overcast	30.07	61	56	5	overcast
12	30.14	64	52	12	very fine	30.13	67	47	20	very fine	30.16	52	47	5	very fine
13	30.23	63	49	14	same	30.20	66	50	16	same	30.20	56	52	4	same
14	30.19	62	57	5	fine	30.13	72	55	15	same	30.08	61	58	3	same
15	29.92	66	59	7	same	29.82	68	65	3	showers	29.73	65	64	1	rain
16	29.66	55	50	5	showers	29.70	56	41	15	fine	29.79	48	39	9	very fine
17	29.97	58	47	11	very fine	30.03	59	45	14	very fine	30.09	52	47	5	overcast
18	30.14	60	52	8	overcast	30.10	63	55	8	fine	30.13	53	51	2	very fine
⊕ 19	30.15	60	40	20	fine	30.14	57	−37	20	very fine	30.18	49	43	6	same
20	30.27	56	46	10	same	30.31	52	45	7	same	30.35	50	45	5	same
21	+30.41	55	47	8	very fine	30.41	56	45	11	same	30.40	50	45	5	same
22	30.39	56	48	8	same	30.30	59	46	13	overcast	30.29	54	49	5	fine
23	30.20	56	47	9	overcast	30.07	59	46	13	same	30.01	56	50	6	overcast
24	29.86	58	52	6	same	29.74	61	47	14	very fine	29.59	53	50	3	very fine
25	29.57	58	57	1	showers	29.51	60	58	2	rain	−29.50	55	55	—	rain
☽ 26	29.52	59	51	8	fine	29.57	58	46	12	very fine	29.63	53	48	5	dull
27	29.57	61	55	6	same	29.61	63	60	3	showers	29.62	60	60	—	rain
28	29.58	65	59	6	showers	29.61	60	58	2	rain	29.59	60	60	—	same
29	29.55	60	60	—	rain	29.62	60	54	6	fine	29.67	58	57	1	showers
30	29.76	63	60	3	showers	29.79	66	62	4	showers	29.79	63	60	3	overcast
Means	29.913	62	53½	8.2		29.909	63	52½	10.2		29.925	56½	53	3½	

1819 September.	Temperature. Max.	Min.	Sun.	Rad.	Wind. Direction.	Force.	Rain.
1	66	46			N W	brisk	
2	68	60			ditto	ditto	
3	+74	52			W	same	0.07
○ 4	70	60			S W	same	0.05
5	66	48			N W	same	0.29
6	66	47			ditto	same	
7	72	61			S W	same	
8	74	58			N W	little	
9	72	59			S E	same	
10	70	52			ditto	ditto	
☾ 11	64	56			N E	ditto	
12	67	45			same	ditto	
13	66	49			S E	calm	
14	72	53			same	little	
15	68	53			S W	ditto	0.06
16	56	43			N	brisk	0.19
17	61	50			same	little	
18	65	46			N W	ditto	
⊕ 19	61	41			N	ditto	
20	60	−40			N E	brisk	
21	61	42			same	ditto	
22	62	50			same	ditto	
23	61	51			same	ditto	
24	64	46			E	ditto	
25	63	50			S W	little	0.32
☽ 26	61	48			W	same	0.13
27	66	57			S W	stormy	0.44
28	66	55			same	same	0.40
29	64	56			same	brisk	0.16
30	66	58			same	ditto	
Means	65	51					2.11

SEPTEMBER.

Mean Temperature	58°	Weight of vapour in a cubic foot.	
—— Pressure	29.915 inches	Mean . . .	4.774 grs.
—— Dew-point	52°	Maximum .	7.187 „
—— Force of vapour . .	0.428 inches	Minimum .	2.861 „
—— Degree of dryness . .	6°		
—— Degree of moisture	813°		

Least observed degree of moisture 503

WINDS.

N. 3 = 43° N.E. 6 = 48° E. 1 = 49° S.E. 4 = 55° S. 0 S.W. 8 = 60°
W. 2 = 55° N.W. 6 = 51°

Amount of rain . . . . 2.11 inches
—— of evaporation 2.94 inches

REMARKS.

The sudden fall of the barometer, at the latter part of the month, was accompanied by high winds.

The westerly winds, which constitute half the amount of the month, average more than their mean quantity of vapour; and it was during their prevalence that the rain fell. The other winds are below the mean.

The weather fine, warm, and seasonable.

The depression of the mercurial column was accompanied by considerable wind, especially during the night.

1819 October.	Morning. Barometer.	Hygrometer.			Weather.	Afternoon. Barometer.	Hygrometer.			Weather.	Night. Barometer.	Hygrometer.			Weather.
1	29.78	67	59	8	fine	29.73	68	63	5	overcast	29.73	62	60	2	very fine
2	29.67	68	59	9	showers	29.71	65	57	8	fine	29.72	60	58	2	fine
○ 3	29.69	65	59	6	overcast	29.64	63	56	7	showers	29.60	57	56	1	showers
4	29.53	62	56	6	same	29.56	56	51	5	fine	29.62	46	42	4	very fine
5	29.88	47	41	6	very fine	29.99	48	33	+15	very fine	30.06	46	39	7	same
6	30.09	53	46	7	overcast	29.95	54	52	2	showers	29.94	53	52	1	showers
7	29.89	56	55	1	same	29.88	60	55	5	same	29.91	57	56	1	overcast
8	30.00	63	59	4	same	30.00	64	61	3	overcast	30.00	59	59	—	rain
9	29.92	60	56	4	same	29.80	61	57	4	showers	29.76	59	57	2	overcast
10	29.78	65	63	2	fine	29.78	70	+66	4	fine	29.79	62	61	1	same
☾ 11	29.82	66	60	6	same	29.84	67	61	6	very fine	29.88	58	57	1	very fine
12	29.96	61	60	1	very fine	29.97	70	63	7	same	29.97	60	60	—	small rain
13	29.93	65	61	4	overcast	29.94	63	59	4	fine	29.99	53	51	2	very fine
14	30.04	56	51	5	dull	30.08	60	56	4	dull	30.11	55	55	—	showers
15	30.23	55	51	4	very fine	30.26	56	45	11	fine	+30.29	49	49	—	fog
16	30.19	51	50	4	dull	30.17	51	39	12	dull	30.07	46	42	4	very fine
17	30.09	45	39	6	very fine	30.12	46	39	7	very fine	30.15	42	39	3	same
18	30.18	46	41	5	same	30.18	49	39	10	same	30.17	39	39	—	fog
⊕ 19	30.09	44	41	3	fog	29.99	49	45	4	fine	29.90	50	49	1	same
20	29.66	56	56	—	showers	29.56	52	52	—	rain	29.56	51	51	—	rain
21	29.50	42	41	1	overcast	29.50	39	37	2	snow	29.38	37	37	—	sleet
22	29.40	35	33	2	fine	29.41	42	42	—	rain	29.43	41	41	—	rain
23	29.33	40	39	1	same	29.30	47	46	1	showers	29.30	41	40	1	very fine
24	-29.30	42	41	1	same	29.32	44	39	5	dull	29.37	39	33	6	dark
25	29.59	41	34	7	same	29.45	42	35	7	fine	29.48	36	35	1	fog
☽ 26	29.57	44	42	2	overcast	29.63	42	35	7	very fine	29.71	35	-32	3	very fine
27	29.75	37	33	4	very fine	29.75	43	43	—	small rain	29.76	38	37	1	dull
28	29.75	39	38	1	fine	29.75	43	41	2	foggy	29.71	34	33	1	fine
29	29.54	36	36	—	rain	29.51	42	42	—	rain	29.50	41	39	2	rain
30	29.52	42	42	—	same	29.53	43	43	—	same	29.61	46	46	—	same
31	29.76	48	48	—	same	29.77	47	47	—	same	29.82	47	47	—	same
Means	29.788	51.6	48	3.5		29.776	53	48.3	4.7		29.783	48.3	46.8	1.5	

1819 October.	Temperature. Max.	Min.	Sun.	Rad.	Wind. Direction.	Force.	Rain.
1	+68	58			S	little	
2	68	56			same	ditto	
○ 3	68	51			S W	brisk	0.06
4	62	38			N W	same	
5	50	39			same	ditto	0.04
6	55	52			W	squally	
7	60	52			W	brisk	
8	65	53			S W	little	0.02
9	61	55			S	brisk	
10	70	57			S W	little	
☾ 11	68	52			S E	same	
12	70	57			same	same	
13	67	47			W	calm	
14	60	46			N W	ditto	0.05
15	59	44			N E	little	
16	55	38		35	N	brisk	
17	50	37		33	same	same	
18	53	32			same	same	
⊕ 19	53	46			W	little	
20	57	40			S W	same	0.35
21	43	30		28	N W	stormy	0.14
22	42	32		25	—	ditto	
23	47	32		27	—	little	
24	44	32		30	N	high	
25	42	30		22	—	little	
☽ 26	44	−27		−17	—	same	
27	43	30		23	N E	same	0.06
28	43	31		25	—	—	
29	43	39		36	—	high	0.30
30	44	44		44	E	same	0.66
31	47	45		43	—	little	0.50
Means	54.8	42.4		29.8			2.18

OCTOBER.

Mean Temperature . . . . 48.6°
—— Pressure . . . . . . 29.782 inches
—— Dew-point . . . . . 45.3°
—— Force of vapour . . 0.344 inches
—— Degree of dryness . . 3.3°
—— Degree of moisture . 900°
Least observed degree of moisture . 595

Weight of vapour in a cubic foot.
Mean . . . 3.869 grs.
Maximum . 7.389 „
Minimum . 2.550 „

WINDS.

N. 6 = 38° N.E. 4 = 40° E. 2 = 45° S.E. 2 = 60° S. 3 = 58° S.W. 4 = 58° W. 4 = 51° N.W. 6 = 42°.

Amount of rain . . . . 2.18 inches
—— of evaporation 1.178 inches

REMARKS.

The weather, during the first half of the month, mostly very fine and warm; but during the remainder, cold and cloudy.

On the 21st it snowed for two hours in the morning, and two or three inches fell during the night. Snow has not fallen so early in the season for seven or eight years past. The appearance of mid-winter, while the leaves were still upon the trees, was very striking. Much damage was done to fruit and forest trees by the great weight. The Aurora Borealis was seen during the month.

1819 November.	Morning. Barometer.	Hygrometer.			Weather.	Afternoon. Barometer.	Hygrometer.			Weather.	Night. Barometer.	Hygrometer.			Weather.
1	29.79	45	45	—	rain	29.70	47	46	1	rain	29.66	42	42	—	overcast
○ 2	29.65	44	44	—	fine	29.65	43	34	9	fine	29.65	39	38	1	fine
3	29.82	40	36	4	very fine	29.88	41	35	6	very fine	29.92	36	35	1	same
4	29.88	46	41	5	fine	29.85	51	46	5	overcast	29.80	46	44	2	same
5	29.68	51	44	7	same	29.57	52	51	1	rain	29.50	49	49	—	fine
6	29.48	47	46	1	same	29.40	48	45	3	fine	29.33	45	41	4	very fine
7	29.39	46	45	1	same	29.45	45	43	2	showers	29.47	38	37	1	fine
8	29.52	40	40	—	same	29.55	42	36	6	overcast	29.72	35	31	4	very fine
☾ 9	29.80	34	31	3	very fine	29.74	40	36	4	very fine	29.60	42	36	6	overcast
10	29.60	47	46	1	overcast	29.32	47	46	1	rain	29.37	44	44	—	rain
11	29.58	46	45	1	rain	29.70	43	43	—	same	29.84	43	41	2	showers
12	29.89	42	39	3	showers	29.88	44	39	5	dark	29.82	42	41	1	dark
13	29.78	44	38	6	overcast	29.73	45	44	1	overcast	29.73	44	44	—	rain
14	29.75	43	43	—	same	29.75	44	42	2	same	29.75	44	42	2	dark
15	29.72	44	44	—	fog	29.65	46	46	—	fog	29.57	47	47	—	rain
16	29.50	41	41	—	rain	29.41	40	40	—	rain	29.40	36	36	—	same
⊕ 17	29.64	42	42	—	small rain	29.73	43	43	—	small rain	29.91	43	43	—	dark
18	29.99	42	34	8	very fine	+30.01	41	32	+9	very fine	30.00	33	31	2	very fine
19	29.91	34	33	1	overcast	29.83	36	31	5	overcast	29.80	36	31	5	overcast
20	29.66	35	26	9	fine	29.40	36	35	1	same	29.12	42	42	—	sleet
21	−29.08	40	39	1	same	29.13	41	41	—	showers	29.33	36	34	2	clearing
22	29.46	32	30	2	very fine	29.49	34	32	2	fine	29.54	30	30	—	very fine
23	29.68	29	27	2	same	29.72	33	30	3	very fine	29.74	29	29	—	same
☽ 24	29.81	27	−25	2	same	29.89	36	33	3	same	29.94	32	32	—	same
25	29.92	31	31	—	same	29.86	34	33	1	same	29.71	31	31	—	fog
26	29.64	36	35	1	rain	29.67	37	37	—	rain	29.76	33	33	—	overcast
27	29.83	36	36	—	overcast	29.83	37	36	1	misty	29.83	29	29	—	fog
28	29.79	31	31	—	fog	29.72	33	33	—	fog	29.66	41	41	—	rain
29	29.61	49	49	—	rain	29.60	50	50	—	rain	29.66	50	50	—	same
30	29.64	51	+51	—	same	29.58	51	51	—	same	29.53	47	47	—	overcast
Means	29.683	40.5	38.5	1.9		29.656	·42	39.6	2.3		29.655	39.4	38.3	1.1	

1819 November.	Temperature. Max.	Min.	Sun.	Rad.	Wind. Direction.	Force.	Rain.
1	47	38		32	E	little	0.02
○ 2	44	33		26	N W	ditto	
3	41	33		26	——	ditto	
4	51	34		30	W	ditto	
5	+52	43		40	S W	ditto	0.17
6	49	38		34	——	brisk	
7	46	34		27	W	little	
8	43	28		21	N	same	
☾ 9	41	38		35	W	brisk	
10	48	43		40	N W	little	0.15
11	46	37		32	N E	high	0.24
12	45	41		38	——	brisk	0.05
13	45	41		38	——	same	
14	45	41		40	——	little	
15	46	41		38	W	ditto	0.10
16	41	36		38	N	little	0.49
⊕ 17	43	40		38	E	brisk	0.09
18	42	31		27	N E	high	
19	37	33		30	——	little	
20	42	37		29	N W	brisk	
21	41	28		22	——	same	
22	35	26		18	——	little	
23	34	−23		−15	——	ditto	
☽ 24	36	28		20	——	ditto	
25	36	28		21	——	ditto	
26	37	32		27	N	ditto	
27	37	29		27	N W	brisk	
28	41	33		31	S W	little	0.57
29	51	48		45	——	brisk	0.15
30	51	46		43	——	ditto	0.12
Means	43.1	35.3		30.9			2.15

NOVEMBER.

Mean Temperature . . . . 39.2°
—— Pressure . . . . . . 29.664 inches
—— Dew-point . . . . . 37.4°
—— Force of vapour . . 0.259 inches
—— Degree of dryness . . 1.8°
—— Degree of moisture . 941

Least degree of observed moisture 739

Weight of vapour in in a cubic foot.
Mean . . . 2.967 grs.
Maximum . 4.684 „
Minimum . 2.019 „

WINDS.

N. 3 = 36° N E. 6 = 38° E. 2 = 43° S.E. 0 S. 0 S.W. 5 = 45°
W. 4 = 41° N.W. 10 = 34°

Amount of rain . . . . 2.150 inches
—— of evaporation 0.480 inches

REMARKS.

The weather, for the most part, uncommonly cold and cloudy, with frequent fogs, and sharp hoar frosts.

The sudden increase of temperature on the 29th, with a south-west wind, is very remarkable.

1819 December.	Morning.					Afternoon					Night.				
	Barometer.	Hygrometer.			Weather.	Barometer.	Hygrometer.			Weather.	Barometer.	Hygrometer.			Weather.
○ 1	29.72	48	47	1	overcast	29.91	46	43	3	fine	29.97	37	37	—	fine
2	29.91	48	48	—	same	29.89	47	47	—	rain	30.00	38	38	—	same
3	30.08	35	33	2	very fine	30.07	39	39	—	fog	+30.09	39	39	—	very fine
4	29.63	49	46	3	fine	29.50	44	44	—	rain	29.52	42	42	—	rain
5	29.73	41	38	3	dark	29.74	40	39	1	dark	29.89	35	35	—	dark
6	29.97	37	37	—	same	29.97	36	36	—	same	29.98	35	33	2	same
7	29.95	33	33	—	sleet	29.90	33	33	—	small rain	29.90	34	34	—	same
8	29.94	29	22	7	clearing	29.96	28	18	+10	very fine	29.97	21	18	3	very fine
☾ 9	29.98	24	23	1	fine	29.96	29	24	5	overcast	29.90	27	27	—	snow
10	29.83	30	29	1	overcast	29.88	30	29	1	same	29.93	21	21	—	fine
11	29.92	18	−15	3	very fine	29.93	23	21	2	fine	29.94	23	21	2	very fine
12	29.89	32	30	2	same	29.83	33	33	—	sleet	29.78	31	31	—	snow
13	29.72	26	25	1	very fine	29.70	31	31	—	very fine	29.70	27	27	—	fine
14	29.66	25	25	—	fine	29.54	33	33	—	rain	29.49	30	30	—	fine
15	29.45	32	32	—	same	29.42	35	34	1	same	29.55	33	33	—	very fine
16	29.79	31	31	—	same	29.89	34	34	—	misty	29.91	33	33	—	rain
⊕ 17	29.66	37	37	—	rain	29.38	44	44	—	rain	29.37	50	50	—	same
18	29.35	51	51	—	dull	29.48	52	52	—	same	29.62	50	49	1	clearing
19	29.77	51	51	—	rain	29.76	53	53	—	rain	29.69	53	53	—	rain
20	29.62	53	+53	—	same	29.60	52	52	—	same	29.66	52	52	—	rain
21	29.86	46	41	5	fine	29.78	42	42	—	same	29.66	51	51	—	same
22	29.64	51	51	—	rain	−29.66	49	48	1	fine	29.46	50	50	—	same
23	29.26	42	42	—	same	29.26	40	35	5	same	29.29	36	36	—	fine
☽ 24	29.30	32	32	—	fine	29.26	33	33	—	fog	29.26	31	31	—	very fine
25	29.31	32	31	1	same	29.33	31	27	4	fine	29.39	29	25	4	same
26	29.42	27	25	2	same	29.43	29	27	2	very fine	29.43	30	30	—	fine
27	29.40	31	31	—	same	29.40	31	29	2	fine	29.42	30	29	1	same
28	29.43	34	34	—	dull	29.43	32	32	—	snow	29.46	31	30	1	overcast
29	29.55	27	27	—	misty	29.57	29	29	—	haze	29.54	23	23	—	fine
30	29.48	25	25	—	fog	29.43	31	31	—	fog	29.37	32	32	—	overcast
○ 31	29.32	27	24	3	fine	29.31	31	26	4	very fine	29.30	21	21	—	very fine
Means	29.662	35.6	34.4	1.1		29.650	36.7	35.4	1.3		29.659	34.6	34.2	0.4	

1819 December.	Temperature. Max.	Min.	Sun.	Rad.	Wind. Direction.	Force.	Rain.
○ 1	48	36		28	N W	little	
2	48	30		25	S W	brisk	
3	39	34		29	——	little	
4	49	39		39	——	brisk	0.13
5	41	35		33	E	high	0.11
6	37	33		31	N E	ditto	0.07
7	34	29		29	——	ditto	0.10
8	29	20		13	E	ditto	
☾ 9	29	27		26	N	ditto	0.21
10	32	−17		−12	——	little	0.20
11	24	20		13	N W	ditto	
12	34	21		15	W	ditto	0.17
13	32	22		15	——	ditto	
14	33	28		22	S W	ditto	
15	35	29		25	W	brisk	0.03
16	35	30		21	——	ditto	
⊕ 17	50	43		40	S	ditto	
18	+53	48		44	S W	ditto	
19	53	51		50	——	little	0.09
20	53	44		42	——	high	0.07
21	51	42		41	——	brisk	
22	51	42		42	——	ditto	
23	42	30		24	N W	ditto	
☽ 24	33	26		19	S E	ditto	
25	32	24		16	N E	ditto	
26	30	29		23	N W	ditto	
27	32	28		25	S E	ditto	
28	34	25		15	E	ditto	
29	29	21		14	S W	ditto	
30	32	25		19	——	ditto	
○ 31	30	17		15	——	little	
Means	38.2	30.4		25.9			1.18

DECEMBER.

Mean Temperature	34.3°	Weight of vapour in a cubic foot.	
—— Pressure	29.657 inches	Mean . .	2.624 grs.
—— Dew-point	32.9°	Maximum .	5.003 „
—— Force of vapour . . .	0.217 inches	Minimum .	1.552 „
—— Degree of dryness . .	1.4°		
—— Degree of moisture .	923°		

Least observed degree of moisture 696

WINDS.

N. 2=25° N.E. 3=32° E. 3=29° S.E. 2=30° S. 1=43° S.W. 12=40° W. 4=31° N.W. 4=31°.

Amount of rain . . . . 1.18 inches
—— of evaporation 0.558 inches

REMARKS.

The first four days were very mild; but with the change of wind, on the fifth, a sharp winter may be said to have commenced. The frost lasted till the 17th, when a thaw took place, and mild (and to the feelings oppressive) weather continued to the 23d. The frost set in again on the 25th, and continued very sharp, with fine weather, to the end of the month.

The Aurora Borealis was observed in the neighbourhood of London on the 14th.

1820 January.	Morning.					Afternoon.					Night.				
	Barometer.	Hygrometer.			Weather.	Barometer.	Hygrometer.			Weather.	Barometer.	Hygrometer.			Weather.
1	29.38	20	19	1	very fine	29.44	27	27	—	fog	29.53	25	25	—	fog
2	29.54	30	30	—	mist	29.50	35	35	—	rain	29.38	37	37	—	rain
3	29.58	30	25	5	fine	29.76	29	25	4	very fine	29.92	26	26	—	very fine
4	29.98	24	24	—	fog	29.97	26	24	2	fog	29.98	21	21	—	misty
5	30.05	20	19	1	mist	30.06	25	25	—	same	30.05	25	25	—	fine
6	30.03	27	27	—	fog	30.04	34	34	—	same	30.06	35	35	—	rain
7	30.18	32	23	9	dull	30.28	28	21	7	dull	30.35	26	20	6	dark
☾ 8	30.42	27	22	5	very fine	30.44	26	21	5	fine	30.52	23	19	4	same
9	+30.59	26	25	1	little snow	30.51	26	23	3	very fine	30.32	28	28	—	snow
10	30.31	25	24	1	fine	30.24	26	26	—	snow	30.11	26	26	—	same
11	29.90	28	28	—	sleet	29.70	32	32	—	same	29.67	33	33	—	dark
12	29.94	25	17	8	clearing	30.03	25	18	7	light snow.	30.10	20	17	3	dull
13	30.05	21	17	4	fine	30.05	28	27	1	mist	30.10	22	17	5	overcast
14	30.12	25	17	8	light snow	30.09	26	17	9	dull	30.01	21	−10	+11	very fine
⊕ 15	29.79	12	10	2	fine	29.71	21	16	5	fine	29.66	28	26	2	dark
16	29.65	22	22	—	same	29.68	25	25	—	mist	29.68	26	25	1	dull
17	29.60	32	32	—	fog	29.57	32	32	—	dull	29.51	32	26	6	snow
18	29.39	29	29	—	snow	29.32	32	32	—	rain	29.02	35	35	—	rain
19	−28.89	45	43	2	mild&damp	28.94	43	41	2	dull	28.98	33	33	—	dull
20	29.50	30	26	4	very fine	29.45	32	24	8	fine	29.13	32	32	—	snow
21	29.12	35	34	1	sleet	29.26	33	33	—	showers	29.63	28	25	3	very fine
☽ 22	29.96	25	22	3	very fine	30.04	28	26	2	very fine	30.07	22	22	—	fog
23	30.02	33	33	—	fog	29.98	36	36	—	fog	29.90	36	36	—	overcast
24	29.79	39	34	5	fine	29.71	41	41	—	showers	29.77	43	43	—	rain
25	29.79	42	42	—	showers	29.63	42	42	—	rain	29.44	42	42	—	same
26	29.47	44	44	—	rain	29.59	47	+46	1	same	29.59	46	46	—	same
27	29.56	47	45	2	very fine	29.50	49	46	3	very fine	29.34	46	46	—	same
28	29.56	43	42	1	dull	29.73	43	37	6	dull	29.94	41	40	1	showers
29	30.09	40	36	4	fine	30.06	42	36	6	fine	30.05	40	37	3	fine
○ 30	29.99	42	42	—	rain	29.99	45	45	—	rain	29.96	44	44	—	rain
31	29.92	42	42	—	very fine	29.91	46	43	3	very fine	29.97	43	43	—	overcast
Means	29.812	31	28.8	2.1		29.812	33.2	30.8	2.4		29.798	31.7	30.3	1.4	

1820 January.	Temperature. Max.	Min.	Sun.	Rad.	Wind. Direction.	Force.	Rain.
1	27	20		12	N W	calm	
2	37	30		29	S W	ditto	
3	30	22		16	N W	little	
4	26	17		14	——	ditto	
5	26	22		16	W	ditto	
6	34	31		31	——	ditto	
7	32	24		21	E	high	
☾ 8	27	21		18	N E	ditto	
9	28	21		19	——	ditto	
10	27	25		25	——	brisk	
11	33	25		25	S W	little	
12	25	−11		− 5	E	brisk	
13	28	22		16	N E	ditto	
14	26	19		14	E	ditto	
⊕ 15	28	19		13	N E	ditto	
16	26	22		12	N W	little	
17	33	27		25	W	ditto	
18	35	32		30	N E	ditto	1.15
19	45	27		21	S W	brisk	
20	32	32		31	S E	ditto	
21	35	22		14	N	ditto	
22	28	19		13	——	ditto	
☽ 23	36	33		29	S W	ditto	
24	43	39		36	——	ditto	0.38
25	43	39		35	S	ditto	0.13
26	48	44		42	S W	ditto	0.07
27	+50	44		40	——	ditto	
28	44	37		33	N E	variable	
29	43	38		37	N W	little	
○ 30	45	39		34	S W	ditto	
31	36	34		28	——	ditto	0.10
Means	34	27.6		23.6			1.83

JANUARY.

Mean Temperature . . . . 30.8°
—— Pressure . . . . . . 29.807 inches
—— Dew-point . . . . . 29.2°
—— Force of vapour . . 0.195 inches
—— Degree of dryness . [illegible]
—— Degree of moisture . 946°
Least observed degree of moisture 671

Weight of vapour in a cubic foot.
Mean . . . 2.285 grs.
Maximum . 4.016 „
Minimum . 1.179 „

WINDS.

N. 2 = 27° N.E. 7 = 29° E. 3 = 17° S.E. 1 = 27° S. 2 = 37° S.W. 9 = 39° W 3 = 28° N.W. 5 = 26°.

Amount of rain, &c. . . . . 1.83 inches
—— of evaporation . . 0.341 inches

REMARKS.

The frost continued unusually severe from the first to the twenty-third.

The weather was generally cloudy and foggy, and there were frequent snow showers.

On the 23d it broke up, and the remaining eight days were mild and cloudy, with light showers of rain.

The feathered tribes suffered very much from cold and hunger, during the frosts. The Thames was much swollen, and was full of floating ice, of great thickness.

The Aurora Borealis was seen during the month.

1820 February.	Morning.					Afternoon.					Night.				
	Barometer.	Hygrometer.			Weather.	Barometer.	Hygrometer.			Weather.	Barometer.	Hygrometer.			Weather.
1	29.89	37	37	—	fine	29.81	40	34	6	very fine	29.77	34	34	—	very fine
2	29.74	32	32	—	fog	29.77	35	32	3	dull	29.81	34	34	—	dull
3	29.87	33	31	2	fine	29.90	34	30	4	same	29.94	33	31	2	same
4	29.95	34	31	3	dull	29.95	37	30	7	same	29.96	33	33	—	same
5	29.89	40	39	1	overcast	29.85	42	42	—	rain	29.84	41	41	—	same
6	29.77	44	44	—	rain	29.90	47	42	5	fine	29.99	45	45	—	overcast
☾ 7	30.06	47	47	—	dull	30.06	49	+49	—	overcast	30.06	46	46	—	same
8	30.06	45	45	—	mist	30.04	46	45	1	same	30.06	45	45	—	same
9	29.95	43	42	1	fine	29.85	46	44	2	very fine	29.79	41	41	—	rain
10	29.80	46	46	—	rain	29.90	47	39	8	same	29.99	36	36	—	very fine
11	30.02	40	38	2	very fine	29.94	45	44	1	fine	29.80	43	43	—	dull
12	29.80	42	42	—	rain	29.88	41	41	—	rain	29.95	40	40	—	same
13	29.95	42	36	6	fine	29.92	42	34	8	fine	29.99	36	36	—	same
⊕ 14	30.12	38	34	4	same	30.13	42	31	+11	same	30.16	36	35	1	same
15	30.20	32	32	—	dull	+30.20	36	32	4	fine	30.20	29	29	—	very fine
16	30.16	28	27	1	very fine	30.10	32	31	1	fog	30.06	28	28	—	same
17	30.00	26	26	—	mist	29.99	32	23	9	very fine	29.99	26	−22	4	same
18	30.00	28	26	2	fine	29.99	33	30	3	fine	30.00	31	29	2	dull
19	30.02	34	30	4	same	29.99	33	31	2	same	29.94	32	32	—	little snow
20	29.80	32	32	—	snow	29.79	32	32	—	snow	29.83	32	32	—	dull
☽ 21	29.78	32	31	1	mist	29.76	34	34	—	sleet	29.69	35	35	—	rain
22	29.75	35	35	—	rain	29.69	42	42	—	fog	29.64	41	41	—	dull
23	29.56	47	47	—	same	29.49	46	46	—	rain	29.46	42	42	—	overcast
24	29.39	39	39	—	rain	−29.30	38	38	—	same	29.30	36	35	1	same
25	29.35	36	36	—	same	29.49	39	39	—	same	29.70	35	35	—	rain
26	29.86	35	35	—	same	29.93	33	30	3	same	30.00	30	28	2	fine
27	30.01	35	26	9	fine	29.97	35	27	8	fine	29.99	29	29	—	very fine
28	29.90	35	30	5	very fine	29.83	36	30	6	very fine	29.83	29	26	3	same
⊙ 29	29.77	29	29	—	mist	29.68	38	33	5	same	29.51	36	35	1	rain
Means	29.874	36.7	35.3	1.4		29.865	39	35.6	3.3		29.870	35.6	35.1	0.5	

1820 February.	Temperature.				Wind.		Rain.
	Max.	Min.	Sun.	Rad.	Direction.	Force.	
1	41	27		23	S	little	
2	35	32		30	E	ditto	
3	37	32		30	N E	ditto	
4	37	32		32	S W	ditto	
5	43	34		34	S	dittc	0.02
6	47	40		37	W	variable	0.04
☾ 7	+49	44		41	S W	little	
8	46	38		31	——	ditto	
9	46	40		35	S	ditto	
10	47	32		26	W	ditto	0.03
11	45	39		36	——	brisk	
12	42	38		36	S	ditto	0.46
13	42	32		29	S E	ditto	
⊕ 14	42	31		29	W	ditto	
15	36	26		23	N E	ditto	
16	32	23		17	——	ditto	
17	32	−21		−16	——	little	
18	33	30		27	——	ditto	
19	34	22		18	——	brisk	
20	32	29		22	S E	ditto	
☽ 21	35	33		33	——	ditto	
22	42	39		38	W	ditto	0.42
23	47	37		36	S E	ditto	
24	39	35		34	S W	ditto	0.10
25	39	34		33	N E	high	0.07
26	35	30		28	——	ditto	0.92
27	35	28		24	——	ditto	
28	36	28		20	——	brisk	
○ 29	38	34		31	S W	ditto	0.02
Means	39.3	32.4		29.2			1.18

FEBRUARY.

Mean Temperature . . . . 35.9°
—— Pressure . . . . . . 29.869 inches
—— Dew-point . . . . . 34°
—— Force of vapour . . 0.232 inches
—— Degree of dryness . . 1.9°
—— Degree of moisture . 939
Least observed degree of moisture . . 684

Weight of vapour in a cubic foot.
Mean . . . 2.699 grs.
Maximum . 4.407 „
Minimum . 1.764 „

WINDS.

N. 0 = N.E 10 = 29° E. 1 = 32° S.E. 4 = 36° S. 4 = 39° S.W 5 = 38°
W. 5 = 39° N.W. 0.

Amount of rain, &c. . . . . . . 1.18 inches
—— of evaporation . . . 0.43 inches

REMARKS.

The weather, though chiefly cloudy, was, for the most part, fair, with hard frost at intervals.

The winter may be considered as having ended with the deep snow on the 20th.

1820 March.	Morning. Barometer.	Hygrometer.			Weather.	Afternoon. Barometer.	Hygrometer.			Weather.	Night. Barometer.	Hygrometer.			Weather.
1	29.49	43	33	10	very fine	29.51	42	30	12	very fine	29.41	38	31	7	dull
2	29.00	35	28	7	snow	29.18	35	20	15	same	29.46	29	26	3	very fine
3	29.68	33	27	6	very fine	29.76	34	27	7	same	29.90	28	25	3	same
4	29.91	35	27	8	same	29.80	34	33	1	same	29.96	29	29	—	fine
5	30.12	33	24	9	same	30.13	32	−19	13	same	30.13	25	21	4	very fine
6	30.08	33	27	6	dull	30.06	33	21	12	fine	30.10	26	24	2	fine
☾ 7	30.09	35	21	14	fine	30.03	35	35	—	sleet	30.09	33	32	1	sleet
8	30.16	36	32	4	same	30.14	38	28	10	very fine	30.16	30	30	—	fine
9	30.09	36	32	4	very fine	30.01	42	34	8	same	29.93	31	31	—	very fine
10	29.79	36	32	4	fine	29.63	41	31	10	very fine	29.59	31	31	—	same
11	29.57	38	36	2	very fine	29.51	48	35	13	same	29.49	35	35	—	same
12	29.40	38	37	1	fine	29.36	42	36	6	same	29.36	38	38	—	dull
13	29.56	45	39	6	very fine	29.68	48	34	14	same	29.85	37	37	—	very fine
⊕ 14	30.01	48	45	3	fine	30.07	53	51	2	overcast	30.13	49	49	—	overcast
15	30.19	58	51	7	same	30.18	59	51	8	very fine	30.22	51	51	—	same
16	+30.29	48	46	2	dull	30.24	49	41	8	same	30.23	39	38	1	very fine
17	30.16	44	42	2	showers	30.16	50	46	4	fine	30.23	41	41	—	sleet
18	30.26	42	36	6	very fine	30.21	45	26	19	very fine	30.20	37	35	2	dull
19	30.14	42	37	5	dull	30.11	44	33	11	fine	30.16	37	29	8	same
20	30.16	39	35	4	fine	30.13	46	37	9	same	30.13	35	35	—	very fine
☽ 21	30.06	45	35	10	same	29.93	48	37	11	same	29.85	41	39	2	showers
22	29.74	47	35	12	very fine	29.71	50	33	17	same	29.53	46	46	—	rain
23	29.30	49	42	7	fine	29.09	50	36	14	same	29.04	41	36	5	very fine
24	29.02	48	36	12	same	−28.87	48	37	11	light show.	28.87	38	38	—	overcast
25	29.21	42	30	12	very fine	29.33	46	23	+23	very fine	29.56	33	32	1	very fine
26	29.67	45	30	15	fine	29.60	45	45	—	light rain	29.64	48	48	—	rain
27	29.74	53	49	4	light show.	29.72	54	50	4	showers	29.76	50	50	—	same
28	29.92	55	49	6	fine	29.94	56	48	8	very fine	29.99	50	50	—	overcast
○ 29	30.03	56	49	7	same	29.99	59	52	7	same	29.92	49	49	—	very fine
30	29.88	53	48	5	same	29.88	59	+54	5	fine	29.91	44	39	5	same
31	29.91	50	42	8	same	29.88	58	37	21	very fine	29.92	47	44	3	same
Means	29.826	43.2	36.5	6.7		29.799	45.9	36.1	9.7		29.829	38.2	36.7	1.5	

1820 March.	Temperature. Max.	Min.	Sun.	Rad.	Wind. Direction.	Force.	Rain.
1	43	32		28	N W	high	
2	35	28		28	——	ditto	0.04
3	34	27		25	N E	ditto	
4	36	26		22	——	brisk	
5	33	25		18	——	ditto	
6	33	−24		−18	——	ditto	
☾ 7	36	33		32	N W	little	0.03
8	39	27		21	N E	ditto	
9	43	29		20	S W	ditto	
10	42	30		26	S E	ditto	
11	50	32		29	S E	brisk	
12	44	35		32	——	ditto	
13	50	34		27	W	ditto	
⊕ 14	56	48		43	S W	ditto	
15	61	44		43	——	ditto	
16	52	36		32	S E	ditto	
17	51	34		30	N E	ditto	0.01
18	47	35		30	——	ditto	
19	45	35		33	——	high	
20	47	35		28	——	little	
☽ 21	49	39		35	N W	ditto	0.01
22	50	44		42	——	brisk	0.02
23	50	38		36	——	ditto	
24	48	33		29	W	ditto	
25	46	28		22	N	ditto	
26	48	44		44	S W	ditto	0.02
27	54	44		41	——	little	0.01
28	57	48		44	W	ditto	
○ 29	+60	41		35	S W	ditto	
30	60	33		27	N W	ditto	
31	60	37		31	——	ditto	
Means	47	34.7		30.6			0.14

MARCH.

Mean Temperature . . . . 40.9°
—— Pressure . . . . . . 29.818 inches
—— Dew-point . . . . . 35.4°
—— Force of vapour . . 0.243 inches
—— Degree of dryness . . 5.5°
—— Degree of moisture . 835°
Least observed degree of moisture . . 449

Weight of vapour in a cubic foot.
Mean . . . 2.790 grs.
Maximum . 5.122 „
Minimum . 1.587 „

WINDS.

N. 1 = 28° N.E. 9 = 30° E 0 = S.E. 4 = 33° S. 0 = S.W. 6 = 45° W. 3 = 40° N.W. 8 = 35°

Amount of rain, &c. . . . . . 0.14 inches
—— of evaporation . . . 1,48 inches

REMARKS.

The weather throughout the principal part of the month very fine and dry.

On the 2d from five to eight A. M. the wind blew a complete hurricane.

There were many sharp frosts during the first fortnight.

1820 April.		Morning.					Afternoon.					Night.				
		Barometer.	Hygrometer.			Weather.	Barometer.	Hygrometer.			Weather.	Barometer.	Hygrometer.			Weather.
	1	30.06	51	42	9	overcast	30.02	56	48	8	overcast	30.07	50	50	—	dull
	2	30.10	55	52	3	dull	30.10	63	53	10	dull	30.12	51	50	1	same
	3	30.21	60	+53	7	same	30.21	60	53	7	same	30.20	47	47	—	fine
	4	30.04	52	40	12	fine	29.96	60	51	9	fine	29.94	45	38	7	very fine
	5	29.85	53	41	12	same	29.78	64	43	21	very fine	29.77	50	42	8	fine
☾	6	29.49	47	45	2	rain	29.46	49	46	3	showers	29.46	36	34	2	very fine
	7	29.52	43	32	11	fine	29.54	46	41	5	fine	29.55	37	37	—	same
	8	29.44	48	37	11	same	29.43	49	41	8	showers	−29.20	44	44	—	rain
	9	29.37	42	40	2	showers	29.44	51	36	15	light show.	29.54	38	38	—	very fine
	10	29.54	49	42	7	dull	29.49	49	49	—	rain	29.49	48	48	—	dull
	11	29.62	50	49	1	same	29.61	55	51	4	showers	29.62	48	48	—	rain
⊕	12	29.89	50	49	1	dull	29.87	55	49	6	rain	30.02	50	48	2	dull
	13	30.01	49	47	2	same	30.00	50	49	1	same	29.84	48	48	—	rain
	14	29.79	50	49	1	same	29.65	50	49	1	same	29.73	46	46	—	same
	15	29.81	55	41	14	very fine	29.87	56	41	15	very fine	30.09	43	39	4	very fine
	16	30.40	58	40	18	same	30.22	61	49	11	same	30.33	51	50	1	same
	17	30.36	64	53	11	same	30.33	65	52	13	same	30.32	53	51	2	same
	18	30.22	58	52	6	same	30.15	62	51	11	same	30.16	48	48	—	same
	19	30.20	62	51	11	same	30.17	67	46	21	same	30.13	54	46	8	same
☽	20	30.19	61	46	15	same	30.23	62	43	19	same	30.28	48	42	6	same
	21	30.30	59	44	15	same	30.29	62	44	18	same	30.33	49	49	—	same
	22	30.50	59	44	15	same	30.38	60	44	16	same	30.50	47	43	4	same
	23	30.45	58	45	13	same	30.45	60	44	16	same	30.50	46	41	5	same
	24	30.49	55	45	10	same	30.51	62	39	23	same	+30.54	45	40	5	same
	25	30.38	52	42	10	same	30.39	60	34	+26	same	30.32	44	42	2	same
	26	30.10	64	44	20	same	29.80	56	44	12	overcast	29.66	47	43	4	overcast
	27	29.71	43	43	—	showers	29.83	42	33	9	same	29.91	39	35	4	dull
○	28	30.01	50	−27	23	fine	30.01	51	33	18	very fine	30.03	40	33	7	very fine
	29	30.10	55	37	18	very fine	30.10	58	40	18	overcast	30.11	48	44	4	overcast
	30	30.18	56	47	9	showers	30.22	56	33	23	very fine	30.27	44	33	11	very fine
Means		30.011	53.6	43.9	9.6		29.983	56.5	44.3	12.2		30.001	46.1	43.2	2.9	

1820 April.	Temperature.				Wind.		Rain.
	Max.	Min.	Sun.	Rad.	Direction.	Force.	
1	59	44		41	W	high	
2	64	49		45	N W	little	
3	63	36		32	——	variable	
4	60	37		34	S E	brisk	
5	66	44		41	W	ditto	
☾ 6	52	30		27	S W	ditto	0.17
7	50	30		25	——	ditto	0.02
8	52	41		41	S	ditto	0.12
9	52	−29		−23	N W	ditto	
10	49	46		46	S	ditto	0.26
11	59	46		40	S W	ditto	
⊕ 12	58	46		43	——	ditto	0.35
13	50	46		45	S E	ditto	0.45
14	50	39		34	——	ditto	0.10
15	57	40		35	N W	little	
16	63	43		39	W	ditto	
17	+67	44		39	N W	ditto	
18	66	43		37	S E	ditto	
19	67	45		41	N W	ditto	
☽ 20	64	41		35	——	ditto	
21	67	41		36	E	variable	
22	63	39		36	——	little	
23	62	40		36	——	ditto	
24	64	38		33	N E	brisk	
25	64	37		29	——	ditto	
26	67	41		38	W	ditto	
27	47	38		36	N E	ditto	
○ 28	53	36		28	N	little	
29	61	41		34	W	ditto	
30	59	36		28	N W	ditto	0.20
Means	59.1	40.2		35.9			1.67

APRIL.

Mean Temperature . . . . 49.6°
—— Pressure . . . . . . 29.998 inches
—— Dew-point . . . . . 42.2°
—— Force of vapour . . 0.306 inches
—— Degree of dryness . . 7.4°
—— Degree of moisture . 774°
Least observed degree of moisture . . 414

Weight of vapour in a cubic foot.
Mean . . . 3.451 grs.
Maximum . 4.934 „
Minimum . 2.063 „

WINDS.

N. 1 = 31° N.E. 3 = 39° E. 3 = 44° S.E. 4 = 47° S. 2 = 43° S.W. 4 = 44° W. 5 = 43° N.W. 8 = 45°

Amount of rain, &c. . . . . . . 1.65 inches
—— of evaporation . . . 2.67 inches

REMARKS.

The first part of the month was remarkably warm and fine, but the weather afterwards became cold, cloudy, and wet. In the middle it again changed, and was fine and seasonable.

The swallows were first seen about the latter end of the month.

1820 May.	Morning.					Afternoon.					Night.				
	Barometer.	Hygrometer.			Weather.	Barometer.	Hygrometer.			Weather.	Barometer.	Hygrometer.			Weather.
1	30.31	56	41	15	very fine	+30.33	56	36	20	very fine	30.33	46	41	5	very fine
2	30.27	58	42	16	fine	30.16	57	41	16	fine	30.14	48	41	7	dull
3	30.15	45	42	3	dull	30.10	50	43	7	dull	30.03	45	40	5	same
4	29.95	49	33	16	very fine	29.88	53	35	18	fine	29.86	42	37	5	same
☽ 5	29.88	51	30	21	fine	29.88	52	−30	22	same	29.88	40	37	3	very fine
6	29.88	59	37	22	very fine	29.79	60	39	21	overcast	29.67	46	43	3	showers
7	29.62	62	43	19	same	29.68	65	41	+24	very fine	29.70	53	43	10	dull
8	29.71	62	53	9	fine	29.69	62	53	9	showers	29.64	54	54	—	showers
9	29.65	64	51	13	same	29.67	63	45	18	very fine	29.67	51	49	2	very fine
10	29.80	68	48	20	very fine	29.88	63	52	11	light show.	29.90	53	52	1	fine
11	29.97	65	50	15	fine	29.96	65	45	20	very fine	30.00	51	46	5	very fine
⊕ 12	30.05	65	46	16	same	30.02	67	47	20	same	30.01	53	53	—	fine
13	29.95	60	53	7	dull	29.90	60	55	5	showers	29.88	53	53	—	same
14	29.88	65	53	12	fine	29.82	65	48	17	overcast	29.82	52	49	3	same
15	29.83	67	47	20	very fine	29.80	66	44	22	very fine	29.74	54	54	—	rain
16	29.70	58	53	5	showers	29.72	61	47	14	showers	29.72	52	52	—	same
17	29.76	65	49	16	fine	29.73	64	45	19	very fine	29.61	53	51	2	dull
18	29.32	52	51	—	rain	−29.16	56	52	4	showers	29.28	50	50	—	rain
19	29.72	69	47	22	very fine	29.81	64	46	18	very fine	29.97	51	46	5	very fine
☾ 20	30.15	68	56	12	fine	30.25	66	54	12	fine	30.30	53	48	5	same
21	30.32	62	50	12	same	30.30	70	53	17	same	30.29	52	49	3	fine
22	30.15	68	53	15	very fine	30.18	77	56	21	very fine	30.16	59	53	6	very fine
23	30.04	69	57	12	same	29.97	72	57	15	same	29.92	59	55	4	same
24	29.81	69	+58	11	same	29.86	70	54	16	same	29.84	58	50	8	dull
25	29.82	58	53	5	fine	29.80	57	54	3	showers	29.80	50	48	2	same
26	29.86	60	47	13	overcast	29.83	59	52	7	same	29.79	56	55	1	showers
○ 27	29.71	55	52	3	showers	29.56	59	55	4	same	29.68	52	47	5	showers
28	29.65	59	48	11	overcast	29.62	56	51	5	rain	29.50	52	51	1	rain
29	29.41	57	43	14	same	29.41	53	50	3	hail&thund.	29.40	46	42	4	fine
30	29.40	55	42	13	same	29.41	58	43	15	light show.	29.42	51	43	8	dull
31	29.45	56	42	14	hail	29.46	60	48	12	showers	29.51	45	42	3	very fine
Means	29.844	60.5	47.4	13		29.826	61.5	47.7	14		29.821	50.9	47.5	3.4	

1820 May.	Temperature.				Wind.		Rain.
	Max.	Min.	Sun.	Rad.	Direction.	Force.	
1	62	40		33	N W	brisk	
2	65	42		39	S E	ditto	
3	51	41		40	E	ditto	
4	54	33		28	——	ditto	
☽ 5	54	−33		−24	N E	ditto	
6	64	45		44	S W	ditto	0.48
7	66	51		46	N W	ditto	
8	67	51		47	S W	little	0.35
9	69	46		42	——	brisk	
10	69	50		47	——	ditto	
11	69	47		41	W	ditto	0.59
⊕ 12	70	43		36	S W	ditto	
13	64	51		49	S E	ditto	
14	70	46		41	S W	ditto	
15	70	49		49	——	little	
16	66	47		46	——	ditto	
17	67	49		46	——	brisk	
18	60	47		47	——	high	
19	70	47		41	——	brisk	
☾ 20	68	44		37	——	ditto	
21	71	47		40	——	ditto	
22	+77	51		46	S E	ditto	
23	72	53		49	——	ditto	
24	72	49		46	S	ditto	
25	62	44		31	S W	ditto	0.50
26	62	52		54	——	little	0.03
○ 27	59	49		44	W	ditto	0.47
28	60	47		43	——	ditto	0.03
29	59	42		37	——	ditto	0.15
30	60	45		40	——	ditto	0.01
31	59	42		37	——	ditto	0.02
Means	64.7	45.9		41.6			2.63

MAY.

Mean Temperature . . . . . 55.3°
—— Pressure . . . . 29.830 inches
—— Dew-point . . . . . 46.8°
—— Force of vapour . . . 0.361 inch
—— Degree of dryness . . 8.5°
—— Degree of moisture . 752°
Least observed degree of moisture . . 444

Weight of vapour in a cubic foot.
Mean . . . 4.084 grs.
Maximum . 5.743 „
Minimum . 2.257 „

WINDS.

N. 0 = N.E. 1 = 32° E. 2 = 38° S.E. 4 = 51° S. 1 = 54° S.W. 15 = 49° W. 6 = 46° N.W. 2 = 40°

Amount of rain . . . . 2.63 inches
—— of evaporation . . 3.68 inches

REMARKS.

During the first part of the month the weather was mostly fine, warm, and dry; but during the latter part cloudy and dull, with frequent falls of rain, accompanied, at intervals, with gales of wind.

1820 June.	Morning.					Afternoon.					Night.				
	Barometer.	Hygrometer.			Weather.	Barometer.	Hygrometer.			Weather.	Barometer.	Hygrometer.			Weather.
1	29.60	56	48	8	showers	−29.60	57	−40	17	fine	29.64	51	43	8	overcast
☽ 2	29.62	56	44	12	fine	29.65	58	50	8	showers	29.71	50	47	3	fine
3	29.80	50	48	2	rain	29.84	56	52	4	rain	29.89	50	46	4	overcast
4	29.95	51	43	8	fine	29.95	51	45	6	fine	29.99	48	45	3	same
5	30.01	56	46	10	dull	30.01	66	51	15	same	29.83	59	56	3	rain
6	29.88	52	50	2	rain	30.02	56	56	—	rain	30.13	49	45	4	fine
7	30.08	59	47	12	overcast	30.05	61	46	15	dull	30.03	50	45	5	same
8	29.97	62	53	9	same	29.94	65	55	10	showers	29.87	58	55	3	dull
9	29.83	60	55	5	same	29.74	61	55	6	same	29.73	50	47	3	same
⊕ 10	29.77	63	47	16	same	29.79	60	41	19	same	29.75	50	45	5	very fine
11	29.66	53	43	10	same	29.62	54	52	2	rain	29.62	50	50	—	rain
12	29.78	58	52	6	fine	29.85	63	50	13	fine	29.93	50	47	3	very fine
13	29.94	58	51	7	same	29.91	57	56	1	thund.storm	29.96	51	51	—	rain
14	30.19	54	50	4	dull	30.12	59	49	10	dull	30.04	53	50	3	same
15	29.88	60	54	6	same	29.96	59	52	7	same	30.00	49	47	2	very fine
16	30.01	61	49	12	fine	29.99	67	51	16	fine	29.98	58	54	4	fine
17	30.01	62	49	13	same	30.01	67	51	16	same	30.05	58	54	4	same
☾ 18	30.11	62	46	16	overcast	30.05	67	49	18	dull	29.95	57	51	6	dull
19	29.79	63	51	12	same	29.80	63	48	15	same	29.80	53	44	9	same
20	29.71	56	51	5	rain	29.82	61	55	6	showers	29.87	54	52	2	showers
21	29.92	62	50	12	very fine	29.99	66	54	12	very fine	30.03	52	51	1	very fine
22	30.09	62	56	6	mist	30.10	75	57	18	same	30.12	60	59	1	fine
23	30.20	66	57	9	very fine	30.20	77	63	14	same	30.23	65	60	5	very fine
24	30.27	74	60	14	same	30.29	84	65	19	same	30.33	68	61	7	same
25	30.40	77	62	15	same	30.40	87	67	20	same	30.43	73	63	10	same
26	30.44	81	67	14	same	30.43	89	+70	19	same	30.44	69	61	8	same
○ 27	30.44	79	66	13	same	+30.46	83	70	13	same	30.34	69	67	2	same
28	30.33	85	67	18	same	30.24	84	64	+20	same	30.23	69	62	7	same
29	30.19	65	60	5	same	30.06	71	63	8	same	30.03	58	58	—	rain
30	29.99	65	62	3	showers	29.94	62	60	2	overcast	30.09	55	52	3	overcast
Means	29.995	62.2	52.8	9.4		29.994	66.2	54.5	11.6		30.001	56.2	52.2	3.8	

1820 June.		Temperature. Max.	Min.	Sun.	Rad.	Wind. Direction.	Force.	Rain.
	1	60	47		42	W	high	0.01
☽	2	60	42		32	——	little	0.32
	3	56	46		41	——	ditto	0.07
	4	55	45		35	N W	ditto	0.01
	5	66	47		47	W	ditto	
	6	59	42		34	N	ditto	0.25
	7	63	46		39	——	brisk	
	8	63	54		52	N.W	ditto	
	9	66	41		33	——	ditto	0.06
⊕	10	63	41		33	——	ditto	
	11	61	48		48	S W	ditto	0.02
	12	64	−40		−30	N	ditto	0.03
	13	61	45		40	S E	ditto	0.63
	14	59	48		31	N W	ditto	
	15	63	51		44	——	ditto	
	16	69	48		40	——	ditto	
	17	69	48		40	——	ditto	
☾	18	68	53		50	W	ditto	0.51
	19	67	49		11	N W	ditto	0.09
	20	63	45		36	——	little	
	21	68	46		38	——	ditto	0.08
	22	76	51		40	S W	ditto	
	23	79	56		46	——	ditto	
	24	85	60		55	——	brisk	
	25	87	63		53	S E	ditto	
	26	+89	62		52	——	ditto	
○	27	86	68		59	E	little	
	28	85	59		56	——	ditto	
	29	73	56		53	S E	ditto	
	30	70	50		45	E	ditto	
Means		68.4	49.9		42.8			2.11

JUNE.

Mean Temperature . . . . 59.1°
—— Pressure . . . . 29.996 inches
—— Dew-point . . . . . 52.2°
—— Force of vapour . . 0.431 inch
—— Degree of dryness . . 6.9°
—— Degree of moisture . 790°
Least observed degree of moisture . . 530

Weight of vapour in a cubic foot.
Mean . . . 5.315 grs.
Maximum . 8.186 „
Minimum . 3.130 „

WINDS.

N. 3 = 48° N.E. 0 E. 3 = 63° S.E. 4 = 60° S. 0 S.W. 4 = 56°
W. 5 = 47° N.W. 11 = 50°

Amount of rain, &c. . . . . . . 2.11 inches
—— of evaporation . . . 3.42 inches

REMARKS.

The first part of this month was cloudy, wet, and unseasonably cold. Rain occurred almost daily. On the thirteenth the storm was very heavy, and accompanied with much thunder and lightning.

Between the 20th and 26th a most extraordinary and rapid increase of temperature took place, and continued till the 28th. During the above period the weather was extremely fine; and from the 23d to the 29th scarcely a cloud obscured the sky.

1820 July.	Morning.					Afternoon.					Night.				
	Barometer.	Hygrometer.			Weather.	Barometer.	Hygrometer.			Weather.	Barometer.	Hygrometer.			Weather.
1	30.21	61	47	14	very fine	30.28	62	49	13	fine	+30.30	53	47	6	fine
☽ 2	30.19	62	52	10	fine	30.09	60	58	2	rain	30.06	52	49	3	clearing
3	30.00	60	51	9	showers	29.89	60	57	3	showers	29.91	55	55	—	overcast
4	29.97	61	52	9	fine	30.01	60	−46	14	fine	30.05	54	51	3	same
5	30.12	60	47	13	same	30.11	58	49	9	dull	30.14	51	48	3	very fine
6	30.17	58	50	8	very fine	30.14	62	52	10	fine	30.16	55	54	1	fine
7	30.13	59	52	7	fine	30.16	64	53	11	same	30.19	55	50	5	same
8	30.18	58	48	10	same	30.17	59	47	12	same	30.19	55	48	7	dull
9	30.19	60	51	9	same	30.15	62	55	7	same	30.16	57	55	2	same
⊕ 10	30.14	61	50	11	overcast	30.10	64	52	12	same	30.09	55	51	4	very fine
11	30.10	65	52	13	fine	30.01	66	55	11	haze	30.01	56	54	2	same
12	29.96	62	55	7	very fine	29.88	64	55	9	overcast	29.86	57	55	2	dull
13	29.83	61	54	7	dull	29.84	61	53	8	same	29.85	55	52	3	same
14	29.87	65	57	8	fine	29.90	68	59	9	fine	29.92	62	61	1	dropsofrain
15	29.97	65	60	5	overcast	30.00	73	63	10	same	30.03	61	60	1	very fine
16	29.99	72	62	10	dark	29.96	75	+67	8	thund.storm	29.92	66	65	1	overcast
17	29.96	68	67	1	rain	29.56	66	66	—	rain	29.52	59	58	1	very fine
☾ 18	29.40	61	61	—	same	−29.39	62	62	—	same	29.41	57	57	—	dull
19	29.45	69	60	9	very fine	29.54	61	61	—	same	29.61	59	59	—	fine
20	29.69	70	60	10	same	29.72	71	60	11	thund.storm	29.81	62	61	1	showers
21	29.89	63	60	3	overcast	29.91	68	60	8	shower	29.94	62	60	2	overcast
22	29.97	70	52	18	very fine	29.97	64	52	12	overcast	29.97	58	52	6	same
23	30.00	58	52	6	fine	30.00	63	54	9	fine	30.02	57	53	4	fine
24	30.03	65	56	9	same	30.01	74	61	13	same	29.99	64	59	5	same
○ 25	29.90	69	55	14	very fine	29.91	65	52	13	same	29.94	57	51	6	same
26	30.05	71	51	+20	same	30.07	71	57	14	very fine	30.08	62	60	2	very fine
27	30.06	72	61	11	same	30.01	74	61	13	same	30.02	62	60	2	same
28	30.07	77	61	16	same	30.03	74	61	13	same	30.07	64	60	4	fine
29	30.10	71	60	11	fine	30.13	78	59	19	same	30.13	67	60	7	very fine
30	30.12	73	61	12	very fine	30.07	75	62	13	heavy	29.90	66	64	2	thund.storm
31	29.90	72	62	10	fine	29.90	81	64	17	fine	29.89	72	66	6	fine
Means	29.987	65.1	55.4	9.6		29.964	66.6	56.8	9.7		29.972	58.9	55.9	2.9	

1820 July.	Temperature.				Wind.		Rain.
	Max.	Min.	Sun.	Rad.	Direction.	Force.	
1	67	−44		−37	N E	little	
☽ 2	68	45		40	W	ditto	0.16
3	66	52		49	N W	ditto	0.12
4	70	53		45	N E	ditto	
5	61	48		39	——	ditto	
6	64	52		45	——	ditto	
7	65	53		47	——	ditto	
8	62	52		50	——	variable	
9	66	54		50	——	ditto	
⊕ 10	68	50		44	S E	little	
11	67	50		44	E	ditto	
12	67	53		52	——	ditto	
13	63	52		49	——	ditto	
14	69	59		57	S E	ditto	
15	78	53		45	——	ditto	
16	76	60		56	——	variable	
17	75	54		47	——	ditto	0.20
☾ 18	74	53		48	S W	little	0.43
19	72	55		48	——	ditto	0.74
20	72	56		52	——	ditto	0.21
21	70	54		46	N W	ditto	0.34
22	70	54		51	——	ditto	0.19
23	64	52		46	——	ditto	
24	75	59		55	——	ditto	
○ 25	73	53		46	——	brisk	0.12
26	75	61		54	——	ditto	0.08
27	76	58		51	W	ditto	0.08
28	78	61		53	——	ditto	0.02
29	79	58		47	——	little	
30	79	58		51	S E	calm	
31	+83	63		57	——	little	
Means	70.7	54.1		48.4			2.69

JULY.

Mean Temperature . . . . 62.4°
—— Pressure . . . . . . 29.974 inches
—— Dew-point . . . . . 55.4°
—— Force of vapour . . 0.482 inch
—— Degree of dryness . . 7°
—— Degree of moisture 799°

Least observed degree of moisture 520

Weight of vapour in a cubic foot.
Mean . . . 5.268 grs.
Maximum . 7.545 „
Minimum . 3.911 „

WINDS.

N. 0 N.E. 7 = 50° E. 3 = 53° S.E. 7 = 60° S. 0 S.W. 3 = 60° W 4 = 58° N.W. 7 = 55°.

Amount of rain, &c. . . . . 2.69 inches
—— of evaporation . . 3.75 inches

REMARKS.

Cloudy, but fine; warm and dry weather prevailed during the greater part of the month. The heavy periodical rains, which are peculiar to this month, commenced on the 16th with a violent thunder storm. Another thunder shower happened on the 20th, but the most violent of all on the 30th. The lightning was vivid in the extreme, and continued, without intermission, from 11 P.M. to 1 A.M., enveloping the atmosphere in one uninterrupted blaze of light the whole time.

1820 August.	Morning.				Afternoon.				Night.			
	Barometer.	Hygrometer.		Weather.	Barometer.	Hygrometer.		Weather.	Barometer.	Hygrometer.		Weather.
☾ 1	29.86	68	59	9	fine	29.88	72	58	14	fine	29.99	61	61	—	very fine
2	30.12	72	55	17	very fine	30.08	71	53	18	very fine	30.08	63	61	2	dull
3	29.99	73	66	7	overcast	29.92	72	+66	6	dull	29.84	66	63	3	same
4	29.75	71	64	7	fine	29.82	71	64	7	overcast	29.86	60	57	3	showers
5	29.88	65	53	12	same	29.90	68	53	15	very fine	29.92	59	57	2	rain
6	29.74	68	52	16	very fine	29.65	68	59	9	rain	29.67	59	57	2	showers
⊕ 7	29.80	65	56	9	fine	29.94	67	53	14	fine	30.00	56	51	5	fine
8	30.02	66	54	12	same	29.88	68	52	16	very fine	29.83	61	55	6	dull
9	29.97	72	53	19	very fine	30.05	71	51	20	same	30.15	58	51	7	very fine
10	+30.26	67	57	10	same	30.23	71	57	14	same	30.23	63	61	2	fine
11	30.26	68	60	8	same	30.26	76	63	13	same	30.25	64	59	5	very fine
12	30.23	65	57	8	same	30.20	74	63	11	same	30.18	53	53	—	same
13	30.15	63	54	9	same	30.10	74	60	14	same	30.10	58	58	—	same
14	30.04	70	60	10	same	30.00	77	48	+29	same	29.97	56	56	—	same
☽ 15	29.90	56	54	2	same	29.80	75	65	10	fine	29.80	65	60	5	dull
16	29.79	71	61	10	overcast	29.83	75	62	13	overcast	29.86	68	65	3	showers
17	29.79	71	60	11	same	29.78	72	65	7	very fine	29.85	60	53	7	very fine
18	29.86	62	52	10	fine	29.86	70	51	19	fine	29.84	59	50	9	fine
19	29.78	63	52	11	very fine	29.79	62	56	6	thunder	29.84	51	48	3	very fine
20	29.92	59	50	9	same	29.93	64	47	17	very fine	29.94	56	48	8	same
21	29.87	58	57	1	rain	29.84	58	58	—	rain	29.84	53	53	—	rain
22	29.82	58	51	7	dull	29.82	57	51	6	overcast	29.99	51	48	3	fine
○ 23	30.09	59	47	12	very fine	30.12	59	−46	13	same	30.14	53	51	2	same
24	30.15	63	46	17	same	30.12	65	48	17	very fine	30.05	54	51	3	very fine
25	29.88	61	55	6	overcast	29.87	65	62	3	dull	29.80	61	61	—	rain
26	29.73	63	52	11	fine	29.73	60	60	—	showers	29.75	55	54	1	fine
27	29.77	58	47	11	fine	29.77	60	60	—	same	29.71	57	54	3	rain
28	−29.55	59	49	10	same	29.60	60	49	11	fine	29.67	49	47	2	fine
29	29.70	59	49	10	same	29.72	63	47	16	very fine	29.79	54	50	4	overcast
☾ 30	29.91	61	48	13	very fine	29.94	58	51	7	fine	30.01	51	50	1	fine
31	30.06	62	54	8	same	30.07	58	51	7	very fine	30.09	50	49	1	very fine
Means	29.923	64.3	54.3	10		29.919	67.1	55.7	11.3		29.936	57.5	54.5	2.9	

1820 August.	Temperature. Max.	Min.	Sun.	Rad.	Wind. Direction.	Force.	Rain.
☾ 1	74	58		53	W	brisk	
2	76	62		55	——	ditto	
3	76	65		63	S W	ditto	
4	74	53		51	——	ditto	
5	71	57		55	W	little	0.70
6	68,	56		52	S W	ditto	0.11
⊕ 7	70	49		41	W	ditto	0.12
8	69	58		51	S W	brisk	
9	73	54		47	W	ditto	
10	73	57		47	——	ditto	
11	76	53		43	——	ditto	
12	75	55		54	N W	little	
13	75	55		45	——	ditto	
14	+77	55		44	——	ditto	
☽ 15	77	64		60	S W	ditto	0.02
16	76	65		60	——	brisk	
17	75	57		49	——	ditto	
18	72	50		42	N W	ditto	
19	68	−42		−35	N E	ditto	
20	65	51		42	N W	ditto	0.21
21	60	50		50	N E	ditto	0.34
22	59	47		42	——	ditto	0.19
○ 23	60	52		48	——	little	
24	67	51		42	S W	ditto	
25	67	53		50	——	ditto	0.12
26	68	47		42	——	ditto	0.08
27	66	51		47	W	brisk	0.08
28	63	43		36	——	ditto	0.02
29	69	48		39	——	ditto	
☾ 30	64	47		39	N W	ditto	
31	64	47		39	N E	ditto	
Means	69.9	53.2		47.1			1.99

AUGUST.

Mean Temperature . . . . 61.8°
—— Pressure . . . . 29.926 inches
—— Dew-point . . . . . 54.4°
—— Force of vapour . . 0.466 inch
—— Degree of dryness . . 7.4°
—— Degree of moisture . 789°
Least observed degree of moisture . . 389

Weight of vapour in a cubic foot.
Mean . . . 5.091 grs.
Maximum . 7.361 „
Minimum . 3.918 „

WINDS.

N. 0 N.E. 5 = 51° E 0 S.E. 0 S. 0 S.W. 10 = 58° W. 10 = 54°
N.W. 6 = 53°

Amount of rain, &c. . . . . . . 1.99 inches
—— of evaporation . . . 3.87 inches

REMARKS.

The weather, for the most part, very fine, warm, and dry.

1820	Morning.					Afternoon.					Night.				
September.	Barometer.	Hygrometer.			Weather.	Barometer.	Hygrometer.			Weather.	Barometer.	Hygrometer.			Weather.
1	30.11	59	52	7	very fine	30.11	67	53	14	very fine	30.12	54	49	5	very fine
2	30.11	60	51	9	same	30.11	64	55	9	overcast	30.11	51	49	2	same
3	30.20	61	54	7	same	30.21	66	55	11	very fine	30.21	46	44	2	same
4	30.20	58	49	9	same	30.20	68	57	11	fine	30.11	53	48	5	same
5	30.16	58	51	7	fine	30.05	61	45	16	very fine	30.07	50	50	—	same
6	30.07	63	55	8	same	30.05	62	47	15	same	30.09	52	52	—	same
⊕ 7	30.12	62	51	11	fine	30.12	64	52	12	same	30.16	55	54	1	fine
8	30.21	66	55	11	very fine	30.11	63	54	9	same	30.14	54	52	2	very fine
9	+30.40	65	53	12	same	30.40	71	54	17	same	30.36	57	52	5	same
10	30.37	60	58	2	fine	30.32	72	52	+20	fine	30.31	50	50	—	same
11	30.34	62	59	3	very fine	30.29	73	+66	7	overcast	30.25	57	52	5	fine
12	30.23	63	60	3	fine	30.21	67	63	4	very fine	30.21	59	59	—	very fine
13	30.21	63	60	3	very fine	30.11	67	58	9	same	30.11	57	57	—	same
14	29.94	66	61	5	same	29.83	69	63	6	fine	29.81	62	62	—	dull
☽ 15	29.68	67	59	8	overcast	29.74	69	61	2	same	29.82	60	59	1	fine
16	30.00	60	49	11	very fine	30.04	63	52	9	overcast	30.01	45	44	1	same
17	30.02	65	54	11	overcast	29.98	64	54	10	same	29.92	54	54	—	rain
18	29.62	52	52	—	rain	29.70	55	48	7	fine	29.78	43	40	3	very fine
19	29.92	49	43	6	very fine	29.97	51	38	13	very fine	30.05	42	−37	5	same
20	29.92	53	41	12	same	29.72	54	50	4	showers	29.44	51	51	—	rain
21	29.44	53	50	3	fine	−29.41	50	49	1	same	29.45	45	44	1	very fine
○ 22	29.72	54	46	8	very fine	29.89	55	51	4	fine	30.00	46	44	2	fine
23	30.02	61	56	5	dull	30.02	67	59	8	same	30.00	59	58	1	same
24	29.79	62	59	3	showers	29.75	64	55	9	same	29.81	48	45	3	very fine
25	29.71	55	45	10	very fine	29.72	55	48	7	very fine	29.76	41	41	—	same
26	29.95	50	42	8	fine	30.02	50	41	9	fine	30.11	42	37	5	same
27	30.11	48	44	4	same	30.03	53	52	1	rain	30.10	50	50	—	overcast
28	30.11	56	52	4	very fine	30.02	58	52	6	fine	30.01	56	55	1	rain
☾ 29	30.05	54	52	2	showers	30.15	58	54	4	showers	30.22	45	44	1	very fine
30	30.19	52	51	1	rain	30.17	60	56	4	same	30.05	57	57	—	rain
Means	30.030	58.5	521.	6.4		30.015	61.7	53.1	8.6		30.019	51.3	49.6	1.7	

1820 September.	Temperature. Max.	Min.	Sun.	Rad.	Wind. Direction.	Force.	Rain.
1	69	48		44	N E	little	
2	66	50		42	N	ditto	
3	67	43		37	——	ditto	
4	70	46		40	N E	ditto	
5	65	46		36	——	ditto	
6	65	47		41	E	variable	
⊕ 7	65	54		44	S E	little	
8	67	47		41	E	ditto	
9	72	48		39	W	ditto	
10	72	50		42	——	ditto	
11	+76	56		50	S E	ditto	
12	76	53		44	——	ditto	
13	69	51		45	——	ditto	
14	74	57		50	S W	ditto	
☽ 15	70	46		37	——	brisk	0.07
16	65	50		41	NW	ditto	
17	67	50		47	W	ditto	0.75
18	64	39		−26	NW	ditto	0.81
19	58	37		28	N	ditto	
20	58	42		33	W	ditto	0.36
21	59	43		34	——	ditto	0.11
○ 22	60	39		34	N W	ditto	
23	69	57		53	S W	ditto	
24	66	45		41	——	ditto	0.10
25	58	37		32	N W	ditto	
26	54	−36		32	——	little	
27	54	43		36	S W	ditto	0.08
28	62	52		50	——	ditto	
☾ 29	60	39		33	N E	ditto	0.03
30	63	48		45	S E	ditto	0.01
Means	65.3	46.6		39.9			2.32

SEPTEMBER.

Mean Temperature . . . . 55.9°
—— Pressure . . . . . . 30.021 inches
—— Dew-point . . . . . 49.8°
—— Force of vapour . . 0.397 inch.
—— Degree of dryness . . 6.1°
—— Degree of moisture . 810°
Least observed degree of moisture . . 520

Weight of vapour in a cubic foot.
Mean . . . 4.481 grs.
Maximum . 7.346 „
Minimum . 2.948 „

WINDS.

N. 3 = 47° N.E 4 = 50° E. 2 = 52° S.E. 5 = 57° S. 0 = S.W. 6 = 55°
W. 5 = 51° N.W. 5 = 45°.

Amount of rain, &c. . . . . . . 2.32 inches
—— of evaporation . . . 2.79 inches

REMARKS.

The weather, during the first sixteen days, was remarkably fine, warm, and dry, and the sky mostly clear. Throughout the remainder of the month, variable, with light falls of rain at intervals.

1820 October.	Morning. Barometer.	Hygrometer.			Weather.	Afternoon. Barometer.	Hygrometer.			Weather.	Night. Barometer.	Hygrometer.			Weather.
1	30.19	54	46	8	very fine	30.26	56	44	12	very fine	30.35	45	44	1	very fine
2	30.43	52	43	9	same	30.48	55	45	10	same	30.51	44	42	2	same
3	30.55	51	46	5	same	+30.61	51	41	10	same	30.61	43	41	2	same
4	30.57	49	46	3	same	30.46	54	46	8	same	30.43	51	51	—	rain
5	30.40	53	51	2	very fine	30.25	55	41	+14	same	30.30	48	45	3	very fine
6	30.26	55	51	4	same	30.25	56	50	6	fine	30.26	47	47	—	same
⊕ 7	30.22	57	51	6	fine	30.17	57	+51	6	very fine	30.17	48	48	—	dull
8	30.21	55	50	5	very fine	30.22	52	48	4	same	30.23	40	40	—	mist
9	30.27	51	45	6	fine	30.21	51	44	7	fine	30.21	48	43	5	dull
10	30.19	51	46	5	dull	30.07	52	46	6	dull	30.05	49	46	3	same
11	30.01	50	45	5	same	29.97	50	44	6	same	30.02	48	45	3	same
12	30.08	45	39	6	very fine	30.06	47	−35	12	very fine	30.06	40	36	4	very fine
13	30.03	44	41	3	dull	30.06	50	47	3	fine	29.99	45	44	1	dull
14	29.78	49	40	9	fine	29.50	49	44	5	dull	29.29	52	52	—	rain
15	29.25	58	57	1	rain	29.12	57	52	5	clearing	29.22	49	46	3	fine
16	29.23	48	43	5	showers	29.20	47	41	6	very fine	29.13	48	43	5	dull
17	28.93	51	47	4	stormy	28.84	51	45	6	fine	28.93	45	42	3	very fine
18	28.96	48	42	6	fine	29.00	49	44	5	very fine	29.10	47	47	—	dull
19	29.30	50	42	8	very fine	29.28	47	42	5	same	29.04	48	48	—	rain
20	28.91	49	44	5	same	29.07	44	39	5	hail	29.22	43	43	—	fine
21	29.43	44	41	3	fine	29.5	46	45	1	ne	29.60	36	36	—	very fine
○ 22	29.02	46	45	1	rain	−28.7	49	48	1	rain	28.82	43	43	—	dull
23	29.11	50	43	7	dull	29.21	49	41	8	clearing	29.11	45	45	—	rain
24	28.89	51	45	6	same	28.78	44	42	2	rain	28.82	42	40	2	lunar bow
25	29.08	46	43	3	same	29.27	47	43	4	fine	29.42	44	42	2	fine
26	29.33	44	40	4	showers	29.05	49	46	3	showers	29.02	49	47	2	same
27	29.07	51	46	5	overcast	29.21	48	44	4	same	29.40	44	41	3	dull
28	29.60	45	42	3	fine	29.64	45	45	—	rain	29.74	37	36	1	fine
29	29.50	45	45	—	rain	29.21	46	46	—	same	29.36	44	43	1	very fine
30	29.61	42	42	—	very fine	29.64	50	41	9	very fine	29.68	40	40	—	same
31	29.52	45	44	1	fine	29.44	46	44	2	same	29.33	45	44	1	fine
Means	29.675	49.3	44.8	4.1		29.637	49.9	44.3	5.6		29.659	45	43.5	1 5	

1820 October.	Temperature.				Wind.		Rain.
	Max.	Min.	Sun.	Rad.	Direction.	Force.	
1	58	41		37	N W	little	
2	58	40		36	—	ditto	
3	58	34		30	N E	ditto	
4	57	47		43	E	ditto	0.02
5	58	46		42	—	brisk	
6	60	45		42	—	ditto	
⊕ 7	60	43		42	N E	little	
8	59	44		38	—	ditto	
9	56	44		43	—	ditto	
10	55	45		44	—	ditto	
11	52	42		37	N	ditto	
12	51	35		31	N	ditto	
13	53	42		40	N W	ditto	
14	53	47		46	S E	ditto	
☾ 15	+63	43		41	S	brisk	
16	53	45		41	S W	ditto	0.05
17	55	41		35	—	high	0.20
18	53	41		35	W	brisk	0.25
19	53	41		41	—	high	0.09
20	52	34		−27	N W	brisk	0.03
21	52	34		27	—	ditto	0.09
○ 22	51	42		39	S W	ditto	0.28
23	55	41		39	W	ditto	0.11
24	51	39		38	S W	ditto	0.23
25	52	37		35	W	ditto	0.19
26	51	45		43	S	ditto	0.07
27	54	37		35	S W	ditto	0.08
☽ 28	50	36		30	N W	ditto	
29	47	−32		29	S	ditto	0.10
30	52	35		33	S E	ditto	
31	49	42		41	—	ditto	0.02
Means	54.2	40.6		37.4			1.81

OCTOBER.

Mean Temperature . . . . 47.4°
—— Pressure . . . . 29.657 inches
—— Dew-point . . . . . 42.4°
—— Force of vapour . . 0.309 inch
—— Degree of dryness . . 5°
—— Degree of moisture . 837°

Least observed degree of moisture . 705

Weight of vapour in a cubic foot.
Mean . . . 3.458 grs.
Maximum . 4.657 „
Minimum . 2.737 „

WINDS.

N. 2=40° N.E. 5=45° E. 3=47° S.E. 3=43° S. 3=46° S.W. 5=43° W. 4=43° N.W 6=42°.

Amount of rain . . . . 1.81 inch
—— of evaporation 1.86 inch

REMARKS.

For the first thirteen days a north-easterly wind prevailed, accompanied with fine, mild, and dry weather. The Barometer was unusually high and very steady.

On the 14th the wind went round to the southward, and the Barometer rapidly fell, in the course of a few days, to 28.74.

The weather, during the remainder of the month, was cloudy and showery.

1820 November.	Morning. Barometer.	Hygrometer.			Weather.	Afternoon Barometer.	Hygrometer.			Weather.	Night. Barometer.	Hygrometer.			Weather.
1	−29.45	42	38	4	fine	29.55	44	38	6	fine	29.67	33	33	—	fine
2	29.78	38	38	—	very fine	29.78	42	42	—	very fine	30.82	37	37	—	very fine
3	29.87	33	33	—	same	29.89	38	37	1	same	29,91	33	33	—	same
4	29.89	37	37	—	fine	29.90	42	42	—	same	29.91	34	34	—	same
5	29.92	32	32	—	dull	29.78	42	42	—	rain	29.69	48	48	—	rain
⊕ 6	29.78	50	50	—	same	29.80	50	50	—	overcast	29.80	48	46	2	dull
7	29.78	52	52	—	same	29.80	54	+54	—	same	29.83	52	52	—	same
8	29.86	52	52	—	fog	29 87	54	50	4	same	29.89	49	47	2	same
9	29.95	48	42	6	fine	29.95	47	41	6	very fine	29.97	46	41	5	same
10	30.00	45	42	3	same	30.00	46	41	5	same	30.06	45	43	2	same
11	30.14	45	42	3	very fine	30.22	45	37	+ 8	very fine	30.24	40	34	6	very fine
12	30.11	41	38	3	dull	29.91	41	40	1	dull	29.77	41	41	—	dull
☾ 13	29.54	38	38	—	rain	29.58	39	39	—	rain	29.66	36	32	4	fine
14	29.79	37	36	1	fine	29.77	37	35	2	sleet	29.77	34	34	—	sleet
15	29.89	38	36	2	very fine	29.97	39	35	4	hail	29.99	35	35	—	very fine
16	29.91	37	36	1	same	29.83	38	35	3	fine	29.80	32	−29	3	same
17	29.66	35	35	—	sleet	29.60	43	43	—	sleet	29.66	43	43	—	rain
18	29.86	35	35	—	fog	29.90	38	38	—	fog	29.94	34	34	—	dense fog
19	29.94	37	37	—	same	29.94	44	44	—	same	30.00	42	42	—	very fine
○ 20	29.99	46	46	—	mist	29.94	49	49	—	rain	29.94	49	49	—	overcast
21	29.91	48	48	—	overcast	29.85	48	45	3	fine	29.83	46	43	3	very fine
22	29.80	46	46	—	rain	29.70	48	48	—	rain	29.70	49	49	—	overcast
23	29.55	47	46	1	same	29.55	47	47	—	same	29.68	40	40	—	very fine
24	29.72	37	37	—	fog	29.70	46	45	1	fog	29.67	44	44	—	dull
25	29.61	44	44	—	mist	29.61	45	45	—	same	29.65	49	47	2	rain
26	29.83	49	49	—	same	29.91	48	48	—	fine	29.92	43	43	—	fine
☽ 27	29.94	43	43	—	fine	29.98	38	38	—	very fine	30.01	36	36	—	very fine
28	30.11	39	38	1	dull	30.15	38	38	—	dull	30.22	38	35	3	dull
29	+30.27	40	36	4	same	30.27	41	36	5	same	30.27	42	40	2	same
30	30.23	42	36	6	fine	30.20	41	36	5	fine	30.20	35	33	2	very fine
Means	29.869	41.7	40.6	1.1		29.863	43.4	41.9	1.8		29.882	41.1	39.9	1.2	

1820 November.	Temperature. Max.	Min.	Sun.	Rad.	Wind. Direction.	Force.	Rain.
1	47	33		30	N W	little	0.03
2	46	31		30	W	ditto	
3	41	31		29	N W	variable	
4	48	30		25	E	ditto	
5	48	32		27	S W	little	0.09
⊕ 6	53	47		44	W	ditto	0.11
7	+57	52		51	S W	ditto	
8	57	46		45	S E	brisk	
9	50	42		40	E	ditto	
10	48	42		40	N	ditto	
11	48	38		32	N E	ditto	
12	42	37		33	W	little	0.12
☾ 13	39	33		30	N	ditto	0.17
14	38	33		29	N E	ditto	0.11
15	42	33	50	28	——	ditto	
16	41	−29	+62	−24	N	ditto	
17	43	33	43	30	S	ditto	0.18
18	39	32	42	30	E	calm	
19	46	37	46	34	S E	little	0.01
○ 20	50	46	50	46	S	ditto	
21	51	46	53	44	——	ditto	0.03
22	49	43	49	39	——	ditto	0.20
23	47	35	51	33	——	variable	0.19
24	45	37	45	37	——	ditto	0.05
25	49	44	49	39	S E	little	0.06
26	51	41	54	38	——	ditto	0.03
☽ 27	42	33	49	27	E	ditto	
28	42	35	42	35	——	ditto	
29	42	40	42	39	——	ditto	
30	42	34	43	31	N	ditto	0.02
Means	46.1	37.5	48.1	34.6			1.40

NOVEMBER.

Mean Temperature . . . . 41.8°
—— Pressure . . . . . . 29.538 inches
—— Dew-point . . . . . 39.7°
—— Force of vapour . . . 0.277 inch
—— Degree of dryness . . 2.1°
—— Degree of moisture . 920°

Weight of vapour in a cubic foot.
Mean . . 3.218 grs.
Maximum . 5.173 „
Minimum . 2.281 „

Least observed degree of moisture 752

WINDS.

N. 4=36° N.E. 3=36° E. 6=38° S.E. 4=45° S. 6=44° S.W. 2=46° W. 3=42° N.W. 2=34°.

Amount of rain . . . . 1.40 inch
—— of evaporation 0.72 inch

REMARKS.

The weather throughout the month dull and cloudy, but, upon the whole, mild for the time of year. Thick fogs occurred frequently, and, at intervals, a few sharp hoar frosts.

A little snow fell on the 14th, being the first of the season.

On the 18th the fog was so dense, that at 10 o'clock, A.M., the coachmen could not see their horses' heads, and, in many instances, they were obliged to be led.

1820 December.	Morning. Barometer.	Morning. Hygrometer.			Morning. Weather.	Afternoon. Barometer.	Afternoon. Hygrometer.			Afternoon. Weather.	Night. Barometer.	Night. Hygrometer.			Night. Weather.
1	30.18	39	39	—	dull	30.09	43	37	6	dull	30.05	40	39	1	dull
2	29.96	40	39	1	same	29.96	42	39	3	fine	30.11	36	36	—	very fine
3	30.13	41	41	—	same	30.03	46	45	1	rain	30.02	48	46	2	dull
4	30.02	51	46	5	fine	30.02	52	48	4	fine	30.00	51	48	3	same
⊕ 5	29.91	50	46	4	same	29.83	52	50	2	same	29.90	51	51	—	rain
6	30.08	46	46	—	rain	30.08	50	50	—	rain	30.08	50	49	1	dull
7	30.08	52	52	—	fine	30.08	53	+52	1	fine	30.13	50	50	—	same
8	30.18	51	51	—	same	30.20	53	52	1	same	30.22	48	48	—	same
9	30.20	49	46	3	same	30·13	48	46	2	very fine	30.13	46	46	—	fine
10	30.05	50	50	—	rain	29.94	51	51	—	rain	29.94	51	51	—	very fine
11	29.89	52	52	—	same	29.89	53	53	—	same	29.89	50	50	—	rain
☾ 12	29.59	51	51	—	same	29.63	53	53	—	same	29.64	50	50	—	same
13	29.56	50	50	—	same	29.73	52	52	—	same	29.80	50	50	—	fine
14	−29.50	34	33	1	very fine	29.70	34	32	2	very fine	29.80	34	32	2	very fine
15	29.90	36	33	3	same	29.89	34	32	2	same	29.88	31	22	9	fine
16	29.71	30	28	2	dull	29.62	38	38	—	snow	29.54	34	34	—	snow
17	29.67	35	35	—	dense fog	29.78	38	38	—	dense fog	29.91	42	42	—	rain
18	30.04	42	42	—	dull	30.12	46	46	—	dull	30.16	46	46	—	fine
○ 19	30.20	47	47	—	same	30.25	49	49	—	rain	30.32	42	42	—	rain
20	+30.32	42	42	—	fog	30.25	48	48	—	same	30.16	50	50	—	same
21	30.12	49	46	3	fine	30.14	49	49	—	fine	30.15	42	42	—	very fine
22	30.15	38	38	—	mist	30.10	45	45	—	same	30.04	43	43	—	rain
23	29.97	40	40	—	very fine	29.90	42	42	—	rain	29.90	41	41	—	same
24	29.87	34	34	—	dull	29.85	35	30	5	dull	29.83	34	29	5	dull
25	29.83	31	29	2	same	29.83	31	28	3	same	29.83	28	24	4	very fine
☽ 26	29.79	31	30	1	same	29.81	32	29	3	same	29.86	31	25	6	dull
27	29.89	31	27	4	same	29.89	32	27	5	same	29.90	32	30	2	same
28	29.90	32	26	6	very fine	29.90	30	23	7	very fine	29.97	29	23	6	same
29	29.97	27	−18	9	dull	29.92	28	19	+9	dull	29.92	28	19	9	same
30	29.95	27	20	7	same	29.94	28	21	7	same	29.95	28	19	9	same
31	29.93	26	22	4	fine	29.89	27	18	9	very fine	29.90	29	20	9	same
Means	29.952	40.4	38.6	1.7		29.948	42.3	40.0	2.3		29.965	40.8	38.6	2.1	

1820 December.	Temperature.				Wind.		Rain.
	Max.	Min.	Sun.	Rad.	Direction.	Force.	
1	43	39	44	38	N W	little	
2	43	35	45	32	—	ditto	
3	51	41	51	41	W	ditto	
4	54	48	55	45	—	ditto	
⊕ 5	54	46	+58	45	—	brisk	
6	52	45	52	45	—	ditto	0.15
7	+55	49	55	48	S W	ditto	
8	52	48	52	48	—	ditto	
9	51	46	51	45	W	ditto	
10	53	50	53	47	S W	ditto	0.13
11	54	49	54	49	—	ditto	
☾ 12	54	49	54	49	—	ditto	0.20
13	52	44	52	43	W	ditto	0.52
14	35	32	38	28	N E	little	0.13
15	35	29	35	24	E	brisk	
16	32	31	35	28	S E	ditto	0.09
17	40	34	42	32	—	ditto	0.14
18	47	42	47	40	—	ditto	0.05
○ 19	50	42	50	39	S	ditto	0.09
20	50	42	50	41	—	little	0.03
21	50	36	50	34	W	ditto	
22	42	39	44	36	—	ditto	
23	43	34	44	33	N	ditto	0.04
24	35	31	35	30	N E	brisk	0.02
25	32	29	32	25	—	ditto	
☽ 26	31	30	32	28	E	ditto	
27	31	30	33	27	—	ditto	
28	32	27	33	25	—	high	
29	28	26	28	26	—	ditto	
30	28	−24	29	22	—	ditto	
31	30	24	32	−19	N E	little	
Means	43.1	37.7	44	35.8			1.59

DECEMBER.

Mean Temperature . . . . 40.4°
—— Pressure . . . . . 29.955 inches
—— Dew-point . . . . . 38.8°
—— Force of vapour . . 0.269 inch
—— Degree of dryness . . 1.6°
—— Degree of moisture 943°
Least observed degree of moisture 718

Weight of vapour in a cubic foot.
Mean . . . 3.143 grs.
Maximum . 4.812 „
Minimum . 1.557 „

WINDS.

N. 1 = 41° N.E. 4 = 27° E. 6 = 24° S.E. 3 = 38° S. 2 = 46° S.W. 5 = 51° W. 8 = 46° N.W. 2 = 36°

Amount of rain . . . . 1.59 inch
—— of evaporation 0.49 inch

REMARKS.

The weather during the greater part of the month remarkably mild and open. On the night of the 24th the frost set in with some severity.

1821 January.	Morning. Barometer.	Hygrometer.			Weather.	Afternoon. Barometer.	Hygrometer.			Weather.	Night. Barometer.	Hygrometer.			Weather.
1	29.88	30	24	6	dull	29.85	30	20	10	overcast	29.81	30	23	7	overcast
2	29.72	26	20	6	fine	29.59	28	19	9	dull	29.50	26	20	6	dull
3	29.34	29	24	9	dull	29.27	32	−13	+19	same	29.28	31	15	16	same
⊕ 4	29.35	27	19	8	overcast	29.38	31	18	13	overcast	29.36	31	22	9	same
5	29.35	31	28	3	same	29.16	32	32	—	snow	29.08	34	34	—	thaw
6	29.14	35	35	—	dull & damp, fog	29.19	39	39	—	damp & dull, fog	29.22	36	36	—	fog
7	29.22	36	36	—	same	29.22	38	38	—	same	29.20	38	38	—	same
8	29.17	40	40	—	same	29.10	39	39	—	same	29.05	38	38	—	same
9	−28.98	38	38	—	same	29.98	41	41	—	same	29.03	42	42	—	same
10	29.11	40	40	—	same	29.11	44	44	—	rain	29.05	44	44	—	rain
☾ 11	29.09	41	41	—	rain	29.11	41	41	—	same	29.19	44	44	—	same
12	29.28	47	47	—	dull & damp	29.28	49	48	1	fine	29.32	48	47	1	fine
13	29.52	46	45	1	fine	29.40	47	47	—	same	29.18	46	46	—	rain
14	29.38	42	42	—	rain	29.76	42	42	—	overcast	30.01	37	36	1	fine
15	30.06	37	37	—	dull	29.84	40	40	—	rain	29.67	48	48	—	rain
16	29.78	47	46	1	fine	29.91	46	41	5	very fine	30.05	39	39	—	very fine
17	30.14	41	41	—	same	30.16	46	46	—	rain	30.19	46	46	—	rain
☉ 18	30.22	47	47	—	dull & damp	30.24	51	+50	1	fine	30.28	50	50	—	dull
19	30.28	48	48	—	rain	30.23	48	48	—	rain	30.23	45	45	—	same
20	30.33	48	46	2	fine	30.40	47	46	1	fine	30.49	41	41	—	dense fog
21	30.64	33	33	—	dense fog	30.64	43	41	2	same	30.64	38	38	—	very fine
22	30.64	40	39	1	fine	30.64	44	44	—	fog	30.66	45	45	—	rain
23	+30.77	38	38	—	same	30.77	39	39	—	very fine	30.77	34	34	—	fog
24	30.77	32	32	—	dense fog	30.66	33	33	—	dense fog	30.66	33	33	—	rain
25	30.62	36	36	—	same	30.62	43	43	—	same	30.60	41	41	—	dull
☽ 26	30.64	39	39	—	mist	30.56	40	35	5	dull	30.56	38	34	4	same
27	30.44	35	34	1	dull	30.33	37	37	—	rain	30.31	37	37	—	rain
28	30.23	34	34	—	same	30.22	36	36	—	mist	30.25	34	31	3	dull
29	30.24	32	32	—	fine	30.20	38	38	—	very fine	30.25	39	39	—	fine
30	30.31	47	47	—	same	30.31	45	45	—	mist	30.35	44	44	—	very fine
31	30.41	48	47	1	same	30.39	48	47	1	same	30.43	46	46	—	dull
Means	29.904	38.3	37.2	1.2		29.887	40.5	38.3	2.1		29.892	39.4	37.9	1.5	

1821 January.	Temperature.				Wind.		Rain.
	Max.	Min.	Sun.	Rad.	Direction.	Force.	
1	30	26	30	24	N E	little	
2	29	−24	30	−23	——	brisk	
3	32	26	32	23	——	high	
⊕ 4	31	28	32	28	——	little	
5	32	30	35	30	E	brisk	0.05
6	37	37	38	32	S E	little	0.26
7	37	35	40	33	N	ditto	0.11
8	38	36	42	34	S E	ditto	0.15
9	39	38	41	34	W	ditto	0.03
10	42	40	44	38	S W	ditto	0.13
☾ 11	47	40	47	39	S.	ditto	0.28
12	50	40	50	37	——	ditto	0.19
13	50	42	51	42	S W	brisk	0.09
14	43	37	45	31	N E	ditto	0.31
15	48	37	49	36	S E	ditto	0.13
16	48	36	48	32	W	ditto	0.09
17	48	40	48	40	S W	little	0.11
○ 18	+52	47	53	47	——	ditto	
19	47	41	49	40	——	ditto	0.08
20	50	33	53	32	——	ditto	
21	43	33	45	31	W	variable	
22	43	38	44	36	——	ditto	
23	41	31	57	28	N E	brisk	
24	36	30	37	30	E	little	
25	43	35	44	35	S W	ditto	
☽ 26	40	35	43	34	N E	ditto	
27	37	34	37	34	S E	ditto	0.03
28	34	32	37	28	——	ditto	0.05
29	47	32	51	30	S	ditto	
30	48	41	+60	40	S W	ditto	
31	51	45	54	44	——	ditto	
Means	41.7	35.4	44	33.7			2.09

JANUARY.

Mean Temperature . . . . 38.5°
—— Pressure . . . . . . 29.894 inches
—— Dew-point . . . . . 36.8°
—— Force of vapour . . 0.253 inch
—— Degree of dryness . . 1.7°
—— Degree of moisture . 944
Least degree of observed moisture 513

Weight of vapour in in a cubic foot.
Mean . . . 2.969 grs.
Maximum . 4.515 „
Minimum . 1.304 „

WINDS.

N. 1 = 37° N E. 7 = 27° E. 2 = 32° S.E. 5 = 37° S. 3 = 41° S.W 9 = 45° W. 4 = 40° N.W. 0.

Amount of rain . . . . 2.09 inches
—— of evaporation 0.46 inch

REMARKS.

The frost only lasted till the fifth, and ended with a snow storm. From this time the temperature rapidly advanced till the 18th, when it reached the unusual amount of 52°. From the 8th to the 18th, rain fell in variable quantities, and with few intermissions.

The height of the barometer on the 23d was extraordinary.

	1821 February.	Morning. Barometer.	Morning. Hygrometer.			Morning. Weather.	Afternoon. Barometer.	Afternoon. Hygrometer.			Afternoon. Weather.	Night. Barometer.	Night. Hygrometer.			Night. Weather.
	1	30.43	49	+49	—	overcast	30.36	48	47	1	very fine	30.27	46	45	1	fine
⊕	2	30.22	47	47	—	rain	30.33	45	40	5	same	30.28	39	36	3	very fine
	3	30.35	40	40	—	very fine	30.23	45	40	5	fine	30.14	44	41	3	dull
	4	29.97	45	45	—	rain	30.05	43	37	6	very fine	30.30	36	32	4	very fine
	5	30.63	34	29	5	very fine	30.77	35	27	8	same	30.78	32	25	7	same
	6	+30.82	35	32	3	fine	30.80	35	30	5	same	30.74	32	32	—	same
	7	30.71	31	31	—	very fine	30.67	39	26	13	same	30.70	35	30	5	same
	8	30.66	38	30	8	same	30.50	41	21	+20	same	30.44	34	33	1	same
	9	30.20	34	34	—	same	30.05	40	37	3	same	30.12	34	33	1	same
☾	10	30.31	41	41	—	rain	30.37	40	35	5	same	30.38	34	34	—	same
	11	30.35	35	34	1	fine	30.27	37	36	1	same	30.27	34	34	—	dull
	12	30.27	34	34	—	dull	30.31	40	38	2	same	30.34	35	35	—	same
	13	30.34	34	34	—	same	30.25	36	34	2	same	30.25	35	33	2	same
	14	30.25	33	29	4	same	30.25	34	29	5	dull	30.32	35	32	3	same
	15	30.38	33	33	—	same	30.42	35	26	9	same	30.51	32	28	4	very fine
	16	30.56	33	29	4	same	30.54	34	30	4	same	30.45	33	32	1	dull
○	17	30.38	31	31	—	dense fog	30.30	32	31	1	same	30.24	32	31	1	same
	18	30.17	33	33	—	dull	30.12	35	35	—	same	30.15	38	38	—	rain
	19	30.32	33	30	3	very fine	30.34	35	28	7	very fine	30.34	31	27	4	very fine
	20	30.27	30	25	5	same	30.11	38	38	—	rain	30.12	38	38	—	fog
	21	30.20	38	36	2	dull	30.21	42	33	9	very fine	30.22	39	35	4	dull
	22	30.33	35	32	3	same	30.33	35	33	2	same	30.33	31	31	—	very fine
	23	30.32	30	30	—	very fine	30.22	36	31	5	same	30.19	31	31	—	fog
	24	30.13	27	27	—	fog	30.06	36	35	1	fog	30.09	31	31	—	very fine
☽	25	30.06	38	33	5	fine	30.02	39	38	1	fine	30.02	39	38	1	overcast
	26	30.09	34	26	8	same	30.06	30	21	9	same	30.06	30	20	10	dull
	27	29.84	29	25	4	very fine	29.71	33	−20	13	very fine	29.60	28	22	6	very fine
	28	29.28	32	30	2	overcast	−29.17	31	31	—	snow	29.27	33	33	—	dull
	Means	30.281	35.3	33.2	2		30.243	37.4	32.3	5		30.247	34.6	32.5	2.1	

1821 February.	Temperature.				Wind.		Rain.
	Max.	Min.	Sun	Rad.	Direction.	Force.	
1	+51	45	54	42	S W	brisk	
⊕ 2	50	34	55	31	W	little	
3	45	40	58	37	——	ditto	
4	44	30	47	25	——	ditto	
5	36	29	60	24	N	ditto	
6	40	29	45	25	S W	ditto	
7	40	32	63	27	——	ditto	
8	42	30	68	25	S	calm	
9	44	33	+80	28	E	variable	
☾ 10	45	32	68	28	——	brisk	
11	40	31	54	27	——	ditto	
12	39	33	59	30	——	ditto	
13	38	32	43	27	N E	little	
14	34	32	35	30	——	variable	
15	36	28	36	23	——	little	
16	36	31	39	26	——	ditto	
○ 17	33	30	35	27	E	ditto	
18	38	31	44	25	N W	ditto	
19	39	27	71	21	N	ditto	
20	39	31	56	27	N W	ditto	
21	43	33	68	26	——	ditto	0.03
22	38	28	50	23	——	variable	
23	38	27	51	21	——	little	
24	38	28	51	21	N	ditto	
☽ 25	42	33	46	30	N W	ditto	
26	34	−25	37	−19	E	brisk	
27	34	27	60	22	S E	little	
28	32	31	38	29	——	ditto	0.10
Means	39.5	31.1	52.5	26.6			0.13

FEBRUARY.

Mean Temperature . . . . 35.3°
—— Pressure . . . . . . 30.257 inches
—— Dew-point . . . . . 31.7°.
—— Force of vapour . . 0.213 inch
—— Degree of dryness . . 3.6°
—— Degree of moisture . 876°
Least observed degree of moisture . . . 500

Weight of vapour in a cubic foot.
Mean . . . 2.520 grs.
Maximum . 4.407 „
Minimum . 1.685 „

WINDS.

N. 3 = 28° N.E. 4 = 30° E. 6 = 32° S.E. 2 = 26° S. 1 = 28° S.W. 3 = 36° W. 3 = 39° N.W. 6 = 33°.

Amount of rain . . . . 0.13 inch
—— of evaporation . . . . 0.84 inch

REMARKS.

The distinguishing feature of the month was the very uncommon deficiency of rain and snow. The atmosphere generally remarkably clear, but sometimes obscured by fogs.

On the 27th the zodiacal light was distinctly seen in the vicinity of London.

1821 March.	Morning. Barometer.	Hygrometer.			Weather.	Afternoon. Barometer.	Hygrometer.			Weather.	Night. Barometer.	Hygrometer.			Weather.
1	29.36	38	38	—	fog	29.41	42	42	—	rain	29.59	42	42	—	dull
2	29.82	42	42	—	very fine	29.81	45	45	—	same	29.77	46	46	—	same
3	29.73	50	50	—	rain	29.61	50	50	—	same	29.61	50	50	—	rain
⊕ 4	29.61	48	48	—	same	29.62	53	+51	2	very fine	29.66	48	48	—	dull
5	29.90	34	34	—	dull	29.95	32	32	—	dull	29.95	33	33	—	rain
6	29.81	35	35	—	same	29.58	46	46	—	rain	29.46	45	45	—	same
7	29.56	46	46	—	fine	29.55	48	45	3	very fine	29.31	48	48	—	same
8	29.23	49	46	3	same	29.23	50	50	—	rain	29.31	48	48	—	dull
9	29.63	50	48	2	very fine	29.56	51	51	—	same	29.53	52	52	—	same
10	29.62	53	52	1	hail & rain	29.62	54	52	2	showers	29.71	46	46	—	very fine
☾ 11	29.80	51	50	1	fine	29.83	50	43	7	very fine	29.90	45	45	—	same
12	29.95	51	50	1	showers	29.95	50	46	4	same	30.00	44	44	—	same
13	30.08	48	46	2	fine	29.03	50	47	3	same	30.03	45	45	—	mist
14	30.15	46	42	4	very fine	30.19	47	31	+16	same	30.31	39	39	—	very fine
15	30.34	40	33	7	same	30.34	44	31	13	same	+30.35	40	31	9	same
16	30.28	39	39	—	mist	30.22	46	31	15	same	30.18	38	32	6	same
17	30.04	35	35	—	fog	29.81	46	41	5	same	29.64	43	43	—	dull
18	29.49	46	37	9	very fine	29.43	45	37	8	showers	29.33	41	35	6	same
○ 19	29.19	42	33	9	same	29.19	41	36	5	very fine	29.21	40	40	—	rain
20	29.27	45	37	8	same	29.30	45	35	10	same	29.30	40	33	7	very fine
21	29.23	43	36	7	same	29.35	42	—30	12	same	29.48	39	36	3	same
22	29.66	41	37	4	dull	29.82	42	32	10	hail	29.95	34	32	2	same
23	30.08	41	30	11	very fine	30.07	43	31	12	very fine	30.05	38	36	2	same
24	29.77	48	40	8	dull	29.60	47	40	7	dull	29.42	42	42	—	rain
25	29.36	46	46	—	rain	29.41	47	44	3	rain	29.56	39	35	4	very fine
☽ 26	29.62	45	41	4	very fine	29.53	47	41	6	showers	29.40	46	36	10	dull
27	29.26	46	36	10	fine	29.35	46	32	14	very fine	29.37	38	35	3	very fine
28	29.09	43	43	—	rain	—29.04	47	46	1	clearing	29.06	46	46	—	rain
29	29.06	47	47	—	same	29.26	47	47	—	rain	29.37	40	40	—	very fine
30	29.67	45	42	3	very fine	29.68	46	31	15	very fine	29.63	39	39	—	same
31	29.38	44	43	1	rain	29.39	45	42	3	showers	29.47	40	38	2	same
Means	29.646	44.4	41.3	3.		29.636	46.2	40.6	5.6		29.642	42.3	40.6	1.7	

1821 March.	Temperature.				Wind.		Rain.
	Max.	Min.	Sun.	Rad.	Direction.	Force.	
1	43	35	46	27	S	little	
2	50	42	50	42	S W	ditto	
3	51	46	52	42	——	ditto	
⊕ 4	+56	34	62	34	——	ditto	0.53
5	35	31	37	30	N E	ditto	
6	46	35	46	32	S E	ditto	
7	52	45	70	41	W	ditto	0.29
8	54	40	63	35	——	brisk	0.28
9	54	48	58	44	S W	ditto	0.05
10	51	41	72	35	——	ditto	0.10
☾ 11	55	38	78	33	——	variable	
12	53	40	81	35	W	little	
13	54	41	75	35	S W	ditto	
14	48	32	+97	25	N E	ditto	0.05
15	47	35	80	26	N	ditto	
16	51	31	81	23	——	variable	
17	49	35	81	30	S W	little	
18	49	37	63	30	N W	high	
○ 19	46	34	67	27	——	ditto	0.20
20	48	35	71	27	——	brisk	
21	45	35	74	27	——	ditto	
22	46	−30	81	−22	N	ditto	0.09
23	47	35	83	28	——	ditto	
24	50	40	54	36	S	high	0.24
25	52	34	60	26	S W	brisk	0.13
☽ 26	51	41	61	33	——	ditto	
27	49	37	56	29	——	ditto	
28	48	43	50	35	S	ditto	0.28
29	49	35	53	27	S W	ditto	0.17
30	50	37	77	31	W	ditto	
31	50	33	69	26	——		0.27
Means	49.3	37.2	66.1	31.3			2.68

MARCH.

Mean Temperature . . . . 43.2°
—— Pressure . . . . . . 29.641 inches
—— Dew-point . . . . . 38.9°
—— Force of vapour . . 0.271 inch
—— Degree of dryness . . 4.3°
—— Degree of moisture 819°

Least observed degree of moisture 571

Weight of vapour in a cubic foot.

Mean . . . 3.127 grs.
Maximum . 4.656 „
Minimum . 2.303 „

WINDS.

N. 4 = 32 N.E. 2 = 35° E. = S.E. 1 = 38° S. 3 = 42° S.W. 12 = 44° W. 5 = 43° N.W. 4 = 35°.

Amount of rain, &c. . . . . 2.68 inches
—— of evaporation . . 1.49 inch

REMARKS.

This period was rainy, and cold to the feelings, although not much below the mean in temperature.

1821 April.	Morning. Barometer.	Hygrometer.			Weather.	Afternoon. Barometer.	Hygrometer.			Weather.	Night. Barometer.	Hygrometer.			Weather.
1	29.63	45	35	10	very fine	29.63	48	41	7	very fine	29.44	44	44	—	rain
⊕ 2	29.31	53	47	6	clearing	−29.20	54	50	4	overcast	29.30	43	40	3	showers
3	29.30	49	38	11	fine	29.30	49	36	13	hail	29.33	43	38	5	very fine
4	29.32	47	40	7	very fine	29.32	50	36	14	very fine	29.36	43	38	5	same
5	29.53	47	34	13	same	29.65	50	35	15	same	29.86	43	38	5	same
6	30.08	46	−31	15	same	30.08	48	34	14	dull	30.06	43	43	—	rain
7	30.03	51	49	2	dull	30.08	58	52	6	same	30.12	52	50	2	very fine
8	+30.15	61	52	9	very fine	30.13	61	54	7	very fine	30.13	53	50	3	same
☾ 9	30.05	54	51	3	same	29.91	58	46	12	same	29.89	52	49	3	same
10	29.83	61	52	9	same	29.80	53	47	6	same	29.72	53	47	6	same
11	29.65	56	48	8	same	29.54	56	43	13	same	29.47	48	46	2	rain
12	29.35	55	41	14	same	29.31	51	47	4	same	29.38	45	40	5	fine
13	29.38	52	42	10	same	29.42	50	48	2	same	29.61	44	40	4	very fine
14	29.61	50	40	10	overcast	29.54	50	48	2	showers	29.45	41	38	3	same
15	29.52	47	44	3	showers	29.52	46	37	9	same	29.52	39	38	1	same
16	29.47	50	40	10	very fine	29.42	52	36	16	very fine	29.44	41	40	1	same
○ 17	29.46	53	42	11	same	29.50	54	36	+18	same	29.55	40	36	4	same
18	29.69	55	45	10	same	29.73	53	42	11	thund.storm	29.78	45	44	1	same
19	29.69	52	50	2	rain	29.54	54	54	—	rain	29.53	51	51	—	rain
20	29.58	54	52	2	dull	29.58	58	56	2	same	29.63	52	52	—	thund.storm
21	29.79	56	56	—	rain	29.88	57	55	2	overcast	29.97	51	51	—	dull
22	29.99	52	49	3	very fine	29.86	53	47	6	very fine	29.75	48	48	—	same
23	29.56	60	55	5	same	29.48	65	58	7	fine	29.37	58	58	—	rain
☽ 24	29.48	65	52	13	same	29.53	65	50	15	very fine	29.63	55	51	4	thunder
25	29.68	68	57	11	same	29.71	70	57	13	same	29.78	59	57	2	very fine
26	29.78	69	+58	11	same	29.65	69	52	17	same	29.65	56	54	2	lightning
27	29.71	64	56	8	thund.storm	29.74	64	52	12	same	29.78	55	52	3	very fine
28	29.84	65	52	13	very fine	29.79	65	52	13	same	29.81	55	52	3	same
29	29.86	58	53	5	same	29.86	58	55	3	same	29.92	50	50	—	dull
30	30.04	51	48	3	overcast	30.04	51	48	3	overcast	30.06	47	43	4	same
Means	29.678	54.8	46.9	7.9		29.658	55.6	46.8	8.8		29.676	48.3	45.9	2.3	

1821 April.	Temperature. Max.	Min.	Sun.	Rad.	Wind. Direction.	Force.	Rain.
1	53	44	82	36	W	brisk	0.11
⊕ 2	59	39	75	30	——	ditto	0.21
3	53	37	66	29	——	ditto	
4	54	37	88	28	N W	ditto	
5	51	−34	98	25	N	ditto	
6	51	43	66	33	N W	ditto	0.19
7	61	49	72	39	W	little	
8	63	47	102	38	——	ditto	
☾ 9	63	49	+110	41	——	ditto	
10	63	49	104	40	——	ditto	
11	60	42	91	32	S W	brisk	0.24
12	56	40	69	30	W	high	0.10
13	55	37	69	27	——	ditto	
14	52	37	54	27	S	ditto	0.13
15	52	35	74	−24	W	brisk	0.18
16	54	36	91	25	——	variable	
○ 17	56	40	94	29	N W	ditto	
18	57	42	92	31	——	brisk	
19	55	50	60	40	S	ditto	0.04
20	60	48	65	37	——	ditto	0.19
21	58	42	80	39	N	ditto	
22	61	47	97	37	E	ditto	
23	68	53	99	43	S E	little	0.09
☽ 24	68	50	103	40	S W	ditto	
25	71	54	104	40	S E	ditto	
26	+74	55	104	48	——	ditto	0.07
27	67	50	97	44	W	brisk	
28	68	49	110	43	N	ditto	
29	61	48	105	46	N E	ditto	
30	53	45	54	44	——	ditto	0.02
Means	59.2	44.2	85.8	35.5			1.57

APRIL.

Mean Temperature . . . . 51.7°
—— Pressure . . . . . . 29.670 inches
—— Dew-point . . . . . 45.5°
—— Force of vapour . . 0.346 inch
—— Degree of dryness . . 6.2°
—— Degree of moisture . 817°
Least observed degree of moisture . . 539

Weight of vapour in a cubic foot.
Mean . . . 3.910 grs.
Maximum . 5.743 ,,
Minimum . 2.376 ,,

WINDS.

N. 3 = 47° N.E. 2 = 48° E 1 = 48° S.E. 3 = 56° S. 3 = 49° S.W. 2 = 48° W. 12 = 45° N.W. 4 = 39°

Amount of rain, &c. . . . . . . 1.57 inch
—— of evaporation . . . 2.31 inches

REMARKS.

The weather during this month was warm, and marked by its usual character of showery.

On the 20th the thunder-storm was accompanied by very heavy rain; the lightning was very vivid and forked, and some flashes descended perpendicularly to the earth. Lightning was very common during the month.

1821 May	Morning.					Afternoon.					Night.				
	Barometer.	Hygrometer.			Weather.	Barometer.	Hygrometer.			Weather.	Barometer.	Hygrometer.			Weather.
⊕ 1	30.03	49	43	6	overcast	29.97	51	48	3	fine	29.92	49	48	1	very fine
2	29.89	57	51	6	same	29.83	59	52	7	very fine	29.83	54	50	4	dull
3	29.76	60	51	9	fine	29.68	65	48	17	same	29.66	56	52	4	same
4	29.71	67	54	13	very fine	29.70	67	54	13	same	29.70	57	50	7	same
5	29.63	64	50	14	same	29.53	63	46	17	same	29.45	54	50	4	overcast
6	29.49	58	46	12	showers	29.55	58	41	17	same	29.68	48	47	1	showers
7	29.86	61	44	17	very fine	29.86	60	53	7	same	29.88	53	53	—	rain
☾ 8	29.94	56	56	—	rain	29.97	62	+56	6	fine	30.06	49	43	6	very fine
9	30.20	60	44	16	very fine	30.22	61	42	19	very fine	30.25	47	43	4	same
10	+30.28	62	44	18	same	30.24	61	45	16	same	30.18	48	43	5	fine
11	30.04	58	54	4	fine	29.97	66	56	10	overcast	29.91	57	56	1	overcast
12	29.89	66	46	20	very fine	29.79	60	40	20	very fine	29.63	46	40	6	very fine
13	29.33	59	43	16	ditto, hail	29.32	54	40	14	ditto, hail	29.32	44	40	4	same
14	29.38	55	40	15	very fine	29.38	53	42	11	very fine	29.35	46	43	3	showers
15	−29.18	51	46	5	showers	29.23	49	47	2	thund.storm	29.49	44	43	1	very fine
16	29.73	60	41	19	hail	29.83	56	35	21	very fine	29.91	44	41	3	same
○ 17	30.00	56	50	6	showers	29.90	52	52	—	rain	29.86	54	54	—	rain
18	29.98	56	46	10	very fine	30.08	56	41	15	very fine	30.15	47	41	6	very fine
19	30.27	59	40	19	same	30.22	56	41	15	same	30.19	54	44	10	dull
20	30.24	48	45	3	rain	30.21	54	36	18	same	30.22	46	43	3	very fine
21	30.18	53	36	17	very fine	30.08	52	39	13	dull	30.08	43	38	5	fine
22	30.03	52	36	16	fine	29.95	51	38	13	very fine	29.92	41	38	3	very fine
23	29.79	48	37	11	dull	29.79	48	43	5	dull	29.89	39	36	3	same
☽ 24	30.00	50	34	16	very fine	30.02	49	34	15	hail	30.02	39	39	—	same
25	29.94	58	37	21	same	29.77	58	42	16	very fine	29.76	41	41	—	rain
26	29.80	51	−28	+23	same	29.83	42	36	6	same	29.88	37	36	1	snow show.
27	29.94	55	38	17	same	29.94	50	36	14	same	29.93	43	41	2	rain
28	29.97	54	39	15	same	30.00	54	46	8	hail	30.01	44	43	1	fine
29	30.11	57	43	14	same	30.14	57	34	23	very fine	30.18	47	45	2	same
30	30.20	56	42	14	same	30.20	56	45	11	same	30.19	46	43	3	very fine
⊕ 31	30.19	56	49	7	fine	30.12	56	46	10	same	30.11	47	46	1	same
Means	29.902	56.5	43.6	12.8		29.881	56	43.6	12.3		29.891	47 2	44.1	3	

1821 May.	Temperature.				Wind.		Rain.
	Max.	Min.	Sun.	Rad.	Direction.	Force.	
⊕ 1	57	44	90	40	N E	brisk	
2	64	53	+114	59	E	ditto	
3	68	51	96	50	W	variable	
4	+72	52	94	50	S W	little	
5	70	48	96	47	S E	brisk	
6	61	44	86	42	W	ditto	0.15
7	63	53	70	53	S W	ditto	0.09
☽ 8	62	42	75	36	——	ditto	0.26
9	64	46	98	40	N W	little	
10	65	48	98	46	——	ditto	
11	65	46	82	40	W	ditto	
12	69	44	95	36	N	brisk	
13	60	40	83	34	——	ditto	0.05
14	60	44	96	40	——	ditto	0.03
15	60	44	65	38	W	ditto	0.15
16	60	39	79	33	——	ditto	
○ 17	57	48	76	47	S	ditto	0.29
18	61	42	102	36	N W	ditto	0.13
19	61	42	114	39	——	ditto	0.02
20	55	38	77	32	E	ditto	
21	56	36	85	−30	——	ditto	0.08
22	54	41	78	36	——	ditto	
23	51	36	54	29	N E	high	0.03
☽ 24	58	38	97	32	N	brisk	0.03
25	61	37	87	37	W	ditto	0.07
26	55	−35	97	32	N	ditto	0.02
27	57	42	82	42	N W	little	0.10
28	58	42	85	38	——	ditto	0.09
29	61	41	101	38	——	ditto	
30	59	41	91	36	E	brisk	
⊕ 31	65	46	103	40	S E	ditto	
Means	60.9	43.3	88.5	39.6			1.59

MAY.

Mean Temperature	52.1°	Weight of vapour in a cubic foot.	
—— Pressure	29.891 inches	Mean . . .	3.631 grs.
—— Dew-point	43.7°	Maximum .	5.511 ,,
—— Force of vapour . .	0.325 inch	Minimum .	2.128 ,,
—— Degree of dryness . .	8.4°		
—— Degree of moisture .	757°		

Least observed degree of moisture . . 454

WINDS.

N. 5 = 38° N.E. 2 = 42° E. 5 = 42° S.E. 2 = 47° S. 1 = 52° S.W. 3 = 51° W. 6 = 45° N.W. 7 = 41°

Amount of rain, &c. . . . . . . 1.59 inches
—— of evaporation . . . 3.22 inch

REMARKS.

The general character of the period may be considered as of unusual asperity for the advanced stage of the season, and particularly remarkable for squally and stormy weather, and chilly showers. There was a snow shower on the 26th, and frequent hail.

1821 June.	Morning. Barometer.	Hygrometer.			Weather.	Afternoon. Barometer.	Hygrometer.			Weather.	Night. Barometer.	Hygrometer.			Weather.
1	30.08	65	56	9	very fine	30.08	65	47	18	very fine	30.08	54	53	1	fine
2	30.04	70	48	22	same	29.96	67	51	16	same	29.94	55	48	7	very fine
3	29.90	65	56	9	fine	29.82	64	47	17	same	29.79	54	53	1	fine
4	29.70	57	54	3	rain	29.64	63	54	9	fine	29.66	56	56	—	mist
5	29.77	67	55	12	very fine	29.81	56	56	—	rain	29.81	56	56	—	rain
6	29.91	67	56	11	same	29.94	67	57	10	very fine	29.88	53	53	—	showers
☾ 7	29.80	64	52	12	same	29.67	63	58	5	same	−29.61	53	53	—	rain
8	29.65	54	52	2	overcast	29.77	53	44	9	dull	29.83	47	43	4	dull
9	29.85	53	42	11	hail	29.84	53	44	9	same	29.80	47	46	1	same
10	29.74	53	48	5	fine	29.78	52	48	4	fine	29.82	47	47	—	showers
11	29.85	52	50	2	showers	29.92	48	48	—	rain	30.02	48	48	—	rain
12	30.16	55	48	7	very fine	30.24	52	47	5	fine	30.25	49	49	—	showers
13	30.16	55	48	7	fine	30.19	51	47	4	very fine	30.19	47	43	4	very fine
14	30.26	55	44	11	very fine	30.25	60	44	16	same	+30.29	49	47	2	same
○ 15	30.30	52	48	4	fine	30.23	64	50	14	same	30.23	52	48	4	same
16	30.20	55	53	2	rain	30.20	56	53	3	overcast	30.22	51	51	—	overcast
17	30.25	53	52	1	dull	30.24	56	53	3	same	30.26	51	51	—	same
18	30.29	51	50	1	fine	30.24	61	52	9	very fine	30.26	48	48	—	very fine
19	30.22	55	50	5	very fine	30.12	60	−35	+25	same	30.12	47	43	4	same
20	30.08	54	50	4	dull	30.04	56	44	12	same	30.03	43	43	—	same
21	30.03	54	44	10	very fine	30.05	56	44	12	same	30.07	42	42	—	same
☽ 22	30.14	55	44	11	same	30.14	55	50	5	dull	30.12	50	50	—	dull
23	30.14	54	49	5	overcast	30.13	55	50	5	same	30.13	49	49	—	very fine
24	30.10	52	47	5	same	30.12	55	50	5	same	30.14	49	49	—	fine
25	30.12	54	50	4	same	30.11	63	46	17	very fine	30.13	52	46	6	dull
26	30.12	57	50	7	overcast	30.06	64	53	11	same	30.08	52	46	6	fine
27	30.09	59	47	12	very fine	30.06	63	46	17	same	30.09	51	48	3	overcast
28	30.15	58	49	9	fine	30.11	63	50	13	same	30.13	54	53	1	very fine
⊕ 29	30.14	64	50	14	very fine	30.06	71	52	19	same	30.03	59	52	7	fine
30	30.00	73	53	20	same	29.84	71	60	11	overcast	29.70	61	+61	—	rain
Means	30.041	57.7	49.8	7.9		30.022	59.4	49.3	10.1		30.023	50.8	49.1	1.7	

1821 June.	Temperature. Max.	Min.	Sun.	Rad.	Wind. Direction.	Force.	Rain.
1	70	46	118	34	E	little	
2	71	50	123	47	N E	brisk	
3	67	51	111	47	——	ditto	
4	68	52	85	46	N W	little	0.21
5	71	51	110	51	W	ditto	0.30
6	69	51	98	50	N W	ditto	0.24
☾ 7	69	52	92	52	S W	brisk	0.41
8	60	41	63	39	N	ditto	0.25
9	69	40	90	40	——	ditto	
10	56	38	90	36	N W	ditto	0.24
11	58	44	97	43	N E	ditto	0.29
12	56	50	91	50	N	ditto	
13	62	42	92	40	——	ditto	
14	61	43	99	45	N E	ditto	
○ 15	66	45	116	46	——	little	
16	65	50	100	50	——	ditto	0.05
17	58	47	86	47	——	ditto	
18	65	45	113	44	——	ditto	
19	64	45	125	43	——	brisk	0.02
20	65	37	108	−31	N	ditto	
21	58	37	92	31	N E	little	
☽ 22	59	48	89	48	N	brisk	
23	56	38	73	33	——	ditto	
24	64	43	80	40	——	ditto	
25	66	49	104	49	N E	ditto	
26	66	43	109	50	——	ditto	
27	65	50	130	47	——	ditto	
28	65	49	+125	45	——	ditto	
⊕ 29	73	53	120	49	——	variable	
30	+78	61	119	59	S W	brisk	0.26
Means	64.6	46.3	101.6	44.4			2.27

JUNE.

Mean Temperature	55.4°	Weight of vapour in a cubic foot.	
—— Pressure	30.028 inches	Mean . . .	4.214 grs.
—— Dew-point	47.8°	Maximum .	6.399 „
—— Force of vapour . . .	0.373 inch	Minimum .	2.666 „
—— Degree of dryness . .	7.6°		
—— Degree of moisture .	772°		

Least observed degree of moisture . . 428

WINDS.

N. 6 = 47° N.E. 15 = 48° E. 1 = 52° S.E. 0 = S. 0 = S.W. 2 = 56°
W. 1 = 55° N.W. 3 = 52°

Amount of rain . . . . 2.27 inches
—— of evaporation . . 3.30 inches

REMARKS.

The weather very cold and unseasonable, with blighting north-east winds.

1821 July.	Morning. Barometer.	Hygrometer.			Weather.	Afternoon. Barometer.	Hygrometer.			Weather.	Night. Barometer.	Hygrometer.			Weather.
1	29.59	66	61	5	overcast	29.61	57	57	—	thund.&hail	29.71	51	51	—	rain
2	29.81	51	51	—	rain	29.80	51	51	—	rain	29.77	51	51	—	same
3	29.82	53	50	3	overcast	29.88	58	46	12	overcast	29.91	48	−41	7	very fine
4	30.08	58	43	15	very fine	30.08	60	43	17	very fine	30.11	54	51	3	same
5	30.14	68	46	12	same	30.12	64	49	15	fine	30.09	57	50	7	dull
6	30.00	63	54	9	same	29.87	61	53	8	showers	29.78	53	52	1	showers
☾ 7	29.79	59	52	7	dull	29.89	54	53	1	same	29.95	50	47	3	clearing
8	30.02	59	49	10	same	30.05	60	51	9	very fine	30.08	52	51	1	dull
9	30.09	70	53	17	very fine	30.08	64	47	17	same	30.07	56	50	6	same
10	30.10	68	52	16	same	30.10	69	49	+20	same	30.11	56	52	4	same
11	30.13	63	55	8	same	30.11	68	50	18	overcast	30.08	54	47	7	same
12	30.08	61	45	16	same	30.03	63	46	17	very fine	30.01	52	46	6	very fine
13	29.97	62	51	11	hazy	29.88	65	50	15	hazy	29.85	56	49	7	dull
14	29.82	63	56	7	very fine	29.77	67	55	12	very fine	29.74	56	51	5	very fine
○ 15	29.68	61	61	—	rain	29.70	60	58	2	rain	29.78	53	53	—	same
16	30.00	66	58	8	very fine	30.03	69	56	13	very fine	30.11	57	55	2	same
17	30.22	67	51	16	fine	30.26	71	51	20	same	+30.28	56	56	—	same
18	30.27	69	49	20	very fine	30.22	69	56	13	same	30.20	59	56	3	same
19	30.01	69	50	19	same	29.90	70	50	20	same	29.83	59	56	3	lightning
20	29.88	70	57	13	overcast	29.81	69	55	14	same	29.79	59	57	2	very fine
21	29.78	71	62	9	showers	29.75	70	62	8	showers	29.75	59	58	1	rain
☽ 22	29.55	70	62	8	same	29.52	68	57	11	very fine	29.56	59	59	—	very fine
23	29.61	68	56	12	very fine	29.64	62	56	6	hail	29.68	56	55	1	dull
24	29.75	66	61	5	showers	29.71	66	54	12	very fine	−29.51	59	59	—	rain
25	29.64	68	60	8	very fine	29.83	67	62	5	showers	29.87	57	57	—	very fine
26	29.91	67	60	7	same	29.91	69	60	9	same	29.94	56	56	—	same
27	29.94	71	53	18	same	29.97	70	52	18	very fine	29.99	56	52	4	same
28	29.94	63	52	11	fine	29.97	64	54	10	same	29.93	54	54	—	showers
⊕ 29	29.91	65	60	5	same	29.90	69	61	8	fine	29.90	58	58	—	very fine
30	29.90	64	64	—	rain	29.88	65	+65	—	rain	29.84	62	62	—	rain
31	29.94	69	64	5	dull	29.92	73	64	9	very fine	29.96	66	64	2	dull
Means	29.913	64.7	54.7	9.6		29.909	64.9	53.9	10.9		29.909	55.8	53.4	2.4	

1821 July.	Temperature. Max.	Min.	Sun.	Rad.	Wind. Direction.	Force.	Rain.
1	75	48	95	50	W	variable	0.39
2	54	48	56	48	E	high	0.25
3	58	−42	65	−41	N E	brisk	
4	68	51	99	49	N	ditto	
5	67	53	104	51	N W	ditto	
6	69	52	70	50	——	ditto	
☾ 7	61	45	70	44	N	ditto	
8	70	48	76	47	——	ditto	0.35
9	71	49	116	52	N W	ditto	
10	72	51	113	48	——	little	
11	71	47	81	45	——	ditto	
12	55	48	120	45	E	ditto	
13	67	49	90	49	S	ditto	
14	71	53	101	50	W	ditto	
○ 15	63	51	67	49	——	ditto	0.26
16	71	52	107	50	N W	ditto	
17	72	52	93	50	——	variable	0.09
18	73	53	100	50	S E	brisk	0.08
19	74	54	+128	52	——	ditto	0.10
20	+76	58	102	55	W	ditto	0.19
21	73	57	113	54	——	ditto	0.16
☽ 22	71	54	105	52	S W	ditto	0.12
23	69	55	91	53	W	ditto	
24	69	59	88	56	S W	high	
25	73	55	101	52	——	ditto	0.05
26	70	55	104	53	S W	brisk	
27	72	54	120	50	W	ditto	
28	72	54	110	50	S E	ditto	
⊕ 29	72	51	101	48	S W	ditto	
30	69	57	69	57	S	little	0.19
31	74	61	95	60	S W	ditto	0.03
Means	69.4	52.1	95.1	50.3			2.26

JULY.

		Weight of vapour in a cubic foot.	
Mean Temperature	60.7°		
—— Pressure	29.910 inches		
—— Dew-point	53°	Mean . . .	4.929 grs.
—— Force of vapour . .	0.444 inch	Maximum .	7.230 ,,
—— Degree of dryness . .	7.7°	Minimum .	3.323 ,,
—— Degree of moisture .	777°		

Least observed degree of moisture . 520

WINDS.

N. 3=48° N.E. 1=45° E. 2=48° S.E. 3=53° S. 2=56° S.W. 6=59° W. 7=56° N.W. 7=51°.

Amount of rain . . . . 2.26 inches
——— of evaporation 3.93 inches

REMARKS.

The weather uncomfortably cold and chilly.

1821 August.	Morning.				Afternoon.					Night.				
	Barometer.	Hygrometer.			Weather.	Barometer.	Hygrometer.			Weather.	Barometer.	Hygrometer.			Weather.
1	29.94	65	62	3	overcast	29.90	72	64	8	overcast	29.95	62	62	—	dull
2	30.07	72	55	17	very fine	30.07	74	54	+20	very fine	30.09	60	60	—	showers
3	30.09	67	57	10	same	30.06	71	54	17	same	30.08	61	61	—	very fine
4	30.09	73	60	13	same	30.01	74	62	12	same	30.00	62	62	—	same
5	29.94	72	64	8	same	29.92	75	62	13	same	29.90	62	62	—	same
☾ 6	28.86	70	65	5	overcast	28.86	71	61	10	same	29.90	59	57	2	overcast
7	29.96	70	60	10	very fine	29.97	71	54	17	same	29.98	56	56	—	very fine
8	29.84	58	56	2	rain	29.62	62	60	2	rain	29.53	61	61	—	rain
9	−29.50	64	59	5	fine	29.50	64	51	13	very fine	29.51	52	50	2	very fine
10	29.50	64	50	14	same	29.50	64	−50	14	same	29.50	55	51	4	same
11	29.57	62	53	9	shower	29.65	62	56	6	shower	29.72	55	55	—	same
12	29.85	63	55	8	very fine	29.89	65	51	14	very fine	29.94	52	50	2	same
○ 13	29.95	65	53	12	same	29.93	63	59	4	shower	29.88	59	59	—	rain
14	29.81	59	59	—	rain	29.67	68	68	—	rain	29.81	56	56	—	very fine
15	29.95	69	59	10	very fine	29.99	70	56	14	very fine	30.01	65	62	3	same
16	30.06	65	61	4	overcast	30.06	74	62	12	same	30.05	62	62	—	dull
17	30.11	68	60	8	very fine	30.06	70	62	8	same	30.01	62	62	—	same
18	29.99	63	60	3	dull	30.05	66	60	6	fine	30.10	54	54	—	very fine
19	30.14	62	60	2	fog	30.13	72	62	10	very fine	30.15	55	54	1	same
☽ 20	+30.16	68	61	7	very fine	30.14	76	58	8	same	30.15	53	52	1	same
21	30.13	67	62	5	same	30.13	77	64	13	same	30.10	58	58	—	same
22	30.11	66	64	2	overcast	30.09	75	63	12	same	30.06	62	62	—	dull
23	30.04	66	64	2	same	29.98	73	65	8	same	29.96	55	55	—	very fine
24	29.92	67	64	3	very fine	29.88	82	+69	13	same	29.89	65	65	—	same
25	29.86	71	68	3	same	29.88	67	64	3	overcast	29.90	65	64	1	rain
26	29.93	58	58	—	rain	29.96	67	63	4	very fine	30.04	59	58	1	dull
⊕ 27	30.09	58	58	—	same	30.06	59	54	5	fine	30.05	56	56	—	rain
28	29.92	57	51	6	overcast	29.89	57	54	3	overcast	29.84	53	53	—	same
29	29.74	57	57	—	rain	29.71	58	58	—	rain	29.70	59	59	—	same
30	29.72	65	62	3	very fine	29.74	66	64	2	same	29.76	54	54	—	same
31	29.80	62	62	—	rain	29.79	64	64	—	same	29.78	61	61	—	thund.storm
Means	29.923	64.9	59.3	5.6		29.906	68.6	59.7	8.7		29.914	58.3	5.78	0.5	

1821 August.	Temperature. Max.	Min.	Sun.	Rad.	Wind. Direction.	Force.	Rain
1	72	56	79	54	S W	little	
2	77	59	121	57	W	ditto	0.11
3	72	59	107	55	——	ditto	
4	77	52	125	50	S	ditto	
5	77	57	125	53	——	ditto	
☾ 6	74	55	106	52	W	ditto	
7	72	52	113	48	——	ditto	0.05
8	63	57	66	54	——	ditto	0.30
9	68	52	94	47	——	ditto	0.16
10	66	54	93	50	——	brisk	
11	68	49	96	43	——	ditto	0.20
12	67	50	94	43	N W	ditto	
○ 13	69	50	102	44	W	ditto	
14	69	50	70	45	——	ditto	0.23
15	73	62	113	60	N	ditto	
16	74	62	106	60	W	ditto	
17	72	60	125	53	——	ditto	
18	68	− 49	98	− 39	——	ditto	
19	73	49	132	40	S E	little	
☽ 20	77	49	+ 144	41	——	ditto	
21	80	56	144	46	——	ditto	
22	75	61	144	61	——	ditto	
23	75	55	144	55	——	ditto	
24	+ 82	57	141	49	S	ditto	
25	73	57	105	57	——	ditto	
26	68	55	92	55	N E	ditto	
⊕ 27	61	56	75	52	E	ditto	
28	58	53	64	53	——	ditto	0.15
29	58	54	58	54	——	stormy	0.22
30	71	54	122	54	S W	brisk	
31	70	54	120	54	——		0.25
Means	70.9	54.6	107.	50.9			1.67

AUGUST.

Mean Temperature	62.7°	Weight of vapour in a cubic foot.	
—— Pressure	29.914 inches		
—— Dew-point	57.1°	Mean . . .	5.617 grs.
—— Force of vapour . .	0.510 inch	Maximum .	7.853 „
—— Degree of dryness . .	5.6°	Minimum .	4.411 „
—— Degree of moisture	837°		

Least observed degree of moisture 524

WINDS.

N. 1 = 59° N.E. 1 = 59° E. 3 = 55° S.E. 5 = 60° S. 4 = 63° S.W. 3 = 61°
W. 13 = 57° N.W. 1 = 52°

Amount of rain . . . . 1.67 inch
—— of evaporation 3.07 inches

REMARKS.

On the 18th the sky was obscured by a haze, through which the sun appeared of a pale blue colour, resembling, in some degree, the flame of sulphur. The phenomenon was observed in several distant places. It was noticed in Essex and Worcestershire, and by several persons in and about London. Mr. Howard saw it in Sussex, where it lasted from nine till near noon, and he describes it as appearing nearly of the colour of watch-spring steel.

1821 September.	Morning.					Afternoon.					Night.				
	Barometer.	Hygrometer.			Weather.	Barometer.	Hygrometer.			Weather.	Barometer.	Hygrometer.			Weather.
1	29.90	62	62	—	rain	29.94	65	62	3	rain	29.97	60	60	—	rain
2	30.03	67	62	5	fine	30.03	67	61	6	very fine	30.03	59	59	—	same
☾ 3	29.93	66	61	5	very fine	29.91	67	64	3	rain	29.89	63	63	—	same
4	29.83	66	+66	—	rain	29.78	68	64	4	same	29.75	62	61	1	dull
5	29.80	63	60	3	very fine	29.83	64	60	4	very fine	29.96	58	58	—	same
6	29.90	66	62	4	same	29.90	70	64	6	showers	29.81	63	63	—	thunder
7	29.61	68	62	6	rain	29.63	67	60	7	very fine	29.65	59	57	2	dull
8	29.61	60	60	—	same	29 65	65	52	+13	same	29.70	53	53	—	very fine
9	29.67	64	59	5	same	29.64	62	59	3	showers	29.55	57	57	—	rain
10	29.66	57	57	—	same	29.74	63	55	8	thunder	29.82	55	55	—	fine
○ 11	29.91	56	54	2	very fine	29.96	63	55	8	very fine	29.96	54	54	—	very fine
12	29.69	60	60	—	rain	29.64	60	60	—	rain	29.83	54	54	—	same
13	29.93	60	54	6	fine	29.95	64	54	10	fine	29.94	54	54	—	rain
14	29.95	56	56	—	rain	30.01	58	57	1	rain	30.09	52	52	—	very fine
15	+30.18	58	55	3	very fine	30.18	62	55	7	very fine	30.18	59	57	2	fine
16	30.17	64	62	2	dull	30.16	69	65	4	dull	30.15	59	59	—	very fine
17	30.11	65	61	4	same	30.08	68	64	4	same	30.06	63	62	1	dull
☽ 18	29.98	62	60	2	showers	29.86	64	58	6	very fine	29.86	56	56	—	very fine
19	29.89	64	52	12	very fine	29.92	62	52	10	same	29.97	51	51	—	same
20	29.96	58	52	6	fine	29.94	62	58	4	rain	29.90	56	56	—	rain
21	29.82	58	57	1	rain	29.78	64	60	4	same	29.72	59	59	—	thund.storm
22	29.76	63	61	2	overcast	29.72	65	62	3	overcast	29.70	55	55	—	very fine
23	29.67	60	60	—	fog	29.62	64	61	3	rain	29.62	54	54	—	fine
24	29.62	60	55	5	fine	29.70	61	58	3	showers	29.85	48	46	2	very fine
25	29.96	55	50	5	same	29.95	59	55	4	very fine	29.96	59	59	—	rain
⊕ 26	29.96	63	63	—	rain	29.96	62	58	4	fine	29.95	59	59	—	same
27	29.85	57	57	—	same	29.90	59	56	3	very fine	29.95	53	51	2	dull
28	29.90	58	52	6	fine	29.70	60	58	2	rain	29.46	56	56	—	rain
29	−29.44	54	54	—	rain	29.60	53	50	3	fine	29.70	47	45	2	very fine
30	29.81	51	−44	7	very fine	29.85	53	44	9	very fine	29.86	51	46	5	same
Means	29.856	60.7	57.6	3		29.851	63	58	4.9		29.861	56.2	55.7	0.5	

1821 September.	Temperature.				Wind.		Rain.
	Max.	Min.	Sun.	Rad.	Direction.	Force.	
1	66	59	67	59	N W	little	0.13
2	71	57	122	45	S W	ditto	
3	72	62	110	59	——	ditto	0.05
☾ 4	71	59	103	51	——	ditto	0.11
5	71	57	112	46	——	ditto	
6	73	63	104	59	S	ditto	
7	+74	57	95	54	——	ditto	
8	66	53	120	47	W	ditto	0.14
9	67	49	87	43	——	variable	0.16
10	65	47	77	46	——	ditto	0.19
○ 11	68	47	120	44	——	ditto	
12	68	47	102	44	——	ditto	0.14
13	65	51	94	52	N W	little	0.13
14	59	−45	70	40	E	ditto	0.25
15	66	53	121	56	N E	ditto	
16	73	59	116	55	W	ditto	
17	66	56	120	54	S W	ditto	
☽ 18	73	60	110	55	W	brisk	
19	66	53	+120	49	N W	ditto	
20	63	54	79	50	——	ditto	
21	66	51	74	59	S W	ditto	
22	67	53	100	50	S E	ditto	0.42
23	65	47	85	42	——	ditto	
24	66	46	100	41	N W	ditto	0.11
25	64	54	87	54	——	ditto	
⊕ 26	68	50	80	59	W	ditto	
27	61	49	92	45	——	ditto	0.13
28	64	51	95	46	——	ditto	0.19
29	60	45	83	−36	——	ditto	
30	59	51	95	48	N W	ditto	
Means	66.7	52.8	98.3	49.6			2.15

SEPTEMBER.

Mean Temperature . . . . 59.7°
—— Pressure . . . . . . 29.856 inches
—— Dew-point . . . . . 55.4°
—— Force of vapour . . 0.483 inch
—— Degree of dryness . . 4.3°
—— Degree of moisture . 868°
Least observed degree of moisture . . . 651

Weight of vapour in a cubic foot.
Mean . . . 5.396 grs.
Maximum . 7.447 „
Minimum . 3.710 „

WINDS.

N. 0 = N.E. 1 = 55° E. 1 = 55° S.E. 2 = 58° S. 2 = 61° S.W. 6 = 60 W. 11 = 56° N.W. 7 = 53°.

Amount of rain . . . . 2.15 inches
—— of evaporation . . 2.19 inches

REMARKS.

Weather very showery and damp. Several thunder-storms, with heavy rain, during the period, were the cause of much damage.

1821 October.	Morning					Afternoon.					Night.				
	Barometer.	Hygrometer.			Weather.	Barometer.	Hygrometer.			Weather.	Barometer.	Hygrometer.			Weather.
1	29.70	59	55	4	fine	29.68	59	47	+12	very fine	29.95	51	47	4	very fine
2	30.04	52	50	2	very fine	30.09	56	51	5	same	30.08	52	50	2	dull
3	29.96	59	56	3	rain	29.86	63	60	3	fine	29.76	61	61	—	rain
☽ 4	29.62	60	60	—	same	29.42	61	61	—	rain	29.39	56	56	—	same
5	29.83	53	46	7	very fine	29.93	53	43	10	very fine	29.99	46	46	—	very fine
6	30.03	55	50	5	same	30.05	58	56	2	same	30.07	56	56	—	overcast
7	30.07	60	56	4	fine	30.05	58	57	1	same	30.00	53	53	—	very fine
8	29.91	57	57	—	rain	30.03	52	51	1	same	30.16	44	43	1	same
9	30.21	36	36	—	very fine	30.19	56	47	9	same	30.15	48	48	—	same
10	30.05	53	47	6	same	29.94	55	50	5	same	29.90	50	47	3	same
○ 11	29.72	53	50	3	same	29.60	55	50	5	same	29.57	50	50	0	rain
12	29.73	52	52	—	overcast	29.91	51	50	1	same	30.06	41	41	—	very fine
13	30.24	49	49	—	mist	30.30	53	50	3	same	+30.32	47	46	1	same
14	30.31	40	46	—	same	30.27	53	50	3	same	30.23	47	46	1	dull
15	30.20	40	40	—	rain	30.20	47	45	2	dull	30.16	37	37	—	very fine
16	30.13	44	42	2	very fine	30.11	47	38	9	very fine	30.13	37	37	—	same
17	30.13	44	42	2	dull	30.06	45	44	1	fine	30.00	44	43	1	dull
☾ 18	29.91	52	49	3	same	29.88	55	54	1	rain	29.86	49	49	—	same
19	29.88	52	49	3	same	29.78	50	45	5	very fine	29.68	49	44	5	same
20	29.19	55	53	2	same	29.10	50	46	4	dull	−29.02	43	43	—	showers
21	29.15	44	40	4	very fine	29.16	47	45	2	same	29.12	44	44	—	rain
22	29.14	46	46	—	fine	29.25	49	47	2	very fine	29.30	46	46	—	same
23	29.32	46	46	—	rain	29.21	50	50	—	rain	29.30	46	46	—	dull
24	29.43	48	48	—	same	29.56	48	48	—	same	29.74	43	43	—	very fine
25	29.94	47	47	—	fine	29.90	52	52	—	same	29.91	51	51	—	dull
⊕ 26	30.03	50	50	—	fog	30.05	57	57	—	fog	30.17	55	55	—	same
27	30.14	56	56	—	rain	30.12	58	+58	—	rain	30.18	55	55	—	rain
28	30.21	56	55	1	same	30.22	51	51	—	very fine	30.24	42	42	—	very fine
29	30.22	39	39	—	fog	30.15	49	46	3	same	30.15	42	42	—	fog
30	30.08	42	42	—	same	29.99	52	52	—	fog	29.95	42	42	—	dense fog
31	29.89	45	45	—	dense fog	29.90	53	50	3	very fine	29.94	50	50	—	showers
Means	29.884	50.	48.3	1.6		29.869	53.	50.	2.0		29.887	47.6	47.	0.5	

1821 October.	Temperature. Max.	Min.	Sun.	Rad.	Wind. Direction.	Force.	Rain
1	64	46	88	40	NW	high	
2	60	47	103	40	W	little	
3	+65	59	82	59	S W	brisk	0.15
☽ 4	61	43	63	42	S	ditto	0.34
5	57	43	88	34	NW	ditto	
6	63	52	87	44	W	ditto	
7	64	51	95	40	S W	ditto	0.09
8	62	37	88	30	N W	ditto	0.21
9	60	41	102	33	W	little	
10	59	42	102	33	S E	ditto	
○ 11	57	48	79	45	——	ditto	0.10
12	56	41	89	30	N W	ditto	
13	60	36	100	30	——	ditto	
14	60	45	100	41	S E	ditto	
15	58	37	84	29	N E	brisk	0.17
16	51	−36	92	−26	N	ditto	
17	58	42	92	36	N W	ditto	
☾ 18	62	47	+104	46	W	ditto	0.21
19	60	46	61	41	N W	ditto	0.09
20	55	40	61	32	W	ditto	0.17
21	58	42	86	38	N W	ditto	
22	54	42	79	32	S W	ditto	0.11
23	52	37	54	31	S E	ditto	0.18
24	52	40	62	30	W	ditto	0.10
25	54	42	69	37	S W	ditto	
⊕ 26	60	49	81	47	——	little	0.24
27	59	54	65	54	——	ditto	0.03
28	62	36	74	30	——	ditto	
29	59	39	79	32	S E	ditto	
30	59	41	87	35	E	ditto	
31	62	48	84	38	S W	ditto	0.04
Means	58.3	43.5	83.2	37.2			2.23

OCTOBER.

Mean Temperature . . . . 50.9°
—— Pressure . . . . 29.880 inches
—— Dew-point . . . . . 46.7°
—— Force of vapour . . 0.361 inch
—— Degree of dryness . . 4.2°
—— Degree of moisture . 874°
Least observed degree of moisture . . 670

Weight of vapour in a cubic foot.
Mean . . . 4.049 grs.
Maximum . 5.868 ,,
Minimum . 2.892 ,,

WINDS.

N. 1 = 39° N.E 1 = 40° E. 1 = 45° S.E. 5 = 47° S. 1 = 59° S.W 8=52° W. 6 = 48° N.W. 8 = 46°.

Amount of rain, &c. . . . . . . 2.23 inches
—— of evaporation . . . 1.61 inch

REMARKS.

The general character of the season mild and dull

1821 November.	Morning					Afternoon.					Night.				
	Barometer.	Hygrometer.			Weather.	Barometer.	Hygrometer.			Weather.	Barometer.	Hygrometer.			Weather.
1	29.91	55	55	—	rain	29.89	57	55	2	fine	29.90	54	54	—	dull
2	29.90	60	+60	—	same	29.87	59	59	—	rain	29.82	58	58	—	rain
☾ 3	29.80	56	55	1	dull	29.69	52	52	—	same	29.43	51	51	—	same
4	29.35	42	40	2	showers	29.40	40	40	—	showers	29.79	36	34	2	very fine
5	30.05	35	−31	4	very fine	30.12	42	37	5	very fine	30.21	33	33	—	same
6	30.25	32	32	—	same	+30.26	39	36	3	same	30.25	35	35	—	same
7	30.18	38	35	3	same	30.14	43	35	8	same	30.13	43	41	2	fine
8	30.11	44	43	1	same	30.09	45	44	1	same	30.08	44	44	—	same
○ 9	30.06	41	40	1	fine	30.05	43	42	1	fine	30.04	42	42	—	same
10	30.03	43	43	—	fog	30.01	50	45	5	same	30.02	54	54	—	rain
11	29.92	55	52	3	dull	29.82	56	56	—	rain	29.74	54	54	—	same
12	29.73	48	48	—	rain	29.94	51	50	1	very fine	29.98	44	44	—	very fine
13	29.94	44	44	—	dense fog	29.79	52	52	—	showers	29.70	54	54	—	rain
14	29.76	57	54	3	fine	29.78	54	54	—	rain	29.76	54	54	—	same
15	29.72	57	57	—	rain	29.55	57	57	—	same	29.50	56	56	—	same
☽ 16	29.42	56	56	—	same	29.41	53	53	—	same	29.38	54	54	—	same
17	29.61	50	47	3	very fine	29.59	52	52	—	same	29.61	52	52	—	same
18	29.76	48	48	—	clearing	30.00	45	45	—	misty	30.05	46	46	—	dull
19	29.95	50	50	—	rain	29.92	46	46	—	very fine	29.96	42	42	—	very fine
20	29.95	44	44	—	overcast	29.82	48	47	1	fine	29.70	46	46	—	dull
21	29.66	50	50	—	rain	29.78	42	37	5	very fine	29.84	35	35	—	fine
22	29.56	46	46	—	same	29.50	54	54	—	rain	29.67	47	47	—	very fine
23	29.70	53	50	3	dull	29.83	44	44	—	fine	29.97	34	34	—	same
⊕ 24	29.73	45	45	—	rain	29.73	47	44	3	showers	29.73	43	43	—	dull
25	29.68	45	42	3	fine	29.75	45	45	—	same	29.55	52	52	—	rain
26	29.36	54	54	—	rain	−29.23	51	51	—	rain	29.90	51	51	—	dull
27	29.45	43	42	1	fine	29.42	51	37	+14	fine	29.82	32	32	—	very fine
28	29.78	39	39	—	rain	29.70	51	51	—	rain	29.69	49	49	—	fine
29	29.55	54	54	—	dull	29.70	48	48	—	fine	29.69	46	46	—	very fine
30	29.81	43	42	1	very fine	29.77	46	46	—	dull	29.35	54	54	—	rain
Means	29.789	47.5	46.6	0.9		29.785	48.7	47.1	1.6		29.808	46.5	46.3	0.1	

1821 November.	Temperature. Max.	Min.	Sun.	Rad.	Wind. Direction.	Force.	Rain.
1	+62	52	67	52	S W	little	
2	62	53	67	52	——	brisk	
☾ 3	57	40	60	34	W	variable	0.10
4	46	31	55	25	——	ditto	0.15
5	46	−28	67	21	——	little	
6	46	32	+72	31	——	ditto	
7	51	38	70	34	S W	ditto	
8	49	37	70	32	E	ditto	
○ 9	48	41	62	38	——	ditto	
10	55	55	57	55	S E	brisk	
11	57	55	57	55	S	ditto	0.23
12	57	40	60	31	W	ditto	0.09
13	57	44	58	44	S E	ditto	0.18
14	58	54	58	54	S W	ditto	0.13
15	60	52	60	52	S	ditto	0.18
☽ 16	56	49	56	49	S W	ditto	0.40
17	56	50	56	50	——	ditto	0.70
18	51	42	68	36	N	ditto	
19	54	38	64	31	S W	little	0.60
20	52	44	52	39	——	ditto	
21	49	35	64	31	N W	ditto	
22	56	37	57	37	W	brisk	
23	56	33	58	28	——	ditto	0.20
⊕ 24	52	42	57	32	——	ditto	0.23
25	52	44	60	41	N W	ditto	0.10
26	55	43	55	39	S W	ditto	
27	46	30	60	−20	N W	ditto	
28	51	45	53	40	W	ditto	0.27
29	54	45	56	38	S W	ditto	
30	55	42	56	42	W	variable	0.09
Means	53.5	42.3	60.4	38.7			3.65

NOVEMBER.

Mean Temperature . . . . 47.9° .
—— Pressure . . . . . . 29.794 inches
—— Dew-point . . . . . 41.7°
—— Force of vapour . . 0.336 inch
—— Degree of dryness . . 3.2°
—— Degree of moisture . 896°

Least observed degree of moisture . 618

Weight of vapour in a cubic foot.

Mean . . . 3.794 grs.
Maximum . 6.222 „
Minimum . 2.431 „

WINDS.

N. 1=46° N.E. 0= E. 2=42° S.E. 2=48° S. 2=55° S.W. 10=50° W. 10=43° N.W. 3=41°.

Amount of rain . . . . 3.65 inches
—— of evaporation 1.17 inch

REMARKS.

Westerly winds prevailed throughout the month, and there was a great fall of rain, accompanied by destructive floods.

On the night of the 3d the wind blew a perfect gale.

The weather was most extraordinarily hot, damp, and stormy.

1821 December.	Morning. Barometer.	Hygrometer.			Weather.	Afternoon. Barometer.	Hygrometer.			Weather.	Night. Barometer	Hygrometer.			Weather.
1	29.61	47	42	5	very fine	29.60	47	47	—	rain	29.74	45	45	—	rain
☾ 2	29.81	45	41	4	fine	29.90	42	41	1	very fine	29.90	42	41	1	very fine
3	29.95	36	36	—	very fine	29.72	50	50	—	stormy	29.95	36	36	—	same
4	29.86	40	41	1	same	29.74	47	47	—	same	29.73	46	46	—	rain
5	29.72	48	48	—	rain	29.81	47	47	—	thund.storm	29.89	41	37	4	very fine
6	30.11	35	35	—	very fine	30.21	39	39	—	fog	30.21	37	37	—	same
7	30.02	41	36	5	dull	29.98	40	40	—	rain	29.77	49	49	—	rain
8	29.96	49	45	4	very fine	30.02	50	47	3	very fine	30.03	46	46	—	overcast
○ 9	30.03	51	51	—	rain	30.02	51	+51	—	rain	30.01	51	51	—	rain
10	30.00	51	51	—	same	29.95	52	51	1	overcast	29.90	51	51	—	same
11	30.15	40	40	—	very fine	30.25	40	40	—	very fine	+30.27	32	32	—	very fine
12	30.21	43	43	—	rain	30.13	47	47	—	rain	30.05	47	47	—	rain
13	29.90	48	45	3	dull	29.90	47	47	—	fine	29.92	42	42	—	showers
14	29.96	41	41	—	fine	29.95	49	49	—	rain	29.90	49	49	—	rain
15	29.89	47	47	—	same	29.87	48	48	—	same	29.87	48	48	—	fine
☽ 16	29.75	50	50	—	very fine	29.70	52	50	2	very fine	29.56	51	45	6	very fine
17	29.46	48	48	—	rain	29.37	48	48	—	rain	29.06	47	47	—	rain
18	29.05	47	47	—	fine	29.08	49	49	—	hail	29.06	47	47	—	same
19	29.08	45	45	—	rain	29.18	46	46	—	rain	29.33	39	39	—	very fine
20	29.40	41	41	—	dull	29.27	43	43	—	dull	28.83	45	45	—	rain
21	28.85	46	46	—	rain	29.23	45	38	+7	very fine	29.36	42	42	—	very fine
22	29.45	42	42	—	fine	29.10	43	43	—	fine	28.90	42	42	—	overcast
23	28.97	39	39	—	rain	29.11	44	40	4	very fine	29.23	38	38	—	very fine
⊕ 24	28.86	44	44	—	same	28.65	44	44	—	rain	−28.12	47	47	—	rain
25	28.23	39	39	—	dull	28.46	38	36	2	very fine	28.55	31	31	—	very fine
26	28.42	39	39	—	rain	28.37	40	40	—	rain	28.75	31	−31	—	same
27	28.90	38	38	—	thund.storm	29.01	40	40	—	hail	29.05	40	40	—	rain
28	28.77	42	42	—	stormy	28.29	46	46	—	rain	28.19	45	45	—	same
29	28.28	45	45	—	rain	28.64	45	45	—	same	28.85	43	43	—	same
30	29.02	42	42	—	overcast	29.25	43	43	—	same	29.54	38	38	—	fine
☾ 31	29.82	36	36	—	very fine	29.94	38	36	2	very fine	29.97	31	31	—	very fine
Means	29.467	43.4	42.7	0.7		29.474	45.1	44.4	0.7		29.467	42.5	42.2	0.3	

1821 December.	Temperature. Max.	Min.	Sun.	Rad.	Wind. Direction.	Force.	Rain.
1	47	39	50	39	W	high	0.20
☾ 2	51	42	60	33	——	brisk	
3	54	33	60	28	——	ditto	0.16
4	48	39	50	38	——	ditto	0.26
5	51	32	55	27	——	ditto	0.41
6	41	32	50	26	N W	ditto	
7	49	40	49	40	S E	ditto	0.10
8	54	45	57	37	W	ditto	
○ 9	52	49	54	49	S W	ditto	
10	+55	40	60	32	S	ditto	0.08
11	45	30	+66	24	N W	ditto	
12	47	42	47	42	S E	ditto	
13	52	36	54	30	——	ditto	0.06
14	52	40	60	35	S	ditto	0.16
15	52	40	62	35	S E	ditto	
☽ 16	54	48	62	42	S W	high	
17	51	48	57	48	——	ditto	0.11
18	51	48	58	47	——	ditto	0.18
19	50	34	50	26	——	ditto	0.13
20	47	40	47	40	——	brisk	0.35
21	47	36	54	32	N W	ditto	
22	44	41	50	40	S W	ditto	
23	46	37	46	31	W	ditto	0.11
⊕ 24	47	37	47	37	S	calm	0.20
25	42	30	49	21	N W	little	
26	42	30	45	25	E	ditto	0.47
27	48	36	50	32	S W	high	0.39
28	47	36	47	36	S E	ditto	0.70
29	46	39	47	34	N W	ditto	0.25
30	46	33	47	30	S E	ditto	0.19
☾ 31	43	−30	61	−19	N W	little	
Means	48.4	38.1	53.2	34.			4.51

DECEMBER.

Mean Temperature . . . . 43.2°
—— Pressure . . . . . . 29.469 inches
—— Dew-point . . . . . 41.2°
—— Force of vapour . . . 0.294 inch
—— Degree of dryness . . 2°
—— Degree of moisture . 924°
Least observed degree of moisture 776

Weight of vapour in a cubic foot.
Mean . . 3.364 grs.
Maximum . 4.684 „
Minimum . 2.451 „

WINDS.

N. 0= N.E. 0= E. 1 = 36° S.E. 6 = 44° S. 3 = 47° S.W. 8 = 45° W. 7 = 42° N.W. 6 = 38°.

Amount of rain . . . . 4.51 inches
—— of evaporation 0.74 inch

REMARKS.

This month was altogether wet, stormy, and oppressive.

The barometer, on the 24th, fell to a point at which it had not been known for 35 years. No storm of wind of any consequence accompanied the great depression which had been coming on for two weeks. A like state of the barometer was extensively observed on the Continent, and very tempestuous weather accompanied it far to the south of this island.

The quantity of rain was nearly double the usual amount.

1822 January.	Morning.					Afternoon.					Night.				
	Barometer.	Hygrometer.			Weather.	Barometer.	Hygrometer.			Weather.	Barometer.	Hygrometer.			Weather.
1	29.73	42	42	—	rain	29.70	44	44	—	rain	29.61	36	33	3	very fine
2	29.60	34	34	—	very fine	29.68	39	37	2	overcast	29.77	36	34	2	same
3	29.77	36	34	2	fine	29.56	37	37	—	rain	29.22	38	38	—	rain
4	−29.06	37	37	—	rain	29.28	37	37	—	same	29.50	35	35	—	stormy
5	29.76	34	32	2	very fine	29.88	34	32	2	very fine	29.91	36	34	2	fine
6	29.95	34	34	—	same	29.97	35	34	1	same	30.00	35	35	—	rain
☾ 7	29.89	33	33	—	snow show.	29.82	35	34	1	fine	29.87	32	32	—	fine
8	30.04	35	34	1	overcast	30.01	38	37	1	overcast	30.02	38	38	—	dull
9	30.12	43	40	3	same	30.14	43	40	3	same	30.14	34	34	—	fine
10	30.10	36	36	—	fog	30.09	41	40	1	fog	30 09	41	41	—	dull
11	30.18	44	44	—	same	30.19	44	44	—	same	30.24	43	43	—	same
12	30.27	44	44	—	same	30.27	45	45	—	mist	30.28	45	45	—	same
13	30.25	46	46	—	mist	30.20	48	+48	—	same	30.22	45	45	—	misty
14	30.23	42	40	2	very fine	30.20	44	40	4	very fine	30.14	40	40	—	fine
○ 15	30.12	40	34	6	same	30.16	37	30	7	same	30.18	31	−25	6	very fine
16	30.11	31	29	2	same	30.05	32	28	4	same	30.02	29	26	3	same
17	30.10	33	32	1	snow	30.10	39	39	—	rain	30.22	33	33	—	same
18	30.27	35	35	—	mist	30.29	39	37	2	very fine	30.32	39	39	—	dull
19	30.33	41	41	—	fog	30.30	41	41	—	rain	30.20	39	39	—	same
20	30.17	40	40	—	dull	30.15	40	40	—	showers	30.26	38	37	1	fine
21	30.33	42	40	2	fine	30.31	44	40	4	fine	30.37	40	40	—	dull
☽ 22	+30.37	44	44	—	fog	30.35	45	45	—	fog	30.29	43	43	—	same
23	30.25	45	45	—	same	30.12	45	45	—	dull	30.05	42	42	—	same
24	29.89	45	45	—	rain	29.84	47	47	—	rain	29.88	46	46	—	same
25	30.00	46	43	3	very fine	30.00	46	41	5	very fine	30.09	42	35	7	very fine
26	30.06	45	44	1	same	30.05	46	38	+8	same	30.15	40	33	7	dull
27	30.27	34	31	3	same	30.25	40	34	6	dull	30.23	40	40	—	same
28	30.14	45	45	—	mist	30.12	45	45	—	mist	30.12	41	41	—	same
⊕ 29	30.15	37	37	—	same	30.15	43	43	—	same	30.22	38	38	—	overcast
30	30.31	34	33	1	very fine	30.31	35	33	2	very fine	30.31	34	34	—	fog
31	30.27	35	34	1	same	30.23	42	34	8	same	30.22	38	34	4	fine
Means	30.067	39.	38.1	0.9		30.057	40.9	39.	1.9		30.069	38.2	37.1	1.1	

1822 January.	Temperature. Max.	Min.	Sun.	Rad.	Wind. Direction	Force.	Rain.
1	45	32		25	S E	little	
2	43	32		26	N W	brisk	
3	39	35		35	W	high	
4	37	33		32	N E	variable	0.20
5	38	31		30	N	brisk	
6	37	28		28	——	ditto	
☾ 7	37	30		28	——	ditto	
8	41	35		34	N W	ditto	
9	45	30		22	N	ditto	
10	45	34		33	N W	ditto	
11	47	40		40	W	little	
12	47	43		40	——	ditto	
13	+49	40		−*32	——	ditto	
14	47	38		*30	N W	ditto	0.06
○ 15	42	30		*20	N	ditto	
16	36	−29		−*19	——	ditto	
17	39	32		*25	W	ditto	0.17
18	41	38		*33	——	ditto	
19	41	38		*37	——	ditto	
20	40	35		*33	——	ditto	
21	48	39		*38	N W	ditto	
☽ 22	46	41		*32	W	ditto	
23	46	41		*35	S W	ditto	
24	48	42		*40	——	ditto	0.10
25	47	40		*32	N W	ditto	
26	47	32		*22	——	ditto	
27	45	35		*33	——	ditto	
28	48	35		*27	W	ditto	
⊕ 29	47	32		*22	——	ditto	
30	39	30		*23	N W	ditto	
	42	35		*28	W	ditto	
Means	43.1	35.		30.1			0.53

JANUARY

Mean Temperature . . . 39°
—— Pressure . . . . . . 30.064 inches
—— Dew-point . . . . . 37°
—— Force of vapour . . . 0.256 inch
—— Degree of dryness . . 2°
—— Degree of moisture . 941°

Least observed degree of moisture . . 759

Weight of vapour in a cubic foot.
Mean . . . 2.967 grs.
Maximum . 4.279 „
Minimum . 2.000 „

WINDS.

N. 6 = 32° N.E. 1 = 36° E. 0 = S.E. 1 = 39 S. 0 = S.W. 2 = 45°
W. 12 = 40° N.W. 9 = 37°

Amount of rain . . . . 0.53 inch
—— of evaporation . . 0.49 inch

REMARKS.

The weather was of the usual character for the time of year, but the temperature was above the mean, and the quantity of rain small.

N.B. The results of the radiating thermometer, marked with an asterisk, were obtained in the metallic reflector, described in the Essay upon Radiation.

1822 February.	Morning. Barometer.	Hygrometer.			Weather.	Afternoon. Barometer.	Hygrometer.			Weather.	Night. Barometer.	Hygrometer.			Weather.
1	30.14	43	40	3	very fine	30.04	40	36	4	very fine	29.90	39	27	2	very fine
2	29.79	47	47	—	rain	29.59	49	+49	—	stormy	29.33	49	49	—	rain
3	29.46	44	43	1	very fine	29.51	43	39	4	very fine	29.64	37	37	—	very fine
4	29.59	44	42	2	showers	29.29	47	47	—	rain	29.27	42	42	—	dull
5	−29.25	49	37	12	fine	29.38	48	46	2	showers	29.83	39	39	—	very fine
○ 6	30.13	35	33	2	very fine	30.10	40	35	5	very fine	29.96	41	41	—	fine
7	29.83	45	45	—	dull	29.81	48	46	2	overcast	29.81	47	47	—	rain
8	29.83	48	41	7	fine	29.90	47	46	1	same	29.90	45	45	—	overcast
9	29.87	48	48	—	overcast	29.81	48	48	—	same	29.85	49	48	1	same
10	29.85	47	44	3	same	29.80	50	46	4	very fine	29.77	47	47	—	rain
11	30.00	45	43	2	very fine	30.05	45	45	—	rain	30.11	45	45	—	same
12	30.20	40	40	—	fog	30.18	41	41	—	fog	30.20	39	39	—	fog
13	30.11	41	41	—	mist	30.03	44	43	1	very fine	30.07	42	42	—	overcast
☽ 14	30.08	44	44	—	dull	30.05	45	44	1	misty	30.04	41	41	—	very fine
15	30.06	49	48	1	very fine	30.14	43	43	—	very fine	30.34	40	40	—	misty
16	30.34	40	40	—	mist	30.33	44	33	11	same	30.36	50	48	2	dull
17	30.39	50	46	4	dull	30.38	52	48	4	same	30.37	42	40	2	very fine
18	30.36	47	46	1	same	30.30	49	44	5	same	30.32	40	40	—	dull
19	30.32	47	47	—	rain	30.26	48	40	8	same	30.28	39	38	1	very fine
20	30.06	44	37	7	dull	29.90	40	40	—	rain	30.04	37	35	2	same
⊕ 21	30.36	37	35	2	very fine	30.44	42	36	6	very fine	30.46	36	32	4	same
22	30.41	41	36	5	dull	30.28	42	34	8	same	30.21	37	32	5	same
23	30.41	41	36	5	same	30.28	42	34	8	same	30.21	37	32	5	same
24	30.30	44	36	8	very fine	30.28	43	−30	+13	same	30.27	40	38	2	same
25	30.21	47	40	7	fine	30.18	50	42	8	same	30.21	48	48	—	dull
26	30.12	49	46	3	overcast	30.09	50	48	2	rain	30.21	45	45	—	clearing
27	30.57	42	33	9	very fine	30.70	44	32	12	very fine	+30.73	35	32	3	very fine
☾ 28	30.68	34	34	—	fog	30.55	43	32	11	same	30.45	35	32	3	same
Means	30.097	44	41	3		30.058	45.2	40.9	4.2		30.076	41.4	40.3	1.1	

1822 February.	Temperature. Max.	Min.	Sun.	Rad.	Wind. Direction.	Force.	Rain.
1	44	35		*31	W	little	
2	51	39		*39	S W	high	0.30
3	47	33		*25	W	variable	
4	47	38		*37	S W	little	0.07
5	51	32		*22	——	ditto	
○ 6	45	35		*29	W	ditto	
7	50	45		*41	S W	brisk	0.07
8	50	42		*36	——	ditto	0.10
9	52	46		*41	——	ditto	
10	52	41		*33	S	ditto	0.07
11	49	40		*34	S W	ditto	0.15
12	44	34		*34	N W	little	
13	46	40		*32	S E	ditto	
☽ 14	48	39		*32	——	ditto	
15	52	38		*32	S	ditto	0.04
16	50	40		*40	W	ditto	0.03
17	+53	40		*33	——	ditto	
18	50	42		*36	——	ditto	
19	51	37		*28	N W	ditto	
20	47	34		*25	W	ditto	0.10
⊕ 21	45	31		*22	N	ditto	
22	46	37		*29	W	ditto	
23	46	37		*29	N W	ditto	
24	47	39		*30	W	ditto	
25	52	46		*43	——	ditto	
26	51	36		*26	S W	ditto	
27	46	31		*22	N W	ditto	
☾ 28	44	− 30		− *20	——	ditto	
Means	48.4	37.7		31.4			0.93

FEBRUARY.

Mean Temperature . . . . 43°
—— Pressure . . . . . . 30.077 inches
—— Dew-point . . . . . 39.3°
—— Force of vapour . . 0.274 inch
—— Degree of dryness . . 3.7°
—— Degree of moisture . 867°
Least observed degree of moisture . . 632

Weight of vapour in a cubic foot.
Mean . . . 3.127 grs.
Maximum . 4.407 „
Minimum . 2.298 „

WINDS.

N. 1 = 34° N.E. 0 = E. 0 = S.E. 2 = 42° S. 2 = 44° S.W. 7 = 45° W. 9 = 39° N.W. 5 = 36°

Amount of rain, &c. . . . . . . 0.93 inch
—— of evaporation . . . 1.17 inch

REMARKS.

Seasonable weather and very mild. The dryness of the atmosphere was great, and the fall of rain remarkably small.

1822 March.	Morning. Barometer.	Hygrometer.			Weather.	Afternoon. Barometer.	Hygrometer.			Weather.	Night. Barometer.	Hygrometer.			Weather.
1	30.28	34	34	—	fog	30.24	43	36	7	very fine	30.24	39	39	—	very fine
2	30.33	46	46	—	overcast	30.31	49	46	3	same	30.31	43	43	—	same
3	30.33	46	46	—	fog	30.24	50	45	5	same	30.22	42	42	—	same
4	30.09	46	46	—	mist	30.00	50	45	5	same	29.98	47	47	—	rain
5	30.00	49	46	3	fine	29.99	48	37	11	same	29.83	47	47	—	same
6	29.56	53	53	—	rain	29.63	52	47	5	dull	29.62	47	47	—	dull
○ 7	29.58	47	41	6	fine	29.49	43	39	4	fine	29.53	39	39	—	very fine
8	29.63	41	41	—	rain	– 29.41	43	43	—	rain	29.51	43	43	—	dull
9	29.75	48	46	2	very fine	29.75	52	52	—	same	29.78	51	50	—	same
10	29.62	53	53	—	rain	29.60	50	45	5	showers	29.66	44	40	4	fine
11	29.87	45	33	12	very fine	30.04	44	32	12	very fine	30.21	37	35	2	very fine
12	30.39	38	38	—	fog	30.38	45	35	10	same	30.36	38	33	5	same
13	30.21	45	40	5	very fine	30.05	50	34	16	same	29.98	45	41	4	same
14	30.00	54	50	4	fine	30.03	55	55	—	showers	30.20	49	49	—	overcast
☽ 15	30.24	42	42	—	fog	30.24	49	31	18	very fine	30.25	43	40	3	very fine
16	30.21	51	51	—	overcast	30.14	53	53	—	rain	30.20	53	53	—	dull
17	30.32	52	49	3	rain	30.17	53	53	—	same	30.05	53	53	—	rain
18	30.33	49	46	3	very fine	30.41	52	44	8	very fine	30.36	46	46	—	overcast
19	30.32	58	54	4	same	30.33	59	57	2	same	30.38	54	54	—	same
20	30.37	54	54	—	overcast	30.31	56	51	5	same	30.35	51	49	2	same
21	30.31	51	49	2	same	30.19	56	53	3	same	30.15	51	50	1	same
22	30.35	50	39	11	very fine	30.38	54	47	7	same	30.34	43	43	—	very fine
⊕ 23	30.26	48	41	7	same	30.06	55	47	8	same	29.85	46	44	2	same
24	29.63	51	50	1	dull	29.70	49	34	15	shower	29.76	40	37	3	same
25	29.80	45	40	5	same	29.75	46	46	—	rain	29.98	38	37	1	same
26	30.08	48	48	—	same	30.15	56	52	4	very fine	30.22	47	46	1	same
27	30.26	52	49	3	same	30.17	56	52	4	same	30.09	48	47	1	same
28	30.01	58	52	6	very fine	30.14	60	+58	2	rain	30.14	52	52	—	same
29	+30.46	53	40	13	same	30.41	57	34	+23	fine	30.25	47	37	10	fine
☾ 30	29.66	52	52	—	rain	29.60	47	47	—	rain	30.03	41	34	7	very fine
31	30.36	41	– 30	11	hail	30.43	40	36	4	hail	30.43	38	30	8	dull
Means	30.084	48.3	45.1	3.2		30.056	50.7	44.7	6.		30.072	45.2	43.4	1.7	

1822 March.	Temperature. Max.	Min.	Sun.	Rad.	Wind. Direction.	Force.	Rain.
1	46	34		*31	W	variable	
2	52	38		*28	S W	little	
3	53	37		*28	——	ditto	
4	52	46		*46	——	ditto	0.03
5	53	46		*46	——	stormy	
6	55	41		*35	S	high	0.25
○ 7	52	36		*29	W	ditto	0.28
8	48	41		*36	——	ditto	0.10
9	53	48		*48	S W	ditto	0.15
10	56	38		*34	——	ditto	0.16
11	47	-33		-*23	N W	brisk	
12	46	37		*28	S W	ditto	
13	54	42		*34	S E	ditto	
14	57	38		*36	S	ditto	0.03
☽ 15	51	40		*32	N	ditto	
16	55	50		*45	S W	little	
17	54	47		*41	——	ditto	0.05
18	58	46		*39	W	ditto	
19	62	51		*46	S	variable	
20	57	45		*40	W	ditto	
21	57	41		*31	——	ditto	
22	55	39		*29	N W	little	
⊕ 23	60	44		*37	W	ditto	
24	53	37		*29	——	ditto	
25	50	36		*27	——	ditto	0.11
26	58	47		*40	——	ditto	
27	58	46		*37	S W	ditto	
28	+66	43		*34	S	ditto	0.07
29	59	47		*40	W	ditto	
☾ 30	56	36		*28	S W	high	0.19
31	45	36		*27	N	ditto	0.08
Means	54.1	41.4		34.9			1.50

MARCH.

Mean Temperature . . . . 47.7°
—— Pressure . . . . . 30.070 inches
—— Dew-point . . . . 43°
—— Force of vapour . . 0.316 inch
—— Degree of dryness . . 4.7°
—— Degree of moisture 849°
Least observed degree of moisture 456

Weight of vapour in a cubic foot.
Mean . . . 3.596 grs.
Maximum . 5.844 ,,
Minimum . 2.297 ,,

WINDS.

N. 2 = 34° N.E. 0 = E. 0 = S.E. 1 = 38° S. 4 = 52° S.W. 11 = 46° W. 11 = 43° N.W. 2 = 36°

Amount of rain . . . . 1.50 inch
—— of evaporation 1.73 inch

REMARKS.

The character of the month was that of fine seasonable weather. The wind, as usual at this period, was often boisterous.

1822 April.	Morning.					Afternoon.					Night.				
	Barometer.	Hygrometer.			Weather.	Barometer	Hygrometer.			Weather.	Barometer.	Hygrometer.			Weather.
1	30.39	42	35	7	very fine	30.37	46	33	13	very fine	30.30	43	43	—	overcast
2	30.38	46	35	11	fine	30 39	46	35	11	same	30.39	40	35	5	very fine
3	+30.41	50	42	8	same	30.35	50	44	6	same	30.28	43	42	1	same
4	30.26	49	44	5	same	30.21	49	45	4	overcast	30.18	47	45	2	overcast
5	30.12	51	42	9	same	30.03	51	42	9	fine	30.01	46	42	4	very fine
○ 6	29.95	52	44	8	very fine	29.90	49	39	10	very fine	29.90	41	35	6	same
7	30.00	45	43	2	fine	30.05	49	38	11	same	30.07	40	35	5	same
8	30.08	45	35	10	very fine	30.06	39	39	—	hail	30.08	35	35	—	same
9	30.11	41	39	2	hail	30.09	44	36	8	same	30.11	35	−34	1	same
10	30.08	42	36	6	fine	30.06	44	40	4	showers	30.08	36	36	—	same
11	30.06	44	36	8	very fine	29.99	44	36	8	very fine	29.91	41	40	1	stormy
12	29.76	41	41	—	rain	29.80	50	50	—	thund.storm	29.85	45	45	—	rain
13	29.93	52	52	—	showers	30.01	55	47	8	very fine	30.07	47	45	2	very fine
☾ 14	30.09	52	45	7	dull	30.06	55	47	8	dull	30.06	50	47	3	dull
15	30.07	57	51	6	fine	30.04	54	54	—	rain	30.02	48	48	—	rain
16	30.04	50	50	—	rain	30.06	54	47	7	fine	30.06	46	43	3	very fine
17	29.94	42	42	—	fog	29.83	48	48	—	rain	29.79	45	45	—	dull
18	29.69	48	48	—	dull	29.68	50	45	5	overcast	29.70	43	40	3	very fine
19	29.76	50	42	8	showers	29.76	50	45	5	thund.&hail	29.75	46	46	—	rain
20	29.76	52	42	10	very fine	29.72	53	40	13	very fine	29.68	48	48	—	same
⊕ 21	29.63	51	45	6	same	29.52	55	50	5	showers	29.46	52	52	—	same
22	29.43	55	47	8	fine	29.40	54	46	8	very fine	29.38	46	46	—	same
23	−29.37	50	50	—	hail	29.49	52	40	12	same	29.55	44	44	—	very fine
24	29.79	56	46	10	very fine	29.79	55	48	7	same	29.66	49	49	—	rain
25	29.57	51	51	—	rain	29.67	53	45	8	same	29.73	45	45	—	very fine
26	29.90	54	43	11	very fine	30.03	55	35	20	same	30.08	48	46	2	fine
27	30.04	51	51	—	rain	30.08	58	55	3	dull	30.18	54	54	—	dull
☽ 28	30.31	56	55	1	haze	30.31	59	55	4	haze	30.29	49	49	—	very fine
29	30.29	51	51	—	dull	30.30	59	51	8	very fine	30.26	49	49	—	same
30	30.30	62	+56	6	very fine	30.32	61	36	+25	same	30.32	50	36	14	same
Means	29.983	49.6	44.6	4 9		29.975	51.3	43.7	7.6		29.973	45	43.3	1.7	

1822 April.	Temperature.				Wind.		Rain.
	Max.	Min.	Sun	Rad.	Direction.	Force.	
1	47	39		*31	N	little	
2	50	39		*31	N E	ditto	
3	54	40		*31	N W	ditto	
4	53	43		*34	——	ditto	
5	53	43		*33	——	ditto	
○ 6	55	39		*30	——	mist	0.11
7	51	36		*27	N	ditto	
8	51	− 32		− *26	N E	ditto	0.19
9	46	33		*28	——	ditto	
10	46	33		*28	——	ditto	
11	46	39		*37	E	high	0.43
12	53	42		*37	——	ditto	0.22
13	58	45		*37	S W	brisk	
☾ 14	57	49		*45	S E	ditto	0.18
15	62	47		*46	E	ditto	0.20
16	55	41		*34	N E	ditto	
17	50	44		*40	S E	ditto	
18	53	41		*31	N	ditto	
19	55	41		*40	N W	variable	0.15
20	56	46		*41	S W	brisk	
⊕ 21	58	48		*47	S E	ditto	0.23
22	58	44		*44	S	ditto	0.08
23	58	42		*36	W	ditto	0.10
24	59	47		*46	——	ditto	
25	58	41		*35	——	ditto	0.19
26	58	46		*40	——	ditto	
27	58	49		*48	S	ditto	0.04
☽ 28	62	46		*40	——	ditto	
29	62	46		*40	E	ditto	
30	+65	45		*38	——	ditto	
Means	54.9	42.2		36.7			2.12

APRIL.

Mean Temperature	48.5°	Weight of vapour in a cubic foot.	
—— Pressure	29.977 inches	Mean	3.593 grs.
—— Dew-point	42.9°	Maximum	5.446 „
—— Force of vapour	0.315 inch	Minimum	2.702 „
—— Degree of dryness	5.6°		
—— Degree of moisture	824°		

Least observed degree of moisture . . . 429

WINDS.

N. 3 = 40° N.E. 5 = 38° E. 5 = 45° S.E. 3 = 46° S. 3 = 50° S.W. 6 = 45° W. 0 = N.W. 5 = 42°.

Amount of rain . . . . 2.12 inches
—— of evaporation . . . 2.01 inches

REMARKS.

This month was cold and rainy, and the frosts in the early part very injurious to vegetation. In the last two or three days a great improvement took place.

1822 May.	Morning. Barometer.	Hygrometer.			Weather.	Afternoon. Barometer.	Hygrometer.			Weather.	Night. Barometer.	Hygrometer.			Weather.
1	+30.38	59	38	21	very fine	30.31	58	38	20	very fine	30.28	46	41	5	very fine
2	30.22	58	48	10	same	30.13	59	46	13	same	30.10	48	48	—	same
3	30.00	55	50	5	same	29.86	60	46	14	same	29.77	53	46	7	fine
4	29.66	62	55	7	fine	29.67	64	56	8	fine	29.70	52	52	—	very fine
5	29.76	57	56	1	showers	29.75	62	60	2	thund.storm	29.78	57	57	—	dull
○ 6	29.80	59	56	3	dull	29.80	67	64	3	showers	29.80	58	58	—	same
7	29.84	59	58	1	rain	29.84	57	57	—	rain	29.91	50	50	—	rain
8	30.02	50	37	13	very fine	30.00	51	38	13	fine	29.96	42	−36	6	fine
9	29.77	50	38	12	same	29.62	49	36	13	dull	29.58	45	45	—	dull
10	29.41	49	49	—	rain	−29.36	52	46	6	fine	29.43	46	46	—	fine
11	29.65	55	51	4	very fine	29.73	58	44	14	very fine	29.77	53	53	—	stormy
12	29.83	49	49	—	rain	29.86	47	46	1	rain	29.88	46	46	—	rain
13	29.88	50	50	—	same	29.84	51	50	1	same	29.90	46	46	—	same
☽ 14	29.90	43	42	1	showers	29.88	45	44	1	showers	29.95	42	42	—	showers
15	29.95	48	47	1	same	29.93	51	50	1	same	29.97	45	45	—	fine
16	29.95	53	43	10	very fine	29.93	63	43	20	very fine	29.94	47	47	—	very fine
17	29.92	59	52	7	fine	29.92	67	58	9	fine	29.93	55	55	—	fine
18	29.95	62	58	4	very fine	29.97	69	56	13	very fine	30.03	57	57	—	very fine
19	30.03	64	52	12	same	30.04	73	50	+23	same	30.06	53	53	—	same
⊕ 20	30.05	58	56	2	same	30.04	72	50	22	same	30.15	58	57	1	same
21	30.21	65	59	6	same	30.24	73	+62	11	same	30.27	55	55	—	same
22	30.31	56	52	4	hazy	30.28	68	62	6	same	30.27	52	50	2	same
23	30.24	56	53	3	very fine	30.16	59	49	10	same	30.10	46	46	—	same
24	30.05	56	54	2	same	30.00	61	46	15	same	30.02	50	50	—	same
25	29.89	63	59	4	same	29.85	60	49	11	thund.storm	29.83	48	48	—	same
26	29.80	60	56	4	dull	29.78	57	57	—	rain	29.97	48	48	—	same
☾ 27	30.08	58	50	8	very fine	30.07	62	48	14	very fine	30.07	55	54	1	rain
28	30.14	59	55	4	dull	30.16	69	51	18	same	30.19	49	49	—	very fine
29	30.20	59	56	3	very fine	30.21	70	59	11	same	30.27	49	49	—	same
30	30.27	58	53	5	fine	30.24	71	55	16	same	30.21	53	50	3	same
31	30.23	64	54	10	very fine	30.23	72	52	20	same	30.25	54	51	—	same
Means	29.980	56.5	51.1	5.3		29.958	61.1	50.5	10.6		29.978	50.2	49.4	0.8	

1822 May.	Temperature.				Wind.		Rain.
	Max.	Min.	Sun.	Rad.	Direction.	Force.	
1	60	41	84	*33	N E	brisk	
2	60	41	95	−*32	E	little	
3	62	52	79	*50	——	ditto	
4	65	47	82	*38	S E	ditto	
5	66	53	90	*52	E	variable	
○ 6	67	53	87	*53	——	ditto	
7	59	45	64	*45	——	brisk	0.67
8	53	−40	70	*38	N E	ditto	
9	52	43	64	*40	——	ditto	
10	56	41	96	*38	——	ditto	0.21
11	60	47	101	*46	S E	ditto	
12	51	46	54	*46	N E	ditto	0.18
13	52	46	62	*44	——	ditto	
☽ 14	48	40	58	*36	N	ditto	
15	52	46	70	*44	N E	ditto	
16	65	45	125	*40	——	ditto	
17	70	51	120	*43	S E	little	
18	70	47	124	*43	——	ditto	
19	74	48	+135	*43	E	ditto	
⊕ 20	+75	49	120	*44	S E	ditto	
21	74	48	119	*42	S	ditto	
22	69	47	131	*41	E	ditto	
23	61	46	123	*40	N	ditto	
24	64	44	122	*37	N E	ditto	
25	65	45	122	*38	N	ditto	0.10
26	60	42	80	*34	S W	ditto	0.13
☽ 27	63	54	102	*54	——	ditto	
28	71	47	120	*40	W	ditto	
29	72	43	121	*35	——	ditto	0.05
30	72	46	113	*40	——	ditto	
31	74	47	120	*42	——	ditto	
Means	63.2	46.1	98.4	41.6			1.34

MAY.

Mean Temperature . . . . 54.6°
—— Pressure . . . . . . 29.974 inches
—— Dew-point . . . . . 48.3°
—— Force of vapour . . 0.380 inch
—— Degree of dryness . . 6.3°
—— Degree of moisture . 810°
Least observed degree of moisture . . 471

Weight of vapour in a cubic foot.
Mean . . . 4.227 grs.
Maximum . 6.436 „
Minimum . 2.856 „

WINDS.

N. 3 = 48° N.E. 9 = 44° E 7 = 53° S.E. 5 = 54° S. 1 = 58° S.W. 2 = 52° W. 4 = 52° N W. 0

Amount of rain, &c. . . . . . . 1.34 inch
—— of evaporation . . . 2.76 inches

REMARKS.

This month was altogether warm and genial, and the weather perfectly delightful.

1822 June.	Morning.					Afternoon.					Night.				
	Barometer.	Hygrometer.			Weather.	Barometer.	Hygrometer.			Weather.	Barometer.	Hygrometer.			Weather.
1	30.19	66	58	8	very fine	30.16	76	53	23	very fine	30.15	57	55	2	very fine
2	30.16	67	60	7	same	30.15	75	63	12	same	30.16	57	56	1	same
3	30.15	70	57	13	same	30.09	77	55	22	same	30.11	64	56	8	same
○ 4	30.13	71	57	14	same	30.10	80	56	+24	same	30.11	65	56	9	same
5	30.08	70	59	11	same	30.07	81	58	23	same	30.10	64	59	5	same
ↄ 6	30.11	71	61	10	same	30.08	81	61	20	same	30.12	65	61	4	same
7	30.10	65	50	15	same	30.00	75	51	24	same	30.03	59	52	7	same
8	29.96	63	60	3	same	29.90	80	+68	12	same	29.89	56	56	—	same
9	29.83	72	64	8	overcast	29.80	78	63	15	shower	29.88	65	64	1	lightning
10	29.88	78	64	14	heavy	29.90	83	63	20	thund.storm	29.94	70	63	7	fine
11	30.04	73	50	23	fine	30.00	75	63	12	very fine	30.00	65	64	1	very fine
☽ 12	30.05	61	54	7	very fine	30.01	68	53	15	same	30.00	56	54	2	same
13	30.00	60	55	5	same	29.98	72	56	16	same	29.99	56	56	—	same
14	30.01	62	60	2	showers	30.00	76	59	17	shower	30.03	58	58	—	same
15	30.00	61	49	12	fine	29.98	70	50	20	same	30.06	56	49	7	rain
16	30.15	62	49	13	very fine	30.12	68	49	19	fine	30.11	56	48	8	fine
17	30.16	60	48	12	same	30.15	64	50	14	very fine	+30.20	52	48	4	very fine
18	30.20	60	51	.9	same	30.09	70	53	17	same	30.04	53	53	—	same
⊕ 19	29.96	65	57	8	same	29.85	73	53	20	same	29.85	58	56	2	fine
20	29.95	60	52	8	fine	30.00	63	54	9	fine	30.06	56	48	8	dull
21	30.17	59	47	12	very fine	30.10	66	−47	19	very fine	30.13	56	52	4	very fine
22	30.12	66	53	13	same	30.02	73	55	18	same	30.02	59	55	4	same
23	29.95	63	62	1	thund. sho.	29.95	69	57	12	same	29.96	62	57	5	dull
24	30.02	67	58	9	fine	30.04	72	60	12	same	30.05	60	55	5	fine
25	30.10	70	59	11	very fine	30.05	73	60	13	same	30.06	61	58	3	very fine
☾ 26	30.01	69	59	10	same	30.01	73	61	12	same	29.99	60	56	4	shower
◡ 27	30.09	66	55	11	same	30.13	69	48	21	same	30.16	55	51	4	very fine
28	30.12	64	54	10	fine	29.99	65	60	5	fine	29.91	60	60	—	rain
29	29.97	56	50	6	overcast	30.00	65	50	15	very fine	30.03	56	51	5	very fine
30	29.96	61	50	11	same	29.83	69	62	7	rain	29.93	55	50	5	same
Means	30.054	65.2	55.4	9.8		30.018	72.6	56.3	16.2		30.035	59.	55.2	3.8	

1822 June.	Temperature.				Wind.		Rain.
	Max.	Min.	Sun.	Rad.	Direction.	Force.	
1	77	52	137	*44	S W	brisk	
2	76	47	135	*40	S E	little	
3	79	43	140	*37	E	ditto	
○ 4	82	49	144	*40	S E	ditto	
5	84	48	+154	*41	N	ditto	
6	87	43	147	*38	N E	ditto	
7	76	42	136	*36	N E	ditto	
8	83	50	130	*44	S E	ditto	
9	84	51	98	*45	S	brisk	
10	+90	56	96	*54	——	ditto	0.23
11	80	50	130	*45	N	little	
☽ 12	72	−38	134	*31	——	ditto	
13	80	39	102	*32	——	ditto	
14	80	54	104	*47	S	ditto	
15	72	49	120	*48	N	ditto	0.51
16	72	40	110	*36	E	variable	
17	70	39	98	−*30	S E	ditto	
18	68	47	112	*35	——	ditto	
⊕ 19	72	52	114	*42	——	ditto	0.05
20	64	44	99	*32	N E	little	
21	67	44	126	*34	——	ditto	
22	74	54	120	*48	S E	ditto	0.01
23	71	58	102	*52	S W	ditto	0.05
24	74	57	104	*49	——	ditto	
25	76	51	124	*50	——	ditto	
☾ 26	77	56	116	*50	——	ditto	0.03
27	70	51	122	*41	N	ditto	
28	69	52	96	*52	W	brisk	0.18
29	65	45	90	*35	N W	ditto	0.05
30	70	50	74	*41	——	ditto	
Means	75.3	48.3	117.1	41.6			1.11

JUNE.

Mean Temperature . . . . 61.8°
—— Pressure . . . . . . 30.035 inches
—— Dew-point . . . . . 52.6°
—— Force of vapour . . 0.441 inch
—— Degree of dryness . . 9.2°
—— Degree of moisture 746°
Least observed degree of moisture 464

Weight of vapour in a cubic foot.
Mean . . . 4.910 grs.
Maximum . 7.719 „
Minimum . 3.997 „

WINDS.

N. 6 = 54 N.E. 4 = 53° E. 2 = 52 S.E. 7 = 55° S. 3 = 62° S.W. 5 = 57° W. 1 = 58° N.W. 2 = 52°.

Amount of rain, &c. . . . . 1.11 inch
—— of evaporation . . 4.50 inches

REMARKS.

Weather unusually dry and hot. The country was very much burnt up from excessive drought.

1822 July.	Morning. Barometer.	Hygrometer.			Weather.	Afternoon. Barometer.	Hygrometer.			Weather.	Night. Barometer.	Hygrometer.			Weather.
1	30.02	60	50	10	very fine	30.08	66	−47	+19	very fine	30.05	56	50	6	very fine
2	29.96	61	51	10	same	29.86	65	58	7	fine	29.78	61	57	4	dull
3	29.92	63	50	13	same	29.95	68	58	10	very fine	29.97	56	56	—	fine
○ 4	29.98	66	62	4	same	29.92	71	60	11	same	29.88	63	60	3	thund.storm
5	29.75	63	63	—	rain	29.67	65	65	—	rain	29.72	58	58	—	rain
6	29.85	57	57	—	fog	29.92	62	56	6	clearing	30.00	54	54	—	fog
7	30.06	62	50	12	very fine	30.05	68	54	14	very fine	30.10	61	54	7	very fine
8	+30.15	59	57	2	overcast	30.14	65	57	8	same	30.10	59	59	—	overcast
9	30.05	63	59	4	very fine	29.98	69	62	7	same	29.94	62	62	—	rain
10	29.83	63	62	1	showers	29.81	71	60	11	very fine	29.85	63	60	3	fine
☾ 11	29.83	64	57	7	overcast	29.63	62	62	—	rain	−29.45	60	60	—	rain
12	29.38	58	56	2	very fine	29.48	65	54	11	very fine	29.71	57	54	3	showers
13	29.86	61	51	10	same	29.90	68	52	16	same	29.95	55	53	2	fine
11	30.01	62	52	11	same	29.97	70	55	15	same	29.97	56	56	—	very fine
15	29.94	59	55	4	same	29.83	66	60	6	overcast	29.80	61	60	1	overcast
16	29.73	62	57	5	same	29.69	64	50	14	very fine	29.69	57	57	—	rain
17	29.66	60	59	1	rain	29.71	67	62	5	showers	29.70	57	57	—	very fine
⊕ 18	29.71	67	+63	4	very fine	29.68	70	57	13	very fine	29.55	63	62	1	lightning
19	29.45	61	61	—	rain	29.51	69	61	8	same	29.46	61	61	—	very fine
20	29.48	64	62	2	very fine	29.46	64	63	1	showers	29.50	61	61	—	same
21	29.54	63	62	1	showers	29.52	62	62	—	rain	29.59	59	59	—	showers
22	29.66	63	60	3	dull	29.79	68	58	10	very fine	29.83	60	59	1	dull
23	29.75	63	61	2	showers	29.63	65	61	4	rain	29.63	61	61	—	rain
24	29.55	63	61	2	same	29.64	67	60	7	very fine	29.64	60	59	1	overcast
☽ 25	29.56	63	62	1	same	29.66	67	56	11	same	29.71	58	58	—	very fine
26	29.73	62	56	6	very fine	29.73	62	62	—	showers	29.73	57	57	—	same
27	29.76	63	53	10	same	29.74	66	58	8	very fine	29.65	58	58	—	rain
28	29.50	61	60	1	showers	29.48	63	61	2	showers	29.46	60	60	—	fine
29	29.48	61	59	2	overcast	29.51	62	61	1	thund.storm	29.55	58	58	—	rain
30	29.52	60	55	5	very fine	29.55	56	54	2	very fine	29.63	51	50	1	very fine
31	29.67	58	52	6	thund. sho.	29.65	59	51	8	same	29.71	51	50	1	same
Means	29.752	67.7	57.2	4.5		29.746	65.5	57.9	7.5		29.751	58.5	57.4	1.	

1822 July.	Temperature. Max.	Min.	Sun.	Rad.	Wind. Direction.	Force.	Rain.
1	68	−46	90	*35	N	little	
2	66	48	100	*40	S W	brisk	
3	70	51	114	*41	W	ditto	
○ 4	+73	56	114	*52	S W	little	
5	65	54	80	*54	S	variable	0.75
6	63	48	70	*41	N	little	
7	69	51	114	*43	N W	ditto	
8	69	57	100	*53	——	ditto	
9	71	57	104	*56	S W	ditto	0.18
10	73	54	106	*45	——	ditto	0.07
☾ 11	66	50	84	*50	——	ditto	0.15
12	66	50	104	*49	N W	brisk	
13	68	46	108	*38	N	ditto	
14	71	47	+121	−*34	N W	ditto	
15	69	57	108	*55	S E	ditto	
16	67	55	110	*55	N	ditto	
17	68	51	92	*44	S W	ditto	0.31
⊕ 18	71	57	104	*57	S	little	
19	69	57	102	*51	——	ditto	0.09
20	69	55	94	*51	——	ditto	0.03
21	67	56	88	*54	——	brisk	0.09
22	70	57	93	*50	W	ditto	0.06
23	66	58	70	*58	S W	ditto	0.10
24	69	57	93	*51	——	ditto	0.02
☽ 25	69	50	83	*43	——	ditto	0.03
26	65	49	84	*38	——	little	0.05
27	68	55	93	*55	——	ditto	0.12
28	66	58	86	*48	——	ditto	0.16
29	68	52	72	*48	W	ditto	0.17
30	62	46	79	*38	N W	ditto	0.03
31	60	46	86	*38	W	ditto	0.19
Means	67.7	52.4	95.	47.2			2.60

JULY.

Mean Temperature	60.0°	Weight of vapour in a cubic foot.	
—— Pressure	29.749 inches		
—— Dew-point	55.1°	Mean	5.289 grs.
—— Force of vapour	0.478 inch	Maximum	6.740 „
—— Degree of dryness	4.9°	Minimum	3.997 „
—— Degree of moisture	853°		

Least observed degree of moisture 536

WINDS.

N. 4 = 52° N E. 0 = E. 0 = S.E. 1 = 61° S. 5 = 61° S.W. 12 = 59° W. 4 = 56° N.W. 5 = 54°.

Amount of rain . . . . 2.60 inches
—— of evaporation 2.54 inches

REMARKS.

The almost daily showers of this month were extremely beneficial, and restored the verdure of vegetation.

1822 August.	Morning. Barometer.	Hygrometer.			Weather.	Afternoon. Barometer.	Hygrometer.			Weather.	Night. Barometer.	Hygrometer.			Weather.
1	29.79	58	52	6	very fine	29.77	63	54	9	very fine	29.75	52	52	—	rain
2	29.70	54	54	—	thund.storm	29.79	61	52	9	hail	29.91	48	48	—	very fine
○ 3	29.98	58	52	6	very fine	29.95	63	53	10	fine	29.91	57	56	1	fine
4	29.83	61	55	6	same	29.74	64	54	10	very fine	29.78	54	54	—	same
5	29.79	54	53	1	fine	29.95	64	54	10	same	29.92	54	54	—	same
6	29.98	59	55	4	very fine	29.98	63	57	6	overcast	30.01	59	57	2	dull
7	30.01	58	56	2	overcast	30.02	65	57	8	very fine	30.06	59	58	1	very fine
8	29.96	59	58	1	fine	29.92	66	59	7	same	29.87	58	58	—	same
9	29.80	60	58	2	very fine	29.73	66	57	9	same	29.76	61	59	2	same
☽ 10	29.78	63	57	6	same	29.79	66	52	14	fine	29.81	62	57	5	dark
11	29.80	64	58	6	same	29.78	68	57	11	very fine	29.79	60	59	1	rain
12	29.81	63	57	6	overcast	29.84	68	62	6	same	29.81	59	59	—	same
13	29.75	65	62	3	rain	29.84	68	61	7	rain	29.93	60	58	2	very fine
14	29.90	65	61	4	same	29.76	66	61	5	overcast	29.66	57	57	—	rain
15	29.69	61	56	5	overcast	29.90	66	54	12	very fine	29.95	54	48	6	very fine
⊕ 16	30.02	58	56	2	very fine	30.10	70	58	12	same	30.10	60	58	2	same
17	+30.18	66	61	5	same	30.18	72	63	9	overcast	30.17	63	63	—	same
18	30.16	64	61	3	haze	30.11	73	62	11	very fine	30.09	63	63	—	same
19	30.12	60	60	—	mist	30.09	73	63	10	same	30.09	59	58	1	same
20	30.08	68	64	4	very fine	30.05	72	64	8	same	30.02	60	60	—	same
21	29.96	68	66	2	same	29.97	81	64	+17	haze	29.91	66	66	—	shower
22	29.86	69	66	3	same	29.82	80	+67	13	very fine	29.86	66	66	—	same
23	29.95	63	62	1	fine	29.92	70	53	17	same	29.90	55	53	2	very fine
☾ 24	29.85	57	54	3	same	29.72	63	56	7	fine	29.60	57	57	—	rain
25	29.70	59	57	2	very fine	29.70	57	57	—	showers	29.68	54	54	—	same
26	29.63	58	56	2	same	29.60	59	54	5	very fine	29.74	54	52	2	very fine
27	29.48	59	58	1	fine	29.50	62	59	3	showers	29.54	58	58	—	rain
28	29.52	60	59	1	dull	29.45	60	60	—	rain	29.42	59	59	—	same
29	−29.35	58	55	3	fine	29.37	64	56	8	showers	29.51	56	55	1	fine
30	29.64	58	55	3	same	29.62	64	57	7	same	29.71	56	55	1	same
31	29.73	60	54	6	very fine	29.73	65	55	10	very fine	29.88	55	55	—	very fine
Means	29.833	60.8	57.6	3.1		29.828	66.5	57.8	8.7		29.843	57.9	56.9	0.9	

1822 August.	Temperature. Max.	Min.	Sun.	Rad.	Wind. Direction.	Force.	Rain.
1	64	50	93	*48	W	little	0.09
2	61	43	100	−*34	N	ditto	0.06
○ 3	65	49	90	*46	N W	ditto	
4	66	47	101	*41	N E	ditto	
5	65	51	106	*42	——	ditto	0.05
6	65	50	96	*41	W	ditto	
7	68	46	104	*39	——	ditto	
8	71	48	110	*39	E	ditto	
9	71	53	109	*47	W	ditto	
☽ 10	68	56	102	*55	——	ditto	
11	69	56	96	*54	S W	ditto	
12	69	57	98	*56	S	ditto	0.05
13	71	56	98	*47	S W	ditto	0.03
14	68	55	88	*51	——	ditto	
15	69	46	103	*39	W	ditto	
⊕ 16	73	52	103	*41	——	ditto	
17	73	54	114	*48	S W	ditto	
18	74	53	114	*41	S E	variable	
19	77	54	117	*50	——	little	
20	75	56	+125	*47	——	ditto	
21	+82	63	123	*60	S	ditto	
22	82	57	118	*50	——	ditto	0.07
23	73	49	120	*37	N	ditto	
☾ 24	64	49	94	*48	W	ditto	0.10
25	64	48	84	*47	——	ditto	0.02
26	70	48	96	*37	N W	ditto	
27	70	49	72	*47	S W	ditto	0.07
28	68	54	70	*50	S E	ditto	0.11
29	67	50	72	*47	W	ditto	0.02
30	68	44	76	*42	——	ditto	0.03
31	69	−41	94	*35	——	ditto	
Means	69.6	51.1	99.5	45.3			0.70

AUGUST.

		Weight of vapour in a cubic foot.	
Mean Temperature	60.3°		
—— Pressure	29.834 inches		
—— Dew-point	54.4°	Mean . . .	5.222 grs.
—— Force of vapour . .	0.467 inch	Maximum .	7.473 ,,
—— Degree of dryness . .	5.9°	Minimum .	4.279 ,,
—— Degree of moisture .	828°		

Least observed degree of moisture . 659

WINDS.

N. 2=53° N.E. 2=54° E. 1=58° S.E. 4=61° S. 3=63° S.W. 5=59° W. 12=55° N.W. 2=53°.

Amount of rain . . . . 0.70 inch
—— of evaporation 3.00 inches

REMARKS.

The weather as favourable as could be wished for all the operations of a very early harvest.

ON THE TRADE WINDS:

CONSIDERED

WITH REFERENCE TO MR. DANIELL'S THEORY OF THE CONSTITUTION OF THE ATMOSPHERE.

In a Letter from Captain BASIL HALL, R.N., F.R.S.

Edinburgh, 19*th January*, 1827.

MY DEAR SIR,

As you wished to have in writing the substance of my remarks during our last conversation, I have thrown them together for you in the following letter; and though they go but a short way to exhaust the subject, I trust they may be of use to you in your very interesting speculations on meteorology.

Many persons have a very distinct, but, as I conceive, a very erroneous conception of the Trade Winds. This idea, which has been acquired at school, subsequent reading will probably have tended to confirm; it is, that in north latitude these winds blow always exactly from N.E., and in south latitude exactly from S.E. Some people, again, have merely a vague recollection that the Trades blow from the eastward: while very few persons, probably, are fully aware of the real state of the fact; and it was this belief which induced me to bring the subject to your notice. It is true that there is hardly any thing stated here which is not somewhere or other adverted to in your Essays, either directly, or by implication; but still I conceive your general doctrines may receive support from the practical illustrations which professional occupations have thrown me in the way of observing.

Professional men, as you well know, are so apt to overrate the importance of their own topic, that I hope you will indulge me with a little rope, in my endeavours to explain one of the most curious, and, at the same time, practically useful phenomena in nature. On the other hand, I am so well aware that it is often difficult for readers to understand subjects which lie much out of the ordinary line of their thoughts, that I shall endeavour to render the whole as simple as possible; indeed my only fear is, that I may be accused of being too elementary

It is a remarkable circumstance in the history of meteorology, that some of the highest authorities should assign a totally erroneous direction to these winds. I may instance in particular the Chart given in Dr. Young's "Natural Philosophy," where some of the most striking facts of the case are altogether misstated. On the other hand, it is curious enough that the best account, not only of the Trades, but of every other wind, is to be found in the works of Dampier, under the express head of an Essay on Winds and Currents. Undoubtedly, the facts which he has so skilfully arranged, might be picked out of the works of Cook and other voyagers; but Dampier, whose means were, beyond all comparison, less than were enjoyed by his successors, had the merit of condensing and separating his information in such a way, as not only to render it available to practical men, but, from the simplicity of the composition, to make his writings agreeable to every class of readers. To persons, therefore, who wish a more detailed account of the different Tropical winds than I can give you here, I can recommend nothing more satisfactory than the Essay alluded to by this prince of all voyagers, though published more than a century ago.

In modern times, by far the most extensive and exact account of the winds, especially of those blowing between the Tropics, is contained in Horsburgh's book of "East India Directions,"—the most valuable gift, perhaps, which well-directed industry has bestowed upon modern navigation. I may be excused, I hope, for using these strong terms, when I mention, that under the sole guidance of this volume, I have sailed over more than a hundred thousand miles of the earth's surface,—sometimes in the dark,—sometimes in stormy weather,—and frequently when not a single soul on board had ever visited the spot. Yet in proportion as my local knowledge became matured, and I could judge of the subject from what I had actually seen, the more unlimited my confidence became in the authority of this admirable navigator.

The only general assertion that can be made with respect to the Trade Winds, as far as their direction is concerned, is, that they blow more or less from the east towards the west. Even this, however, is not universally true. Neither are they alike on both sides of the Equator; nor do they exhibit the same aspect, at different seasons of the year, on the same spot.

I shall first glance at the received ideas upon the subject, and then describe the actual state of the facts as I have observed them; after which, I shall endeavour to give the laws, which regulate these singular phenomena, a place in your imagination, by a theoretical consideration of their cause.

I may mention that these views having suggested themselves to my mind before I met with your book, I had intended to publish some notice respecting them: on seeing, however, the complete manner in which you had exhausted the subject, I abandoned my intention; and I only resume it now, as you seem to think the corroboration of a practical man's opinions may help to substantiate the truths which your sagacity and industry have deduced theoretically.

A few words will serve to describe the common notions upon the subject.—The north-east Trade Wind is conceived to blow from the exact north-east point, nearly to the Equator, when it takes a graceful bend, and blows more and more from the east point, till at length it becomes parallel to it; that is, blows from due east. The south-east Trade, in like manner, is supposed to blow at first precisely at south-east, or at an angle of 45° with the meridian, and at last to assume an exact parallelism with the Equinoctial Line*. This, however, is altogether erroneous. The real state of

* See the Chart in Dr. Young's "Natural Philosophy."

things is as follows. The Trade Winds in the Atlantic and Pacific ocean extend to about twenty-eight degrees of latitude on each side of the Equator, sometimes a degree or two farther; so that a ship, after passing the latitude of thirty degrees, may expect every day to enter them. It will perhaps assist the apprehension of the subject, to suppose ourselves actually making a voyage to the Cape, first outwards, and then homewards; by which means we shall have to cross each of these winds twice. Shortly after leaving Madeira, which is in $32\frac{1}{2}°$, we get into the Trades, and instead of finding the wind blowing from N.E., as the accounts would lead us to suppose, we shall find it blowing from east, or even sometimes a little southerly. You are seaman enough to be aware that, with the wind at east, a south course can readily be steered, first towards the Canaries, and then to the Cape de Verd islands. It is the most approved practice, I think, to pass just within sight of these islands to the westward of them; that is to say, leaving them on the left hand. As the ship advances to the southward, she finds the Trade Wind drawing round gradually from east to north-east, and finally to north-north-east, and even north at the southern verge of the north-east Trade. This last-named or northern direction, it will be observed, is at right angles to that usually assigned to it—due east, near the Line. The southern limit to the north-east Trade Wind varies with the season of the year, reaching at one time to within three or four degrees of north latitude, and at other times, not approaching it nearer than ten or twelve degrees; but it never crosses the Equator and enters the southern latitudes. It will aid the memory in this matter, to bear in mind that the line, which limits or marks the termination of this Trade Wind, follows the sun. In July and August it recedes from the Equator, in pursuit, as it were, of the sun; while in December and January, when the sun has high southern declination, it reaches almost to the Line.

The great difficulty of the outward-bound voyage commences after the ship is deserted by the N.E. Trade, as she has then to fight across a considerable range of calms, and of what are called the "variables," where the wind has generally more or less southing in it. At certain seasons it blows freshly from the S.S.W., and greatly perplexes the young navigator, who, from trusting to published accounts, expects to find the wind, not from south, but from east. This troublesome range varies in width from 150 to 550 miles; is widest in September, and narrowest in December or January. I speak now of what takes place in the Atlantic; for it is not quite the same far at sea in the Pacific Ocean, where fewer modifying circumstances interfere with the regular course of the phenomena, than in the comparatively narrow neck formed by the protuberances of Africa and South America.

I may remark in passing, that it is upon a knowledge of these deviations from the general rule, which we are pleased to call *irregularities*, that much of the success of tropical navigation depends. A seaman, who trusts to theory alone, will, in all probability, make a bad passage; while another, who relies solely upon past experience, will probably, if the season happens to be different, ao quite as badly. The judicious navigator will endeavour to unite the two; and having attentively studied the theory of his subject, and sought to reduce every case to its principles, checking these from time to time by fresh experience, may be able, when occasions arrive where his own knowledge or that of others entirely fails him, to take that course which, all things considered, is most likely to serve the purpose he has in view.

I knew an officer who was ordered to cruize in the Mozambique channel, between Africa and Madagascar, until a certain day, and then to proceed to the Isle of France. At the time appointed he sailed to the northward; but, though he proceeded nearly to the Line in search of a N.W. wind, he could not make a bit of easting; and, after six weeks of ineffectual struggle between the north end of Madagascar and the Equator, he was obliged, for want of water, to run for a port in Africa, where he lost the half of his crew by sickness, and was compelled to bear up at last for the Cape of Good Hope, and the whole object of his mission was defeated. Unfortunately, while he knew nothing of the theory of these subjects, he had heard, in a general way, that north westerly winds occasionally blew in that quarter, between the Trade Winds and the Equator. He was right, indeed, as to the fact, but wrong as to the season. A very slight knowledge, however, of the principles which regulate the winds, might have taught him that the *irregularity*, of which he hoped to take advantage, would probably not have occurred at that particular season, and that he ought to have gone, not to the north, but, in the opposite direction, to the southward, where he would certainly have found a fair wind. Had he done so, a fortnight or three weeks would have placed him in the port he wished to reach.

But I am forgetting our voyager. We had reached that spot where the N.E. Trade Wind left us rolling about in a dead calm, or with only an occasional violent squall, accompanied by deluges of rain, in a climate so hot that the slightest cat's-paw of wind is hailed with the utmost delight. In process of time, the ship, by taking advantage of every such puff of wind, gets across this troublesome stage of her journey, and meets the S.E. Trade. It is very material to remark that this wind does not blow from the east, as the navigator is led to expect, or in a direction parallel to the Equator, and which would be to him a fair wind; but it meets him, as it is emphatically termed, *smack in the teeth*. Instead, therefore, of steering away S., or S.S.E. for the Cape of Good Hope, he is obliged to keep his wind as closely as possible, and he may think himself fortunate, in a dull sailer, if he can clear the coast of Brazil without making a tack. As he proceeds on, however, the wind gradually hauls to the south-eastward, then to E.S.E., and at last E., at the southern limit of the Trade Winds properly so called. Here, after a little baffling weather, he is almost certain of finding westerly winds, which prevail in the latitudes beyond the Trades in both hemispheres.

Such are the phenomena most generally observed

with respect to the regular Trade Winds outward-bound. We shall now, in order to make things quite clear, invert the order of the voyage, and suppose the ship, after having reached the Cape of Good Hope, to turn back again. At first she may be plagued with westerly and north-westerly winds; but she will generally be able to stretch into the Trades, where she will at first find the wind hanging far to the east, and it may even have some northing in it at first. As she proceeds onwards to St. Helena, which lies directly in the track of homeward-bound ships, the wind will draw to the east, —east-south-east,—south-east, — and, eventually, to south-south-east. At crossing the Equator, it will probably be blowing from due south, and not (I must again beg you to take particular notice) from due east, as we are generally led to suppose. After reaching three or four degrees of north latitude, the ship will lose the south-east Trade, and re-enter the "variables," where, when it is not calm, she will generally find light southerly winds, and, at one period of the year, namely, about July and August, blowing briskly from the south-west, as far as ten or twelve degrees of north latitude. At other seasons, especially when the sun is near the Line, a ship may expect light winds from all quarters of the compass, long calms, and now and then a furious squall, with deluges of rain. But, at every season of the year, the homeward-bound passage, or that from the southward, is much easier made than the reverse.

On reaching the southern limit of the N.E. Trade Wind, the seaman finds the wind blowing in his face from the north, (exactly as he formerly met the S.E. Trade, blowing, not from east, but from the south Pole,) and is obliged to stretch away to the W.N.W. at first, and then N.W, as if he were going to the United States of America—not to Europe. As he sails on, and gets more into the Trade, it draws round gradually to N.E. and E.N.E., which allows of his "coming up" more and more every day, till at length he can steer north,—and even north-east; so that he is enabled frequently to "look up" for the Azores or Western Islands. By-and-by he bids adieu to the N.E. Trade, in about 28 or 29 degrees of north latitude, as he formerly did of the other Trade, in the correspondent degree south. In like manner also he will now almost always meet with westerly winds, which will carry him to the Channel. It may be remarked by the way, that these westerly winds are not so regular as they are in the southern hemisphere, owing probably to the comparative absence of land, which enables the general principle, by which the winds are produced, to act there with greater uniformity*.

If these descriptions have been rendered suffi-

* The number of days required by the packets between Liverpool and New York, to make the passage outwards and homewards, places this in a striking point of view:

The average of the whole of the passages made by the packets, in six years, from Liverpool to New York, that is, from east to west, is .... 40 days.

The average, during the same period, of the same vessels from New York to Liverpool, or from west to east, is .... 23 days.

—See *Hodgson's Letters on North America*, vol. ii. p. 345.

ciently intelligible to a person who has not before considered the subject, I think he will be in a situation to comprehend the theory; and when that is duly fixed in his imagination, he will find it useful to go back again to the facts stated above, with sharper powers of observation, and a judgment more fitted to arrange and generalize these materials to good purpose. To persons indeed already acquainted with your admirable "Meteorological Essays," much of what follows will appear superfluous; but I prefer giving my own views, without reference to your book, (which I had not seen at the time I conceived the following explanations,) in order to keep up the connexion between my experience and the theory actually suggested thereby at the time.

It may be right to state, however, that I by no means pretend to assert that these ideas have any claim to originality; for there are some treatises in which parts of the theory are to be found, though I am not acquainted with any work, antecedent to yours, which accounts for the direction and force of the winds on principles applicable to practice.

The most elaborate work I am acquainted with on these subjects, is Col. Capper's "Observations on the Winds and Monsoons," printed in 1801. But he never once, as it appears to me, throughout his work, assigns the right cause for the phenomena, which, in most cases, he describes extremely well. Dr. Halley's theory of the wind following the course of the sun in his diurnal motion to the west, (which Col. Capper quotes,) is equally unsatisfactory. In describing the trade winds of the

Atlantic, Colonel Capper errs essentially in several particulars, especially where he says that the north-east perennial extends sometimes to four or five degrees south of the Equator; which, I believe, it never does*.

If air, at any particular spot, be heated, it becomes specifically lighter than the adjacent cooler parts, and consequently rises; while its place is speedily occupied by the contiguous less rarefied or colder air. Now the region of the globe lying between the Tropics, or we may say between thirty degrees on each side of the Equator, being exposed to the most direct rays of the sun, becomes heated; and the air in contact with this belt or zone becoming rarefied, rises with more or less rapidity, according to the circumstances under which the earth is situated. Where an open ocean is found, the incumbent air will be less heated, as in the Pacific, than where districts of dry earth are found, as in Mexico for instance. The partial vacuum thus formed will, in both hemispheres, be supplied by the adjacent air lying, we shall suppose, between the latitudes of thirty and fifty degrees. If this be admitted, most of the phenomena of the Trade Winds will, I conceive, be readily explained. It must be granted, however, before proceeding farther, that a volume of air put into motion is, like every other body, possessed with a momentum, which will continue that motion till stopped by its friction against the fluid through which it is propelled, or by that of the surface of a

* Capper, page 36.

solid body along which it may be impelled. Any one who has observed the ring of smoke sometimes projected from the mouth of a cannon will understand this ; or the familiar experiment of blowing out a candle by means of the air forced from an uncharged gun, by means of one of the copper priming caps, affords ample illustration that a mass of air once put in motion, will retain that motion like any other portion of matter.

The velocity of the earth's rotation at the Equator is, in round numbers, 1000 miles an hour : at latitude 30° it is about 860, or about 140 miles an hour slower. The average velocity of the earth's easterly motion, in the space between the Equator and latitude 30°, may be stated at 950 miles an hour; while that of the belt lying between thirty and forty degrees, is not much above 800 miles an hour*.

The superincumbent air at these places respectively, *supposing no difference of temperature to exist*, would of course partake of the earth's velocity, and

* If the Equator be supposed to move at the rate of 1000 miles an hour, the different parallels of latitude will move at the following rates, which are to 1000, as the cosine of the latitude to radius.

Latitude.	Velocity per Hour.	Latitude.	Velocity per Hour.
°		°	
0	1000	50	643
10	985	60	500
20	940	70	342
30	866	80	174
40	766	90	

there would be an universal calm. But, if we suppose the Tropical region to be heated, the air over it will instantly ascend, and take its station above the cold ; while the colder and more dense air lying beyond the Tropics will rush in to occupy its place, below that which has been heated. This hardly needs illustration ; but, as I have more than once met with people who did not immediately see the consequences which follow from placing two fluids of different density side by side, I may suggest the experiment of a trough, divided, by a sluice in the centre, into two spaces, one of which may be filled with water, the other with quicksilver ; both fluids will of course be at rest until the sluice be drawn up, when the heavier fluid will instantly rush in beneath the lighter, and the lighter will flow along above the quicksilver. If, instead of these fluids, we substitute hot and cold water, the same thing will take place, the cold always flowing under the hot, towards the place formerly occupied by the lower strata of the heated fluid ; while the heated portion flows along over the cold, towards the place formerly occupied by the upper strata of the cold fluid. Exactly the same thing will take place if two portions of air, at different temperatures, be the contiguous fluids ; though the phenomena will not now strike the senses so strongly.

It would not be difficult, I conceive, to have a globe fitted with a contrivance which should represent the operation of the Trade Winds ; and perhaps a description of such an apparatus will be as ready a method as any other of explaining my views of this theory. Having taken a common globe, I would inclose its tropical region from thirty degrees north to thirty degrees south, in a glass zone or coating concentric with the globe, and also each of the belts lying between the latitudes of thirty and fifty degrees in like manner, with distinct cases placed respectively in close contact with the tropical glass coating, and divided from it by partitions removeable at pleasure; I would fill the tropical case with hot water, and the middle latitude cases, or those embracing the space contained between the latitudes of thirty and fifty degrees in both hemispheres, with cold water; or, which would represent the actual fact still better, a broad ring of heated iron might be fixed round the equator to represent the torrid zone, while the middle or temperate latitudes, both north and south, should be encircled with rings of ice. The water might also be coloured in order to render the effects visible. Things being arranged as above described, and the globe being supposed *for the present* at rest, if the division between the hot and the cold fluids were removed, the cold water would gradually slide along *under* the hot towards the equator, while the heated water would be carried *over* the cold towards the poles; and, if nothing else were done, that is to say, if the globe were allowed to remain at rest, a mere circular interchange would take place. The temperate portions of the fluid, on coming into contact with the torrid zone of the globe, and being thereby heated and rendered specifically lighter, would necessarily rise; while the hot portion, on flowing towards the cooling substance in latitudes farther from the equator, would descend to occupy the place of the cold water drawn off to supply the place of the lighter heated water at the equator. A steady current would in this way be produced, running below towards the equator, and at right angles to it, and above towards the poles; this would evidently be the only motion impressed on the fluid as long as the globe stood still.

It is material to remark here, that this motion would be less and less obvious as the currents approached the equator, where the cold fluid would gradually become heated, and have a tendency to rise as well as to flow along, so that their course would be checked, till at length, at the equator, the opposite currents would meet and produce a calm.

While things are supposed to be in this situation, let the globe be put into rapid motion from west to east, we shall say, for the sake of illustration, at the rate of one thousand feet in a minute, while all the circumstances as to temperature remain as before. The cold water would continue to flow just as before, under the hot, towards the equator, where the rarefying cause existed, but it would now come to the equatorial regions, possessed, not only with a motion directly towards the equator, but with the easterly velocity due to that circle of latitude which it had left, or about eight hundred feet in a minute; and if we suppose these equatorial regions to be moving to the eastward at the average rate of nine hundred and fifty feet in the same interval, the cold water moving at the slower rate would inevitably at its first arrival there be left behind; or, which is the same thing, the surface of the globe would go faster to the eastward than the superincumbent water, and this, in effect, would produce an apparent or relative motion of the water from east to west; or, if the fluid in question were air, we should there have what we call an easterly wind.

This, in its most general sense, is what really takes place with the Trade Winds, and, if what I have said be well understood, all the modifications which they undergo will be readily seen to follow.

The cold air, however, (it must be carefully observed,) which comes towards the equator, is acted upon by two forces, or, in other words, is influenced by two sources of motion; first, by that which has been impressed upon it, in a due easterly direction, by the rotation of the earth in the temperate latitudes it has left: and, secondly, by a motion, in the direction of the meridian, towards the equator, and at right angles to it. This last is caused by the air rushing in to fill up the space left by that which has been rarefied by the heat of the torrid zone, as shown in the first experiment where the globe stood still; in which case, it will be remembered, this was the only motion to which the fluid was exposed. The combined effect of these two motions is to produce the south-east Trade Wind in south latitude, and the north-east trade on the other side of the equator.

When the comparatively slow moving air of the temperate zone, caused by the rotatory motion of the earth to the east, first comes into contact with the quick moving or tropical belt of the globe, the difference of their velocities is great compared with the other motion of the air above described, or that directly towards the equator; and consequently the wind blows at the extreme edge of the Trades nearly from the east point. As this cool air, however, is drawn nearer to the equator, and comes successively in contact with parallels of latitude moving faster and faster, this constant action of the earth's rapid easterly motion gradually imparts to the superincumbent air the rotatory velocity due to the equatorial regions which it has now reached; that is to say, there will be less and less difference at every moment between the easterly motion of the earth and the easterly motion of the air in question; while, at the same time, the other motion of the same air, or that which has a tendency to carry it straight towards the equator, having been exposed merely to the friction along the surface without meeting any such powerful counteracting influence as the earth's rotation, will remain nearly unchecked in its velocity. Thus, as I conceive, the Trade Wind must gradually lose the eastern character which it had on first quitting the temperate for the tropical region, in consequence of its acquiring more and more that of the rotatory motion of the earth due to the equatorial regions it has now reached. While this cause operates, therefore, to destroy the easterly direction of the Trades, their meridional motion, as it may be called, or that towards the equator, by remaining constant or nearly so, will become more and more apparent, till at length, when the friction of the earth in its rotatory motion has reduced the velocity of the cool air to the tropical rate, there will be left only this motion towards the equator, which is found invariably to characterize the equatorial limits of both trade winds. This velocity, also, is at length checked, first, by its friction on the surface of the earth; secondly, by the air becoming heated, which causes it rather to rise than to flow along the surface; and thirdly, by the meeting of the two opposite currents—one from the north, the other from the south.

In confirmation of these doctrines, I may state that, in the Trade Winds, the higher clouds are very seldom, if ever, observed to go in the same direction as the wind below. In general they are seen to move nearly in the contrary direction; and I find it noted in my journal, that on the top of the peak of Teneriffe, the wind was blowing from the south-west, directly in the opposite direction to the Trade Wind below.

In what has been said above, the quickest moving or equatorial belt of the earth is assumed as being also the hottest and consequently that over which the air has the greatest tendency to rise. This, however, is not the case universally; and where variations in this respect occur, effects very different from those described are the result. The most striking examples, with which I am personally acquainted, of this deviation from the general law of the Trade Winds, or that which would obtain, were the earth a uniform mass of water, or land, occur in India and Mexico. That portion of the Pacific Ocean, which stretches from the Isthmus of Panama to the Peninsula of California, lies between eight and twenty-two degrees of north latitude. Now, the sun's rays strike directly upon the adjacent great territory of Mexico, and, by heating the land violently, cause the air to rise over it. But the vacuum is filled up not only from the northward, but by the comparatively cold air of the equatorial regions in the neighbourhood. This air coming from that part of the globe which revolves quickest, to one which moves more slowly, produces not an easterly, but westerly and south-westerly winds;—so that the navigator, who works by what is called the rule of thumb, and takes things for granted, instead of inquiring into them, will be very apt to make sad blunders in his navigation. I confess that I once laid myself open to an accusation little short of this, for which I had less excuse, perhaps, than another man, since, from having long speculated upon these topics, I had in a great measure satisfied myself of the truth of these theories. Yet when I was sent to visit the south-west coast of Mexico alluded to, and was left to my own choice as to the manner of performing the voyage, I miscalculated the probable effect of so vast a heater as Mexico, and expected to find the winds from east or north-east; and therefore began my voyage at Panama. I soon learned, however, to my cost, that instead of being to windward of my port, I was dead to leeward of it, and I had to beat against westerly winds for many weeks.

After all, however, it is by this union of theory and experience, (which is not the worse for being dearly bought), that effectual knowledge can be obtained; and the disasters into which we are led by ignorance must be serious indeed, if they be not more essentially profitable, than mere unobservant success would have been. I mean that our finding things as we expected them is not always a proof that we have reasoned correctly,—for had I visited this coast at another season of the year, and found an east wind blowing, I might have called it the north-east trade, perhaps, and brought away none of the local knowledge, which is now, I trust, well engraved on my mind by the laborious process of rectifying my original error.

The monsoons in India, in like manner, are striking illustrations of this modified part of the theory. When the sun has great northern declination, the Peninsula of Hindostan, the north of India, and China, being heated, the quick-moving equatorial air rushes to the northward to fill up the slow-moving rarefied space, and this supply being possessed not only with a rapid eastern velocity, but with a motion from the south, produces the south-west monsoon in the Indian Ocean, Bay of Bengal, and in the China Sea. When the sun, on the other hand, goes to the south, the same seas are occupied by air which, coming from regions beyond the northern tropic, possesses less easterly velocity than the space they are drawn to, which gives them an easterly character; and this combined with their proper motion, if I may so call it, from the north, produces the north-east monsoon.

There are numberless other less striking modifications of these principles, which give a high degree of interest to the science of navigation, particularly between the tropics;—but which it is needless to enter into just now. It may however be useful to mention one important case which occurs in the Atlantic, when the sun has high northern declination, and the north of Africa is much heated; the equatorial air is then invited to the north, and a brisk south-west or south-south-west wind blows in the space between the equator and the southern limit of the north-east Trade Wind, which lies then in ten or twelve degrees of latitude, greatly to the astonishment of the inexperienced navigator, who, trusting to his books, expects a wind directly the reverse.

The same reasoning, precisely, will serve to account not only for the direction but for the degree of strength with which the winds blow between the trades and the polar regions,—that is from 30° to 60°. The heated air which rises over the tropical belt, is carried towards the poles, till it is sufficiently cooled, when it descends, and, by encountering a part of the globe going to the eastward at a much slower rate, produces westerly winds. It must be observed also that, as the lower or cold air of this range proceeds towards the equator, it encounters, at every stage of its course along the surface, parallels of latitude moving faster and faster to the eastward, and consequently is exposed to more and more friction, by which means the relative difference between its velocity and that of the earth becomes at every moment less and less, till it subsides at length into a calm. But the equatorial air, on the contrary, in its progress towards the middle latitudes, comes constantly to regions of the globe moving with less and less velocity, so that it descends from the high regions of the atmosphere, along which it has passed with less friction to check its easterly motion, than the lower or cold current must have had to contend with, in its passage along the earth's surface. This equatorial air, therefore, comes, with scarcely any diminution of its original velocity, into contact with a part of the earth moving more than a hundred miles more slowly to the eastward than itself. Consequently we have furious westerly gales as far as Madeira, on the one side, and the Cape of Good Hope on the other, which lie just beyond the north-east and south-east trade winds, in the opposite hemispheres.

There are many other modifications of this theory of the winds with which it is not at present my purpose to trouble you, but I may mention, before closing my letter, that I do not remember to have met with a single circumstance connected with the winds, in any part of the globe, that, when I succeeded in understanding it, was inconsistent with those philosophical reasonings which you have the undoubted honour of having first brought distinctly before the public.

I remain, my dear Sir,

Most truly yours,

Basil Hall.

To J. F. Daniell, Esq.

ON EVAPORATION,

AS CONNECTED WITH ATMOSPHERIC PHENOMENA.

THE subject of evaporation has occupied, at various times, much of the attention of natural philosophers, and many accurate and interesting observations have been recorded of the formation and diffusion of elastic fluids, from various kinds of liquids. The circumstances, especially, attending the rise and precipitation of aqueous steam in the atmosphere, are acknowledged to be important in the highest degree, as upon their silent influence depends the adjustment of many of those important meteorological phenomena, with which is connected the welfare of the organized creation. The labours of De Luc, De Saussure, and particularly of Mr. Dalton, have thrown considerable light upon this never-ceasing process; but something appears to be still wanting to complete the investigation, and to follow up the results to their ultimate consequences. The following observations, however inadequate to fulfil this desirable purpose, may possibly attract some attention to the subject, and may be the means of indicating the points which most require elucidation.

It is a well-known fact that water, under all circumstances, is endued with the power of emitting vapour, of an elastic force proportioned to its temperature. It is also well understood, that the gaseous atmosphere of the earth, in some degree, opposes the diffusion, and retards the formation of this vapour ; not, as Mr. Dalton has shown, by its weight or pressure, but by its *vis inertiæ*. What is the amount of this opposition, and by what progression it is connected with the varying circumstances of density and elasticity, have never yet been experimentally explained.

It may facilitate the comprehension of the subject, to distinguish three cases with regard to the evaporating fluid : the first, when its temperature is such as to give rise to vapour equivalent in elasticity to the gaseous medium, and when it is said to boil; the second, when the temperature is above that of the surrounding air, but below the boiling point; and the third, when the temperature is below that of the atmosphere.

With regard to the first, all the phenomena have been accurately appreciated. The quantity evaporated from any surface, under any given pressure, is governed, in some measure, by the intensity of the source of heat, and is in no way affected by the motions of the aerial fluid. The elasticity of the vapour is exactly equivalent to that of the air, which yields *en masse* to its lightest impulse. When disengaged, it is immediately precipitated in the form of cloud, giving out its latent caloric to the ambient medium ; and under that form is again exposed to the process of evaporation, according to the laws of the third division of the process. All the phenomena attending the process of boiling, have been ably investigated by Gay-Lussac, Dalton, Ure, and Archdeacon Wollaston ; but, as they have but little connexion with the atmospheric relations, which are the particular object of the present paper, I shall proceed to the second case of evaporation.

When the evaporating fluid is of a higher temperature than the surrounding air, but not so high as to emit vapour of equal elasticity to it, the exhalation is proportionate to the difference of temperature. The gaseous fluid, in contact with the surface, becomes lighter by the abstraction of portions of the excess of heat, and, rising up, carries with it, in its ascent, the entangled steam. This, as in the former case, is precipitated, and, in the form of cloud, exposed to the third species of evaporation. This process is not only proportioned to the difference of temperature, and the elasticity of the vapour, but is also governed by the motion of the air. A current of wind tends to keep up that inequality of heat upon which it depends, and prevents that equalization which would gradually take place in a stagnant air. Such is the evaporation which often takes place in this climate, in Autumn, from rivers, lakes, and seas, and which is indicated by the fogs and mists which hang over their surfaces.

It is, however, the third modification of circumstances, which is the most interesting in the point of view which I have suggested, and from which I have merely distinguished the preceding, to free the subject from ambiguity. When the temperature of water is below that of the atmosphere, it still exhales steam from its surface ; but, in this case, the vapour, neither having the force necessary to displace the gaseous fluid, nor heat enough to cause a circulation, which would raise it in its course, is obliged to filter its way slowly through its interstices; and the nature of the resistance it meets with in this course is the first object of investigation.

The force of vapour, at different temperatures, has been determined with great accuracy, and the amount of evaporation has been shown to be, *cæteris paribus*, always in direct proportion to this force. The quantity is also known to depend upon the atmospheric pressure, but I know of no experiments which establish the exact relation between the two powers. I attempted to elucidate the point as follows:—

By inclosing in a glass receiver, upon the plate of an air-pump, a vessel with sulphuric acid, and another with water, and by properly adjusting the surfaces of the two, it is easy to maintain, in the included atmosphere of permanently-elastic fluid, an atmosphere of vapour of any required force; or, in the usual mode of expressing the same fact, the air may be kept at any required degree of dryness. The density of the air, in such an arrangement, may, of course, be varied and measured at pleasure. Now there are three methods of estimating the progress of evaporation in such an atmosphere: the first, and most direct, is to weigh the loss sustained by the water in a given time; the second, to measure, by a thermometer, the depression of temperature of an evaporating surface; and the third, to ascertain the dew point, by means of the hygrometer.

Experiment 1.

The receiver, which I made use of, was of large capacity, and fitted with a hygrometer. I placed under it a flat glass dish, of 7½ inches diameter, the bottom of which I covered with strong sulphuric acid. The glass bell but just passed over it, so that the base of the included column of air rested everywhere upon the acid. In the centre of the dish, was a stand with glass feet, which supported a light glass vessel of 2·7 inches diameter, and 1·3 inches depth. Water to the height of an inch was poured into the latter, the surface of which stood just three inches above that of the acid. A very delicate thermometer rested in the water, upon the bottom of the glass, and another was suspended in the air. It may be necessary to observe, that the sides of the vessel were perpendicular to its bottom, which was perfectly flat. The height of the barometer was 29·6, and the temperature of the water 56°. In twenty minutes from the beginning of the experiment, the hygrometer was examined, and no deposition of moisture was obtained at 26°.

This being the greatest degree of cold which could be conveniently produced by the affusion of ether, the experiment was repeated, with a contrivance which admitted of the application of a mixture of pounded ice and muriate of lime, to the exterior ball of the hygrometer. In this manner the interior ball was cooled to 0°, without the appearance of any dew. The temperature of the water and air was, in this instance, 58°, and the pressure of the atmosphere 30·5.

From this experiment it appears, that in the arrangement above described, the surface of water was not adequate to maintain an atmosphere of the small elasticity of ·068 inch; in other words, the degree of moisture in the interior of the receiver could not have exceeded 129, the point of saturation being reckoned 1000. How much less it was than this, or whether steam of any less degree of elasticity existed, the experiment, of course, did not determine. We may reckon, however, without any danger of error in our reasoning, that the sulphuric acid, under these circumstances, maintained the air in a state of almost perfect dryness.

Experiment 2.

The same trial was made with atmospheres variously rarefied, by means of the pump. No deposition of moisture was, in any case, perceived, with the utmost depression of temperature which it was possible to produce; and the state of dryness was as great, in the most highly attenuated air, as it was in the most dense. In the higher degrees of rarefaction, the water however became frozen.

Experiment 3.

The water, which had been previously exposed to the vacuum of the pump to free it from any air in solution, was weighed in a very sensible balance, before it was exposed to the action of the sulphuric acid under the receiver. Its temperature was 45°, and the height of the barometer 30·4. In half an hour's time, it was again weighed, and the loss by evaporation was found to be 1·24 grains. It was replaced, and the air was rarefied till the gauge of the pump stood at 15·2; in the same interval of time it was re-weighed, and the loss was 2·72, but its temperature was reduced to 43°. The loss from evaporation, in equal intervals, with a pressure constantly diminishing one-half, was found to be as follows:—

Pressure	Temperature. Beginning	Temperature. End	Loss Grains
30·4	45	45	1·24
15·2	45	43	2·87
7·6	45	43	5·49
3·8	45	43	8·80
1·9	45	41	14·80
·95	44	37	24·16
·47	45	31	39·40

When the exhaustion was pushed to the utmost, the gauge stood at 0·07, and the evaporation in the half hour was 87·22 grains. During this last experiment, the water was frozen in about eight minutes, while the thermometer under the ice denoted a temperature of 37.

Now, before we infer from these experiments the state of evaporation from different degrees of atmospheric pressure, it is necessary to apply to the results a correction for the variation of temperature which took place during their progress. The quantity of evaporation having been determined to be in exact proportion to the elasticity of the vapour, we must estimate the latter from the mean of the temperatures before and after the experiments, and calculate the amount for any fixed temperature accordingly. This will, doubtless, give us a near approximation, although, from the last experiment, we perceive that the method of estimating the temperature of the surface water cannot be absolutely correct. The following table presents us with the former results so corrected for the temperature of 45°.

Pressure.	Grains.
30·4	1·24
15·2	2·97
7·6	5·68
3·8	9·12
1·9	15·92
·95	29 33
47	50·74
·07	112·32

Notwithstanding the slight irregularity of the above series, we can, I think, run no risk in drawing from it the conclusion, that the amount of evaporation is, *cæteris paribus*, in exact inverse proportion to the elasticity of the incumbent air; and that De Saussure was misled by his hygrometer, when he inferred from its indications, that a diminution of one-third the density doubled the rate.

Before we proceed, it is necessary to say a few words upon the apparent discrepancy between the results of Mr. Dalton's experiments and mine, as to the amount of evaporation, at the full pressure of the atmosphere. He found, upon the supposition of no previous vapour existing in the air, that the full evaporating force of water, of the temperature of 45°, would be 1·26 grains per minute, from a vessel of six inches in diameter. This amount, reduced in proportion to the squares of the diameters of the two vessels, would give 7·65 grains in half an hour, from the glass of 2·7 inches diameter which I employed. It must, however, be recollected, that Mr. Dalton's calculations were founded upon experiments made at a temperature very considerably above that of the surrounding medium, and that consequently a current must have been established in the latter which greatly accelerated the progress. It is true, that he afterwards subjected his calculations to the test of experience, at common atmospheric temperatures; but then he expressly states, that "when any experiment, designed as a test of the theory, was made, a quantity of water was put into one of them (vessels), and the whole was weighed to a grain; then *it was placed in an open window, or other exposed situation*, for ten or fifteen minutes, and again weighed, to ascertain the loss by evaporation." In this way he ascertained that, with the same evaporating force, a strong wind would double the effect. The difference, however, even after these considerations, is still very striking; but, from several repetitions of the experiment, I have no doubt of its exactness.

Experiment 4.

The arrangement described in the last experiment, having been found adequate to maintain in the receiver a state approaching to that of complete dryness, I had no opportunity of judging whether the elasticity of the vapour, as it rose from the surface of the water, varied in any degree with the pressure of the air, or whether any part of the increase of evaporation were dependent upon such variation. To determine this point, I placed the sulphuric acid in a glass, of the diameter of 2·8 inches, so that its surface was very little more than equal to that of the water. The vessels were placed, side by side, upon the plate of the air-pump, and covered with the receiver. The temperature of the water and air was 52°, and the height of the barometer 29·8. The following table shows the dew point, which was obtained, at intervals of half an hour, at different degrees of atmospheric pressure:—

Barom.	Temp. of Water and Air	Dew Point
29·8	52	36
14·9	53	37
7·45	52	35
3·72	53	36
1·86	52	34
·93	52	36
·15	52	36

The differences of these results are so extremely small, and are moreover so little connected with the variations of density, that there can be no difficulty in regarding them as errors of observation, and we may conclude, that the elasticity of vapour, given off by water of the same temperature, is not influenced by differences of atmospheric pressure. The equal surfaces of sulphuric acid and water here made use of, maintained, at the temperature of 52°, a degree of saturation equal to 570. I repeated the experiment, at the temperature of 61°, and the following are the results:—

Barom.	Temp. of Water and Air	Dew Point
29·6	61	48
14·8	61	49
7·4	60	48
3·7	61	50
1·85	61	48
·92	60	48
·15	61	48

Under these circumstances, the amount of saturation was 651; an increase evidently dependent upon the force of the vapour, but not in exact proportion to its augmentation.

Experiment 5.

Being now desirous of ascertaining in what degree the temperature of an evaporating surface would be influenced by differences in the density of the air, I made the following disposition of the apparatus:—To a brass wire, sliding through a collar of leathers, in a ground brass plate, I attached a very delicate mercurial thermometer; this was fixed, air-tight, upon the top of a large glass receiver, which covered a surface of sulphuric acid of nearly equal dimensions with its base. Upon a tripod of glass, standing in the acid, was placed a vessel containing a little water, into which the thermometer could be dipped and withdrawn by means of the sliding wire. The bulb of the thermometer was covered with filtering paper. At the commencement of the experiment, the barometer was at 30·2 inches, and the temperature of the air 50°. Upon withdrawing the thermometer from the water, it began to fall very rapidly, and in a few minutes reached its maximum of depression. The following table presents the results of the experiment, for different degrees of the air's density; the intervals were each of twenty minutes:—

Barom.	Temp. of Air	Temp. of wet Ther.	Difference
30·2	50	41	9
15·1	49	37	12
7·5	49	34	15
3·7	49·5	31·5	18

Barom.	Temp. of Air.	Temp. of wet Ther.	Difference.
1·8	49·5	28·5	21
·9	49	24·5	24·5
·4	49	23	26

Here, in an atmosphere which a former experiment has proved to be in a state of almost perfect dryness, we find that, at the full atmospheric pressure, the wet surface of the thermometer was reduced 9°. It is worthy of remark, also, how small a quantity of water is required to produce this effect. It has been previously shown, that a surface of 2·7 inches diameter, only lost 1·24 grains in half an hour. This would have been 1·41 grains at the temperature of 49°. The surface of the wet thermometer could not have exceeded $\frac{1}{50}$th of that of the evaporating vessel, and the maximum effect was produced in ten minutes, or $\frac{1}{3}$d of the time, so that the weight of water evaporated in this case was not more than (·0094 grains) one-hundredth of a grain. It will be seen, that the depression increased with the rarefaction of the air, but in the proportion only of the terms of an arithmetical progression to those of a geometrical. The increase is attributable, not to the augmented quantity of the evaporation, but to the decreased heating power of the atmosphere. MM. Du Long and Petit, in their experiments upon the cooling power of air, determined it to be nearly as the square root of the elasticity; but whether the heat which it is capable of communicating to a cold body, follow the same progression, the experiments above detailed are not sufficient to determine with precision. We may, however, certainly conclude from them, that the temperature of an evaporating surface is not affected by the mere quantity of evaporation.

It is right to remark that, in the last experiment, care was always taken to station the evaporating thermometer in the same place in the receiver, for I found that, when the air was highly rarefied, a greater degree of cold could be produced by approximating the wet bulb to the surface of the acid. No difference, however, could be perceived from such a change at the full atmospheric pressure. I also ascertained that no change of relative position in the surfaces of the acid and water produced any alteration in the dew point under any circumstances.

The few simple facts above determined appear to me to be intimately connected with the solution of some very important atmospheric phenomena, and I shall endeavour briefly to indicate their relation.

The aqueous fluid is so abundantly spread over the face of the earth, that there can be no doubt that the permanently-elastic atmosphere, which surrounds it, would very speedily be saturated with its steam, did not some cause, analogous to the sulphuric acid in the receiver, prevent its universal diffusion. This never-failing cause is inequality of temperature. As in the small experiment we found that the degree of dryness was proportioned to the energy of the absorbent mass, and that the existing vapour was equally diffused between it and the exhaling surface; so, in the larger operations of nature, we shall find that the state of saturation is dependent upon the point of precipitation, and that the aqueous atmosphere is nearly uniform between it and the source of steam.

Now, it is well understood that the temperature of the gaseous atmosphere in its natural state must decrease with its density as we ascend to its upper parts; so that a great degree of cold is at all times to be found within a very moderate distance from the surface of the waters. It is this low temperature which determines the tension of the aqueous atmosphere; and it is evident that the evaporation which is thus caused at the base of the aërial fluid, must be accompanied by a simultaneous and equal precipitation above. What then becomes of the precipitated moisture? Let us endeavour to trace the order of this phenomenon. We will first suppose a calm state of the atmosphere, a temperature of eighty degrees, and the barometer at thirty at the surface of the earth. By a calm state of the atmosphere is here meant, one that is free from any lateral wind, and in which, the only currents being in an ascending and descending direction, evaporation would proceed at the rate exhibited in the first column of Mr. Dalton's table. The dew-point at the surface of the earth is sixty-four degrees, and this is determined by the temperature at the height of about 5000 feet, where the barometric column would maintain itself at twenty-four inches. The degree of saturation below would therefore be 600, and the amount of evaporation 1·74 grains per minute from a surface of six inches diameter. This quantity we therefore suppose condensed at the height before named. But the state of saturation in the atmosphere, above this point of precipitation, is again diminished; for we may suppose the force of the vapour to be determined by a temperature of thirty-one degrees at a height of 15,000 feet, where the barometer would stand about sixteen inches. The force of evaporation would, therefore, be 1·71 grains per minute, at the full atmospheric pressure; and this amount, increasing as the pressure diminishes, would give 2·13 grains per minute; so that the power of evaporation at this stage exceeds the supply of moisture, and no cloud could possibly be formed. Above the second point of condensation let us now suppose the force of the vapour to be determined, in still loftier regions, by a temperature of twelve degrees. The force of evaporation would then be 0·44 grains, increased in the proportion of sixteen inches to thirty, or 0·82 grains. Here, then, the power of evaporation would be insufficient to diffuse in the upper regions the whole of the moisture supplied from the surface of the earth, and a cloud, it might be supposed, must consequently result. But another modification of the process now ensues; the precipitated moisture has a tendency to fall back into the warm air below it, and consequently would again assume the elastic form with a rapidity proportioned to the rarefaction of the stratum in which it is diffused. There is, I think, no difficulty in supposing that no visible cloud, or one of extreme tenuity, would be formed during this double process of evaporation. A very important re-action, however, must take place upon the strata of vapour beneath; the elastic force being increased above, enables the water below to maintain an atmosphere of a higher degree, and the quantity of evaporation must decrease as the point of saturation rises. A different arrangement of the points of precipitation would ensue in the progress of these effects.

An important distinction must here be drawn between the ultimate effects of the superior and inferior evaporation denoted above. In the first, the whole weight of water is condensed and simultaneously exhaled; and although it constitutes steam of an inferior degree of force, there is little or no difference in the quantity of its latent heat, and no effect is therefore produced upon the temperature of that portion of the atmosphere in which the change takes place. But in the second, the condensation happens at one spot, and the vaporization at another inferior to it; the latent heat is therefore evolved at the former and communicated to the air, while at the latter the process is reversed, and the air is cooled. The process of this operation would, therefore, tend to equalize the temperature of the atmosphere.

We will next imagine that the surface of the earth is swept by a high wind, and that the atmosphere, instead of resting calmly upon its base, moves laterally with great velocity. Under these circumstances experience has shown that the amount of evaporation will be nearly doubled; but the force of evaporation is not altered in the upper regions. The inferior exhaling surface being immoveable, the motion of the air perpetually changes, and renews the points of contact, and prevents accumulation at any one place; but in the heights of the atmosphere the exhaling surface of the cloud is borne upon the wind, and their relative situations never change.

The progress of precipitation must, therefore, necessarily, under these circumstances, outstrip that of evaporation, and the disturbance of the atmospheric temperature will be greatly accelerated.

There is another cause which would also quicken evaporation below, without equally increasing its power of diffusion at any given height above; and that is a decrease in the density of the air at the surface of the earth. Under the circumstances of our first supposition, imagine the barometer to fall to twenty-eight inches, the evaporation would be increased from 1·74 grains per minute, to 1·86 grains; but this decline of two inches at the surface would indicate a contemporaneous fall of little more than one inch at the height of 15,000 feet, and the rate of diffusion would vary accordingly. When it is considered that great falls of the barometer are generally accompanied by high winds, and that this disparity is multiplied by the force of the current, it is easy to appreciate the influence of this local increase of the power of evaporation.

The facility of evaporation in the rarer regions of the atmosphere will also go far to account for the state of saturation in which the air of mountainous countries is generally found, and many minor meteorological phenomena might probably meet with their explanation from variations of the same cause; such as the fogs which frequently accompany a very high degree of atmospheric pressure, and that peculiar transparency of the air which often precedes rain, and is accompanied by a falling barometer. But to return again to the more general and extended influence of the vapour upon the boundless strata of the atmosphere:—that the phenomena of evaporation and condensation, as we have been contemplating their progress, have not been described with any bias to theoretical considerations, but are in strict accordance with facts and observations, any one might easily convince himself with less difficulty than would at first be supposed. To prove the assertion, I shall extract the following passages from the works of De Luc, who was probably one of the most accurate observers of nature that ever existed, and who seldom, indeed, allowed any hypothetical considerations to warp his description of what he had observed. They will afford a complete illustration of the preceding remarks, although they were penned by him to support a very different hypothesis.

"Si l'on ne fait qu'une légère attention à la surface de ces brouillards vus des montagnes pour en jouir comme d'un beau spectacle, on peut penser qu'ils sont permanens ; que l'évaporation est arrivée à son maximum à la surface des eaux, parce que l'air est parvenu à l'humidité extrême ; et que les vapeurs vésiculaires qui troublent la transparence de cet air restent les mêmes durant des semaines ou même des mois ; c'est-à-dire, tant que le brouillard se conserve à une même hauteur. Mais le phénomène diffère beaucoup de cette première apparence : l'évaporation continue à la surface des eaux, les vapeurs vésiculaires qui s'en forment montent sans cesse, et une nouvelle évaporation a lieu à la surface des brouillards. C'est un spectacle aussi amusant qu'instructif, que celui que fournit cette surface, vue d'un lieu peu élevé au-dessus d'elle, et dans une grande vallée où l'on ait, à quelque distance, des montagnes rembrunies par des fôrets de sapins. Une telle vallée éclairée par les rayons du soleil semble être comblée de coton, filé dans toute sa surface par des êtres invisibles en fils invisibles: il s'y fait par-tout des tumeurs, semblables à celle que produit une fileuse sur sa quenouille en tirant le coton pour former son fil, et elles disparoissent successivement en se dissipant dans l'air. Quelquefois ces tumeurs s'allongent et se séparent de la masse en tendant à monter; on les voit alors s'étendre comme un paquet de gaze qui se déploie et peu à peu elles disparoissent. Les brouillards se forment donc constamment à la surface des eaux et du sol; mais constamment aussi ils se dissipent dans l'air supérieur: et cependant on n'apperçoit point que l'humidité y augmente."—*Idées sur la Météorologie*, tom. xi. p. 78.

"Depuis que mes idées ont changé sur la cause de la pluie, j'ai fort souvent fixé mon attention sur les nuages, et j'ai reconnu très évidemment, qu'ils s'évaporent même tandis qu'ils grossissent. Si l'on fixe ses regards sur leur bord découpé qui, lorsqu'il a pour fond l'azur du ciel, présente mille figures singulières, celles que l'imagination leur prête alors, peut aider à l'examen dont je parle, en rendant leurs changemens plus frappans. Il arrive souvent, que la partie sur laquelle on fixe son attention se dissipe au lieu même où l'on a commencé à l'observer: souvent aussi on la voit s'étendre, sans que la totalité du nuage se meuve, et elle ne se dissipe pas moins durant cette extension. Quelquefois, tandis que l'un des festons du nuage se dissipe on en voit d'autres se former, s'étendre, produire eux-mêmes de nouveaux festons, par où le nuage grossit: d'autres fois il diminue; et alors tous ses festons s'évaporent successivement et il n'en acquiert de nouveaux, que parce qu'il se découpe: on apperçoit en même tems, qu'il devient plus mince, et il disparoît enfin totalement.

" C'est ce qui m'a conduit à penser qu'il y a en effet dans l'air, une source générale de vapeurs qui en fournit en certaines circonstances; que ces vapeurs sont produites au lieu même où se forme un nuage; que c'est par la durée de cette production de vapeurs, que les nuages subsistent, s'aggrandissent même, quoiqu'en s'évaporant tout le tour; et que lorsqu'ils se dissipent c'est que leur évaporation n'est plus réparée par la formation de nouvelles vapeurs."—*Idées sur la Météorologie,* tom. xi. p. 117.

I shall now conclude this essay with an observation which is intimately connected with the subject of the preceding pages. It has been argued that the quantity of heat which would be communicated to the air by the condensation of atmospheric vapour would be trifling, and inadequate to produce those expansions in the aërial currents to which, in my essay upon the constitution of the atmosphere, I have ascribed the fluctuations of the barometer. Now, I have therein shown how the gradual spread of a small increase of temperature, through a considerable stratum, is sufficient for the purpose;

and a very little consideration will, I think, convince any one that the evolution of caloric is by no means so small as has been supposed.

The following rough calculation will place the facts in a striking point of view:—The latent heat of steam has been proved to be somewhere about 970°, and it is known that, whatever be its density, or the temperature at which it is produced, the amount will differ but little from this estimate. The condensation, therefore, of a pound of steam of any degree of elasticity would be adequate to raise a pound of water 970°. The capacity of atmospheric air, of mean density, for heat, compared to that of water, is as ·2669 to 1; therefore the same quantity of heat which would raise a pound of water 1°, would raise a pound of air 3°·7. The condensation of a pound of steam would, therefore, elevate the same weight of air to 3589°. A pound of air is equal to about eleven cubic feet, so that the evolution of heat from the condensation of a pound of steam, would be sufficient to raise the temperature of 3657 cubic feet of air 10°.

When we now look to the depth of water which falls upon the surface of the earth, and recollect that this is not the sole measure of the effect we are endeavouring to estimate, but that the unceasing precipitation and exhalation of the clouds is perpetually extending this influence to the most inaccessible heights, we shall, perhaps, have a juster notion of the prodigious power of atmospheric vapour; and it will, I think, be granted that I have not over-rated the impulse which it is calculated to impart.

ON CLIMATE:

CONSIDERED WITH REGARD TO HORTICULTURE.

[The following Essay was read before the Horticultural Society, August 17th, 1824, who honoured it by the presentation of their Medal. It is here reprinted from their "Transactions," with the leave of the Council.]

THE following observations were committed to paper, and submitted to the consideration of the Horticultural Society, at the particular request of their Secretary. The author would scarcely have thought them novel or important enough for such a destination, but he defers to his judgment, and shall, at all events, have had the pleasure of complying with his wishes.

Horticulture differs from Agriculture in one very material respect. The latter has for its object the fertilization of the soil by manures, and the different processes of cultivation, in the manner best adapted to the peculiarities of any given climate: it concerns itself only with the growth and nourishment of such plants as are indigenous, or, by a long course of treatment, have become inured to the vicissitudes of weather incidental to a particular latitude. The former occupies a much wider field of research; it not only seeks to be conversant with the constitution of soils, but, as it aspires to the preservation and propagation of exotic vegetation, it necessarily embraces the consideration of varieties of climate : and it labours, by art, to assimilate the confined space of its operations to that constitution of atmosphere which is most congenial to its charge, or to protect them at different periods of their growth from sudden changes of weather which would be detrimental to their health. Experience has anticipated theoretical knowledge in suggesting various artifices, by which these ends may be effected; a connected view of which has never, I believe, been attempted, but may prove to be not without interest and utility. The suggestions of experience may probably enlarge the conclusions of theory, while it is not impossible that the improved state of the latter may be found to furnish some assistance to the former.

The science of Horticulture, with regard to climate, will be best considered in two divisions: the first comprises the methods of mitigating the extremes, or exalting the energies, of the natural climate in the open air; the second embraces the more difficult means of composing and maintaining a confined atmosphere, whose properties may assimilate with those of the natural atmosphere in intertropical latitudes. I shall commence my observations with the former.

The basis of the atmosphere has been proved to be of the same chemical composition in all the regions of the globe. All the varieties of climate will therefore be found to depend upon the modifications impressed upon it by light, heat, and moisture ; and over these, art has obtained, even in the open air, a greater influence than at first sight would appear to be possible. By judicious management, the climate of our gardens is rendered congenial to the luxurious productions of more favoured regions, and flowers and fruits from the confines of the tropics, flourishing in the open air, daily prove the triumphs of knowledge and industry.

For the complete understanding of the subject in all its bearings, and to enable us to derive all the practical advantages which such an understanding would certainly afford, it would be necessary to have a full knowledge of the peculiarities of the climate of every region of the earth; a knowledge which we are very far from yet possessing, but to which rapid advances are daily making. But above all, it seems necessary that we should understand the atmospheric variations of our own situation. These, though not constituting the greatest range with which we are acquainted, are great, and oftentimes sudden. The range of the thermometer in the shade is from 0° to 90° of Fahrenheit's scale; but under favourable circumstances the heat of the sun's rays reaches 135°: the changes of moisture extend from 1.000, or saturation, to 389. Now the great object of the Horticulturist is to stretch, as it were, his climate to the south, where these extremes of drought and cold never occur; and not only to guard against the injurious effects of the ultimate severity of weather, but to ward off the sudden changes which are liable to recur in the different seasons of the year. To enable us to understand the methods of effecting this end, it will be necessary to consider the means by which these changes are brought about in the general course of nature. The principal of these will be found to be, wind and radiation.

The amount of evaporation from the soil, and of exhalation from the foliage of the vegetable kingdom, depends upon two circumstances,—the saturation of the air with moisture, and the velocity of its motion. They are in inverse proportion to the former, and in direct proportion to the latter.

When the air is dry, vapour ascends in it with great rapidity from every surface capable of affording it, and the energy of this action is greatly promoted by wind, which removes it from the exhaling body as fast as it is formed, and prevents that accumulation which would otherwise arrest the process.

Over the state of saturation, the Horticulturist has little or no control in the open air, but over its velocity he has some command. He can break the force of the blast by artificial means, such as walls, palings, hedges, or other screens; or he may find natural shelter in situations upon the acclivities of hills. Excessive exhalation is very injurious to many of the processes of vegetation, and no small proportion of what is commonly called *blight* may be attributed to this cause. Evaporation increases in a prodigiously rapid ratio with the velocity of the wind, and anything which retards the motion of the latter, is very efficacious in diminishing the amount of the former; the same surface, which in a calm state of the air would exhale 100 parts of moisture, would yield 125 in a moderate breeze, and 150 in a high wind. The dryness of the atmosphere in spring renders the effect most injurious to the tender shoots of this season of the year, and the easterly winds especially are most to be opposed in their course. The moisture of the air flowing from any point between N. E. and S. E. inclusive is to that of the air from the other quarter of the compass, in the proportion of 814 to 907 upon an average of the whole year: and it is no uncommon thing in spring for the dew-point to be more than 20 degrees below the temperature of the atmosphere in the shade, and I have even seen the difference amount to 30 degrees. The effect of such a degree of dryness is parching in the extreme, and if accompanied with wind is destructive to the blossoms of tender plants. The use of high walls, especially upon the northern and eastern sides of a garden, in checking this evil, cannot be doubtful, and in the case of tender fruit-trees, such screens should not be too far apart.

And here theory would suggest another precaution, which, I believe, has never yet been adopted, but which would be well worthy of a trial. When trees are trained upon a wall with a southern aspect, they have the advantage of a greatly exalted temperature; but this temperature, in spring, differs from the warmth of a more advanced period of the year, or of a more southern climate, in not being accompanied by an increase of moisture. In the extremely dry state of the atmosphere to which I am now alluding, the enormous exhalation from the blossoms of tender fruit trees, which must thus be induced, cannot fail of being extremely detrimental; the effect of shading the plants from the direct rays of the sun should therefore be ascertained. The state of the weather to which I refer, often occurs in April, May, and June, but seldom lasts many hours. Great mischief, however, may arise in a very small interval of time, and the disadvantage of a partial loss of light cannot be put in comparison with the probable effect which I have pointed out.

During the time in which I kept a register of the weather, I have seen in the month of May the thermometer in the sun at 101°, while the dew-point was only 34°; the state of saturation of the air, upon a south wall, consequently only amounted to 120, a state of dryness which is certainly not surpassed by an African Harmattan. The shelter of a mat on such occasions, would often prevent the sudden injury which so frequently arises at this period of the year.

Some of the present practices of gardening are founded upon experience of similar effects; and it is well known that cuttings of plants succeed best in a border with a northern aspect protected from the wind: or, if otherwise situated, they require to be screened from the force of the noon-day sun. If these precautions be unattended to, they speedily droop and die. For the same reason, the autumn is selected for placing them in the ground, as well as for transplanting trees; the atmosphere at that season being saturated with moisture, is not found to exhaust the plant before it has become rooted in the soil.

Over the absolute state of vapour in the air we are wholly powerless, and by no system of watering can we affect the dew-point in the free atmosphere.

This is determined in the upper regions; it is only therefore by these indirect methods, and by the selection of proper seasons, that we can preserve the more tender shoots of the vegetable kingdom from the injurious effects of excessive exhalation.

Radiation, the second cause which I have mentioned as producing a sudden and injurious influence upon the tender products of the garden, is one that has been little understood, till of late years, by the natural philosopher; and even to this day has not been rendered familiar to the practical gardener; who, although he has been taught by experience to guard against some of its effects, is totally unacquainted with the theory of his practice. Dr. Wells, to whose admirable *Essay upon Dew* we are so much indebted for our present knowledge upon this important subject, thus candidly remarks upon this anticipation of science: "I had often, in the pride of half-knowledge, smiled at the means frequently employed by gardeners to protect tender plants from cold; as it appeared to me impossible that a thin mat or any such flimsy substance could prevent them from attaining the temperature of the atmosphere, by which alone I thought them liable to be injured. But when I had learned that bodies on the surface of the earth become, during a still and serene night, colder than the atmosphere, by radiating their heat to the heavens, I perceived immediately a just reason for the practice which I had before deemed useless."

The power of emitting heat in straight lines in every direction, independently of contact, may be regarded as a property common to all matter; but differing in degree in different kinds of matter. Co-existing with it, in the same degrees, may be regarded the power of absorbing heat so emitted from other bodies. Polished metals, and the fibres of vegetables may be considered as placed at the two extremities of the scale upon which these properties in different substances may be measured. If a body be so situated that it may receive just as much radiant heat as itself projects, its temperature remains the same; if the surrounding bodies emit heat of greater intensity than the same body, its temperature rises, till the quantity which it receives exactly balances its expenditure; at which point it again becomes stationary: and if the power of radiation be exerted under circumstances which prevent a return, the temperature of the body declines. Thus, if a thermometer be placed in the focus of a concave metallic mirror, and turned towards any clear portion of the sky, at any period of the day, it will fall many degrees below the temperature of another thermometer placed near it, out of the mirror: the power of radiation is exerted in both thermometers, but to the first all return of radiant heat is cut off, while the other receives as much from the surrounding bodies, as itself projects. This interchange amongst bodies takes place in transparent *media* as well as in *vacuo*; but in the former case the effect is modified by the equalizing power of the medium.

Any portion of the surface of the globe which is fully turned towards the sun receives more radiant heat than it projects, and becomes heated; but when, by the revolution of the axis, this portion is turned from the source of heat, the radiation into space still continues, and being uncompensated, the temperature declines. In consequence of the different degrees in which different bodies possess this power of radiation, two contiguous portions of the system of the earth will become of different temperatures; and if on a clear night we place a thermometer upon a grass-plat, and another upon a gravel-walk or the bare soil, we shall find the temperature of the former many degrees below that of the latter: the fibrous texture of the grass is favourable to the emission of the heat, but the dense surfaces of the gravel seem to retain and fix it. But this unequal effect will only be perceived when the atmosphere is unclouded, and a free passage is open into space; for even a light mist will arrest the radiant matter in its course, and return as much to the radiating body as it emits. The intervention of more substantial obstacles will of course equally prevent the result, and the balance of temperature will not be disturbed in any substance which is not placed in the clear aspect of the sky. A portion of a grass-plat under the protection of a tree or hedge, will generally be found, on a clear night, to be eight or ten degrees warmer than surrounding unsheltered parts, and it is well known to gardeners that less dew and frost are to be found in such situations than in those which are wholly exposed.

There are many independent circumstances which modify the effects of this action, such as the state of the radiating body, its power of conducting heat, &c. If, for instance, the body be in a liquid or aëriform state, although the process may go on freely, as in water, the cold produced by it will not accumulate upon the surface, but will be dispersed by known laws throughout the mass; and if a solid body be a good radiator but a bad conductor of heat, the frigorific effect will be condensed upon the face which is exposed. So, upon the surface of the earth, absolute stillness of the atmosphere is necessary for the accumulation of cold upon the radiating body; for if the air be in motion, it disperses and equalizes the effect, with a rapidity proportioned to its velocity.

It is upon these principles that Dr. Wells has satisfactorily explained all the phenomena connected with dew or hoar frost. This deposition of moisture is owing to the cold produced in bodies by radiation, which condenses the atmospheric vapour upon their surfaces. It takes place upon vegetables, but not upon the naked soil. The fibres of short grass are particularly favourable to its formation. It is not produced either in cloudy or in windy weather, or in situations which are not perfectly open to the sky. It is never formed upon the good conducting surfaces of metals, but is rapidly deposited upon the badly conducting surfaces of filamentous bodies, such as cotton, wool, &c.

In remarking that dew is never formed upon metals, it is necessary to distinguish a secondary effect, which often causes a deposition of moisture upon every kind of surface indiscriminately. The cold which is produced upon the surface of the radiating body, is communicated by slow degrees to the surrounding atmosphere; and, if the effect be great and of sufficient continuance, moisture is not only deposited upon the solid body, but is precipated in the air itself; from which it slowly subsides, and settles upon everything within its range.

The formation of dew is one of the circumstances which modify and check the refrigerating effect of radiation; for, as the vapour is condensed, it gives out the latent heat with which it was combined in its elastic form, and thus, no doubt, prevents an excess of depression which might in many cases prove injurious to vegetation. A compensating arrangement is thus established, which, while it produces all the advantages of this gentle effusion of moisture, guards against the injurious concentration of the cause by which it is produced.

The effects of radiation come under the consideration of the Horticulturist in two points of view: the first regards the primary influence upon vegetables exposed to it; the second the modifications produced by it upon the atmosphere of particular situations. To vegetables growing in the climates for which they were originally designed by nature, there can be no doubt that the action of radiation is particularly beneficial, from the deposition of moisture which it determines upon their foliage: but to tender plants artificially trained to resist the rigours of an unnatural situation, this extra degree of cold may prove highly prejudicial. It also appears probable, from observation, that the intensity of this action increases with the distance from the equator to the poles; as the lowest depression of the thermometer which has been registered between the tropics, from this cause, is 12°, whereas in the latitude of London, it not unfrequently amounts to 17°. But, however this may be, it is certain that vegetation in this country is liable to be affected at night from the influence of radiation, by a temperature below the freezing point of water, ten months in the year; and even in the two months July and August, which are the only exceptions, a thermometer covered with wool will sometimes fall to 35°. It is, however, only low vegetation upon the ground which is exposed to the full rigour of this effect. In such a situation the air which is cooled by the process, lies upon the surface of the plants, and from its weight cannot make its escape; but from the foliage of a tree or shrub, it glides off and settles upon the ground.

Anything which obstructs the free aspect of the sky arrests, in proportion, the progress of this refrigeration, and the slightest covering of cloth or matting annihilates it altogether. Trees trained upon a wall or paling, or plants sown under their protection, are at once cut off from a large portion of this evil; and are still further protected, if within a moderate distance of another opposing screen. The most perfect combination for the growth of exotic fruits in the open air would be a number of parallel walls within a short distance of one another, facing the south-east quarter of the heavens: the spaces between each should be gravelled, except a narrow border on each side, which should be kept free from weeds and other short vegetables. On the southern sides of these walls, peaches, nectarines, figs, &c. might be trained to advantage, and on their northern sides many hardier kinds of fruit would be very advantageously situated. Tender exotic trees would thus derive all the benefit of the early morning sun, which would at the earliest moment dissipate the greatest accumulation of cold which immediately precedes its rise, and the injurious influence of nocturnal radiation would be almost entirely prevented. Upon trees so trained, the absolute perpendicular impression could have little effect, and this little might even be prevented by a moderate coping.

Mats or canvass, upon rollers to draw down occasionally in front of the trees, at the distance of a foot or two from their foliage, would, I have no doubt, be a great advantage in certain dry states of the atmosphere, before alluded to, and in the case of walls which are not opposed to others, would be a good substitute for the protection of the latter.

Experience has taught gardeners the advantages of warding off the effects of frost from tender vegetables, by loose straw or other litter, but the system of matting does not appear to be carried to that extent which its simplicity and efficacy would suggest. Neither does the manner of fixing the screen exhibit a proper acquaintance with the principle upon which it is resorted to: it is generally bound tight round the tree which it is required to protect, or nailed in close contact with its foliage.

Now it should be borne in mind that the radiation is only transferred fram the tree to the mat, and the cold of the latter will be conducted to the former in every point where it touches. Contact should therefore be prevented by hoops or other means properly applied, and the stratum of air which is enclosed will by its low conducting power effectually secure the plant. With their foliage thus protected, and their roots well covered with litter, many evergreens might doubtless be brought to survive the rigour of our winters, which are now confined to the stunted growth of the greenhouse and conservatory.

The secondary effect which radiation has upon the climate of particular situations, is a point which is less frequently considered than the primary one which we have been investigating; but which requires, perhaps, still more attention. The utmost concentration of cold can only take place in a perfectly still atmosphere: a very slight motion of the air is sufficient to disperse it. A low mist is often formed in meadows in particular situations, which is the consequence of the slow extension of this cold in the air, as before described: the agitation of merely walking through this condensation is frequently sufficient to disperse and melt it. A valley surrounded by low hills, is more liable to the effects of radiation than the tops and sides of the hills themselves; and it is a well known fact that dew and hoar frost are always more abundant in the former than in the latter situations. It is not meant to include in this observation, places surrounded by lofty and precipitous hills which obstruct the aspect of the sky, for in such, the contrary effect would be produced. Gentle slopes, which break the undulations of the air, without naturally circumscribing the heavens, are more efficient in promoting this action; and it is worthy of remark and consideration, that by walls and other fences we may artificially combine circumstances which may produce the same injurious effect.

But the influence of hills upon the nightly temperature of the valleys which they surround is not confined to this insulation; radiation goes on upon their declivities, and the air which is condensed by the cold rolls down and lodges at their feet.

Their sides are thus protected from the chill, and a double portion falls upon, what many are apt to consider, the more sheltered situation. Experience amply confirms these theoretical considerations. It is a very old remark, that the injurious effects of cold occur chiefly in hollow places, and that frosts are less severe upon hills than in neighbouring plains. It is consistent with my own observations that the leaves of the vine, the walnut-tree, and the succulent shoots of dahlias and potatoes, are often destroyed by frost in sheltered valleys, on nights when they are perfectly untouched upon the surrounding eminences; and I have seen a difference of 30 degrees on the same night between two thermometers placed in the two situations, in favour of the latter.

The advantages of placing a garden upon a gentle slope must be hence very apparent: a running stream at its foot would secure the further benefit of a contiguous surface, not liable to refrigeration, and would prevent any injurious stagnation of the air. Few situations are likely to fulfil all the conditions which theory would suggest for the most perfect mitigation of the climate in the open air; but the preceding remarks may not be without their use in pointing out localities, which, with this view, are most to be avoided.

Little is in the power of the Horticulturist to effect in the way of exalting the powers of the climate in the open air; except by choice of situation with regard to the sun, and the concentration of its rays upon walls and other screens. The natural reverberation from these and the subjacent soil, is however very effective, and few of the productions of the tropical regions are exposed to a greater heat than a well-trained tree upon a wall in summer. Indeed, it would appear from experiment, that the power of radiation from the sun, like that of radiation from the earth, increases with the distance from the equator; and there is a greater difference between a thermometer placed in the shade, and another in the solar rays in this country, than in Sierra Leone or Jamaica. The observations of the President of this Society upon the growth of pine-apples is in exact accordance with this idea; for he has remarked that this species of plant, though extremely patient of a high temperature, is not by any means so patient of the action of very continued bright light as many other plants, and much less so than the fig and orange tree; and he is inclined to think that on this account they may be found to ripen their fruit better in the spring than in the middle of the summer *. This energy of the sun is at times so great that it often becomes necessary to shade delicate flowers from its influence;

* See Horticultural Transactions, vol. iv. page 548.

and I have already pointed out a case in which it would be desirable to try the same precaution with the early blossom of certain fruit-trees. The greatest power is put forth in this country in June, while the greatest temperature of the air does not take place till July. The temperature of summer may thus be anticipated a month, in well-secured situations.

The greatest disadvantage to which Horticulture is subject in this climate, is the uncertainty of clear weather; a circumstance which art has, of course, no means to control; no artificial warmth is capable of supplying the deficiency when it occurs, and without the solar beams fruits lose their flavour and flowers the brightness of their tints. It has been attempted to communicate warmth to walls by means of fire and flues, but without the assistance of glass no great success has attended the trial.

It is well known that solar heat is absorbed by different substances with various degrees of facility dependant upon their colours, and that black is the most efficacious in this respect. It has therefore been proposed to paint garden walls of this colour; but no great benefit is likely to arise from this suggestion. It is probable that in the spring, when the trees are devoid of foliage, the wood may thus be forced to throw out its blossom somewhat earlier than it otherwise would; but this would be rather a disadvantage, as the flower would become exposed to the vicissitudes of an early spring. It is more desirable to check than to force this delicate and important process of vegetation, as much injury may arise from its premature development. When the tree has put forth its foliage, the colour of its protecting support can have no influence in any way: the leaves cover the surface, and absorb the rays by their own inherent powers. The only known advantage which can be taken of this peculiar power in dark substances, is in the case of covering up fruits, to preserve them from the ravages of flies; grapes which are enclosed in bags of black crape ripen better than those in white; but I believe that it is admitted that neither do so well as those which are freely exposed.

I come now to the consideration of a confined atmosphere; the management of which being entirely dependent upon art, requires in the Horticulturist a more extended acquaintance with the laws of nature, with regard to climate, and greater skill and experience in the application of his means. The plants which require this protection are in the most artificial state which it is possible to conceive; for, not only are their stems and foliage subject to the vicissitudes of the air in which they are immersed, but, in most cases, their roots also. The soil in which they are set to vegetate is generally contained in porous pots of earthernware, to the interior surface of which the tender fibres quickly penetrate and spread in every direction; they are thus exposed to every change of temperature and humidity, and are liable to great chills from any sudden increase of evaporation. This part of the subject naturally divides itself into two branches. The first regards the treatment of such exotics as are wholly dependent upon the artificial atmosphere of hot-houses: the second refers to the management of those hardier plants which only require to be preserved in green-houses part of the year, but during the summer months are exposed to the changes of the open air. I shall offer a few remarks first on the atmosphere of a hot-house.

The principal considerations which generally guide the management of gardeners in this delicate department are those of temperature; but there are others, regarding moisture, which are, I conceive, of at least equal importance. The inhabitants of the hot-house are all natives of the torrid zone, and the climate of this region is not only distinguished by an unvarying high degree of heat, but also by a very vaporous atmosphere Captain Sabine, in his meteorological researches between the tropics, rarely found, at the hottest period of the day, so great a difference as ten degrees between the temperature of the air and the dew-point; making the degree of saturation about 730, but most frequently 5° or 850; and the mean saturation of the air could not have been below 910. Now I believe, that if the hygrometer were consulted, it would be no uncommon thing to find in hot-houses, as at present managed, a difference of 20° between the point of condensation and the air, or a degree of moisture falling short of 500. The danger of over-watering most of the plants, especially at particular periods of their growth, is in general very justly appreciated; and in consequence the earth at their roots is kept in a state comparatively dry; the only supply of moisture being commonly derived from the pots, the exhalations of the leaves is not enough to saturate the air, and the consequence is a prodigious power of evaporation. This is injurious to the plants in two ways: in the first place, if the pots be at all moist, and not protected by tan or other litter, it produces a considerable degree of cold upon their surface, and communicates a chill to the tender fibres with which they are lined. The danger of such a chill is carefully guarded against in the case of watering, for it is one of the commonest precautions not to use any water of a temperature at all inferior to that of the hot air of the house; inattention to this point is quickly followed by disastrous consequences. The danger is quite as great from a moist flower-pot placed in a very dry atmosphere.

The custom of lowering the temperature of fluids in hot climates, by placing them in coolers of wet porous earthenware, is well known, and the common garden-pot is as good a cooler for this purpose as can be made. Under the common circumstances of the atmosphere of a hot-house, a depression of temperature amounting to 15 or 20 degrees, may easily be produced upon such an evaporating surface. But the greatest mischief will arise from the increased exhalations of the plants so circumstanced, and the consequent exhaustion of the powers of vegetation. The flowers of the torrid zone are many of them of a very succulent nature, argely supplied with cuticular pores, and their tender buds are unprovided with those integuments and other wonderful provisions by which nature guards her first embryo productions in more uncertain climates. Comparatively speaking, they shoot naked into the world, and are suited only to that enchanting mildness of the atmosphere, for which the whole system of their organization is adapted. In the tropical climates the sap never ceases to flow, and sudden checks or accelerations of its progress are as injurious to its healthy functions as they are necessary in the plants of more variable climates to the formation of those *hybernacula* which are provided for the preservation of the shoots in the winter season. Some idea may be formed of the prodigiously increased drain upon the functions of a plant arising from an increase of dryness in the air from the following consideration. If we suppose the amount of its perspiration, in a given time, to be 57 grains, the temperature of the air being 75°, and the dew-point 70, or the saturation of the air being 849, the amount would be increased to 120 grains in the same time if the dew-point were to remain stationary, and the temperature were to rise to 80°; or in other words, if the saturation of the air were to fall to 726.

Besides this power of transpiration, the leaves of vegetables exercise also an absorbent function, which must be no less disarranged by any deficiency of moisture. Some plants derive the greatest portion of their nutriment from the vaporous atmosphere, and all are more or less dependent upon the same source. The *Nepenthes distillatoria* lays up a store of water in the cup formed at the end of its leaves, which is probably secreted from the air, and applied to the exigences of the plant when exposed to drought, and the quantity, which is known to vary in the hot-house, is no doubt connected with the state of moisture of the atmosphere.

These considerations must be sufficient, I imagine, to place in a strong light the necessity of a strict attention to the atmosphere of vapour in our artificial climates, and to enforce as absolute an imitation as possible of the example of nature. The means of effecting this is the next object of our inquiry.

Tropical plants require to be watered at the root with great caution, and it is impossible that a sufficient supply of vapour can be kept up from this source alone. There can however be no difficulty in keeping the floor of the house, and the flues, continually wet, and an atmosphere of great elasticity may thus be maintained in a way perfectly analogous to the natural process. Where steam is employed as the means of communicating heat, an occasional injection of it into the air may also be had recourse to: but this method would require much attention on the part of the superintendent, whereas the first cannot easily be carried to excess.

It is true that damp air or floating moisture of long continuance would also be detrimental to the health of the plants, for it is absolutely necessary that the process of transpiration should proceed; but there is no danger that the high temperature of the hot-house should ever attain the point of saturation by spontaneous evaporation. The temperature of the external air will always keep down the force of the vapour; for as in the natural atmosphere the dew-point at the surface of the earth is regulated by the cold of the upper regions, so in a house the point of deposition is governed by the temperature of the glass with which it is in contact. In a well-ventilated hot-house, by watering the floor in summer, we may bring the dew-point within four or five degrees of the temperature of the air, and the glass will be perfectly free from moisture; by closing the ventilators we shall probably raise the heat 10 or 15 degrees, but the degree of saturation will remain nearly the same, and a copious dew will quickly form upon the glass, and will shortly run down in streams. A process of distillation is thus established, which prevents the vapour from attaining the full elasticity of the temperature.

This action is beneficial within certain limits, and at particular seasons of the year; but when the external air is very cold, or radiation proceeds very rapidly, it may become excessive and prejudicial. It is a well-known fact, but one which I believe has never yet been properly explained, that by attempting to keep up in a hot-house the same degree of heat at night as during the day, the plants become scorched; from what has been premised, it will be evident that this is owing to the low temperature of the glass, and the consequent low dew-point in the house, which occasions a degree of dryness which quickly exhausts the juices.

Much of this evil might be prevented by such simple and cheap means as an external covering of mats or canvass.

The heat of the glass of a hot-house at night does

not probably exceed the mean of the external and internal air, and taking these at 80° and 40°, 20 degrees of dryness are kept up in the interior; or a degree of saturation not exceeding 528. To this, in a clear night, we may add at least 6° for the effects of radiation, to which the glass is particularly exposed, which would reduce the saturation to 434°, and this is a degree of drought which must be nearly destructive. It will be allowed that the case which I have selected is by no means extreme, and it is one which is liable to occur even in the summer months. Now by an external covering of mats, &c. the effects of radiation would be at once annihilated, and a thin stratum of air would be kept in contact with the glass which would become warmed, and consequently tend to prevent the dissipation of the heat. But no means would of course be so effective as double glass including a stratum of air: indeed, such a precaution in winter seems almost essential to any great degree of perfection in this branch of Horticulture. When it is considered that a temperature at night of 20° is no very unfrequent occurrence in this country, the saturation of the air may upon such occasions fall to 120°, and such an evil can only at present be guarded against by diminishing the interior heat in proportion.

By materially lowering the temperature we communicate a check which is totally inconsistent with the welfare of tropical vegetation. The chill which is instantaneously communicated to the glass by a fall of rain and snow, and the consequent evaporation from its surface, must also precipitate the internal vapour, and dry the included air to a very considerable amount, and the effect should be closely watched. I do not conceive that the diminution of light which would be occasioned by the double panes, would be sufficient to occasion any serious objection to the plan. The difference would not probably amount to as much as that between hot-houses with wooden rafters and lights, and those constructed with curvilinear iron bars, two of which have been erected in the garden of the Horticultural Society. It might also possibly occasion a greater expansion of the foliage; for it is known that in houses with a northern aspect, the leaves grow to a larger size than in houses which front the south. Nature thus makes an effort to counteract the deficiency of light by increasing the surface upon which it is destined to act.

The present method of ventilating hot-houses is also objectionable, upon the same principles which I have been endeavouring to explain. A communication is at once opened with the external air, while the hot and vaporous atmosphere is allowed to escape at the roof; the consequence is, that the dry external air rushes in with considerable velocity, and becoming heated in its course, rapidly abstracts the moisture from the pots and foliage. This is the more dangerous, in as much as it acts with a rapidity proportioned in a very high degree to its motion. I would suggest, as a matter of easy experiment, whether great benefit might not arise from warming the air to a certain extent, and making it traverse a wet surface before it is allowed to enter the house.

There is one practice universally adopted by gardeners, which is confirmatory of these theoretical speculations, namely, that of planting tender cuttings of plants in a hot-bed, and covering them with a double glass. Experience has shown them that many kinds will not succeed under any other treatment. The end of this is obviously to preserve a saturated atmosphere; and it affords a parallel case to that of Dr. Wells of the anticipation of theory by practice.

The effect of keeping the floor of the hot-house continually wet has been already tried at the Society's garden, at my suggestion, and it has been found that the plants have grown with unprecedented vigour: indeed their luxuriance must strike the most superficial observer.

To the human feelings the impression of an atmosphere so saturated with moisture is very different from one heated to the same degree without this precaution; and any one coming out of a house heated in the common way, into one well charged with vapour, cannot fail to be struck with the difference. Those who are used to hot climates have declared that the feel and smell of the latter exactly assimilate to those of the tropical regions.

But there is a danger attending the very success of this experiment, which cannot be too carefully guarded against. The trial has been made in the summer months, when the temperature of the external air has not been low, nor the change from day to night very great. In proportion to the luxuriance of the vegetation, will be the danger of any sudden check; and it is much to be feared, that, unless proper precautions are adopted, the cold, long nights of winter may produce irreparable mischief.

I am aware that a great objection attaches to my plan of the double glass, on account of the expense; but I think that this may appear greater at first sight than it may afterwards be found to be in practice. It is however, at all events, I submit, a point worthy of the Horticultural Society to determine; and if the suggestion should be found to be effective, the lights of many frames, which are not commonly in use in winter, might, without much trouble, be fitted to slide over the hot-houses during the severe season; and in the spring, when they are wanted for other purposes, their places might be supplied at night by mats or canvass.

The principles which I have been endeavouring to illustrate should be doubtless extended to the pinery and the melon-frame, in the latter of which a saturated atmosphere might be maintained by shallow pans of water. An increase in the size of the fruit might be anticipated from this treatment, without that loss of flavour which would attend the communication of water to the roots of the plants.

I have but few additional observations to offer upon the artificial climate of a green-house. The remarks which have been made upon the atmosphere of the hot-house are applicable to it, though not to the same extent. The plants which are subject to this culture seldom require an artificial temperature greater than 45° or 50°, and few of them would receive injury from a temperature so low as 35°. When in the house they are effectually sheltered from the effects of direct radiation, which cannot take place through glass: but the glass itself radiates very freely, and thus communicates a chill to the air, which might effectually be prevented by rolling mats. With this precaution, fire would be but rarely wanted in a good situation, to communicate warmth. But in this damp climate it may be required to dissipate moisture. The state of the air should be as carefully watched with this view, as where a high temperature is necessary, to guard against the contrary extreme. Free transpiration, as I have before remarked, is necessary to the healthy progress of vegetation, and when any mouldiness or damp appears upon the plants, the temperature of the air should be moderately raised, and free ventilation allowed. When the pots in the proper season are moved into the open air, it would contribute greatly to their health and preserve them from the effects of too great evaporation, to imbed them well in moss or litter: as a substitute for this precaution, the plants are generally exposed to a northern or eastern aspect, where the influence of the sun but rarely reaches them, but which would be very beneficial if their roots were properly protected. The advantage of such a protection may be seen when the pots are plunged into the soil, a method which communicates the greatest luxuriance to the plants, but unfits them to resume their winter stations.

When a green-house is made use of, as it often is, after the removal of the pots, to force the vine, the same precautions should be attended to as in the management of the hot-house, and the elasticity of the vapour should be maintained by wetting the floor; but after a certain period a great degree of dryness should be allowed to prevail, to enable the tree to ripen its wood, and form the winter protection for its buds. In this its treatment differs from that of the tropical plants, which require no such change, and to which, on the contrary, it would be highly detrimental. The same observation applies to forcing-houses for peaches, and other similar kinds of trees. As soon as the fruit is all matured they should be freely exposed to the changes of the weather.

Upon an attentive consideration and review of the subject, it appears to me certain that a frequent consultation of the indications of the hygrometer is quite as necessary to the Horticulturist as of those of the thermometer; and it is not unworthy of the consideration of the Horticultural Society whether correct registers of the state of the climate, both in their houses and out of doors, and a connected series of experiments upon the modifications of which it is susceptible, might not contribute something to the perfection of that art, which they are making such honourable exertions to perfect and communicate.

To me it will be a source of great satisfaction if any observations which I have made, or may make, upon the subject of climate, should prove at all instrumental in forwarding their important views.

ON THE

OSCILLATIONS OF THE BAROMETER.

AMONGST the phenomena for which I have endeavoured to account in my first essay, I slightly alluded to the coincidence which was known often to occur in the movements of barometers situated at great distances from each other; and mentioning the observation that this unison of action extends further in the direction of the latitude than in that of the longitude, I remarked that the fact confirmed the theory: for, as the grand currents of the atmosphere flow nearly in the direction of the meridians, any irregularity in their courses would most readily be propagated in the same line. Further consideration convinced me that this argument should have been greatly extended; and I perceived that a strict adherence to the legitimate conclusions of the hypothesis would establish, not only a partial and frequent coincidence of the aërial undulations in different latitudes, but a wide-extending and constant agreement. The want of combination in the meteorological observers of the present day made me despair of being able to bring the idea to the test of sufficient experiments, till accident threw in my way a register of inestimable value.

My friend Mr. Howard some time ago put into

my hands, and obliged me with the loan of, some of the volumes of the Ephemerides of the Meteorological Society of the Palatinate: a work which, if it had been continued with its original spirit to the present time, would probably have left little to be desired in the way of observation; and which, even in its present state, would be found by the diligent inquirer to contain more *data* for a correct history of European weather than all other works upon the same subject taken together. During a tour which I made in Germany, I succeeded in obtaining a complete copy of these transactions, from their commencement in 1781, to their termination in 1792. As this record is very little known in this country, and in its complete state very scarce, I shall be excused for giving a short account of its origin, and that of a society which might, undoubtedly, afford the most perfect model of a similar institution at the present day for promoting the Science of Meteorology. I shall hereafter, probably, if time should allow me to complete the task which I have commenced, draw up a memoir upon the climate of Europe founded upon these careful documents. The labour will not be trifling, but cannot be thrown away in a case where such full reliance can be placed upon the experimental basis of the inductions.

The Meteorological Society of the Palatinate was established in the year 1780, under the auspices of the Elector Charles Theodore, who not only gave it the support of his public patronage, but entered with spirit and ability into its pursuits, and furnished it with the means of defraying the expense of instruments of the best construction, which were gratuitously distributed to all parts of Europe, and even to America. One of the first acts of the Association was, to write to all the principal universities, scientific academies, and colleges, soliciting their co-operation, and offering to present them with all the necessary instruments properly verified by standards, and free of expense. The offer was immediately accepted by thirty societies; and the list of distinguished men who undertook to make the observations shows the importance which was attached to the plan, and the zeal with which it was promoted in every part of the Continent. Amongst those who, for the good of science, undertook and executed this daily drudgery, we find the names of Hemmer, Weis, Planer, Senebier, Bugge, Van Swinden, König, Cotte, Egel, Pictet, Toaldo, and Euler. The Secretary Hemmer appears to have been indefatigable in his exertions to perfect this truly princely plan of operations; and, even now, but little could be added to the precautions taken in the preparation of the instruments which he describes, or to the ample instructions for their use, which he transmitted with them. Some idea may be formed of the comprehensive scale of the register, when it is known that it contains observations, three times in the day, of the barometer, thermometer in the shade and in the sun, hygrometer, magnetic needle, direction and force of the wind, quantity of rain and of evaporation, the height of any neighbouring water, the changes of the moon, the appearance of the sky, and the occurrence of meteors and of the Aurora Borealis. To these must be added, in some places, observations upon the electrical state of the atmosphere, upon the progress of vegetation, the prevalence of disease, changes of population, and migration of animals. The field of observation extended from the Ural Mountains in the east, to Cambridge in the United States in the west; and from Greenland and Norway in the north, to Rome in the south. This range included also stations upon three high mountains in Bavaria, and upon the summit of St. Gothard. The observations of each year are summed up, and compared with those which precede, in copious and most laborious tables of mean and extreme results; and many very interesting essays upon various branches of Meteorology are interspersed throughout the volumes. Unfortunately for science, the Secretary Hemmer died in the month of May, 1790, and from that time the Society appears to have languished, and finally to have become extinct amidst the troubles and the wars of the French Revolution. Would that another Hemmer could be found in this age to direct the uncombined efforts of meteorologists to one common purpose! And would that the scientific men of the present day, laying aside all petty and degrading jealousies, might see the advantage of uniting in a system such labours as lose the greatest part of their value from wanting unity of purpose!

Amongst other valuable suggestions, in these volumes, upon the proper uses of meteorological observations, I have found the first exemplification of the method of representing the oscillations of the barometer by a curved line upon a scale—a method which I think will appear, from the subsequent part of this essay, to be of the utmost consequence in connecting detached observations, and exhibiting their mutual relations. The instances of this application in the volumes before me are very few, and for very short intervals; but it has been employed, upon a small scale, to show the accordance of great changes of the mercurial column at distant points. It is by an extension of this plan that I shall now proceed to show, that, within certain limits, the movements of the barometer coincide by some general law over large portions of the surface of the globe. I shall endeavour to trace, as far as the observations will allow, the limits of this coincidence, the particular direction in which it occurs, and the circumstances, if any, which modify its regularity. I shall speak of the facts first; and I shall afterwards endeavour to apply them in illustration of that theory by which, in reality, I was guided to their discovery. By this method of proceeding, whatever success I may be thought to obtain in establishing the latter, I shall at least have the satisfaction of fixing *data*, upon which others may found their own reasoning, but which must hereafter claim an explanation in every theory of the phenomena of the atmosphere.

Plates III. and IV. represent the oscillations of the barometer at eighteen stations on the continent of Europe for one twelvemonth; they comprehend a space of nearly 18° of latitude, and 14° of longitude. The observations are laid down twice in the day, *viz.*, at 7 A. M., and 9 P. M., upon a scale of English inches, which, however, has been reduced in the engraving; each perpendicular division representing a tenth of an inch, and each horizontal division comprehending a day. The curves have been arranged in order from north to south, commencing with Spidberg, in Norway, and ending with Rome and Padua. In selecting the stations, I have endeavoured to confine one set, as much as possible, to the same meridian; while I have chosen a second nearly approaching one another in latitude, but differing widely in longitude.

Spidberg, the first station, is a small parish in Norway, situated between Christiana and Frederickshall, within a short distance of the North Sea on its western side, and of the Baltic on its eastern. The longitude of the place of observation, which was the church and residence of the minister, is represented in the Transactions as 9° 4′ E.; but there must be some mistake in this, as from the locality it cannot be less than 10° 50′ E. The latitude is 59° 30′ N., and the altitude above the level of the sea about 426 feet.

Stockholm, the second station, is situated in about the same degree of latitude, but nearly 8° apart in the longitude. The place of observation was the Observatory close to the shore of the Baltic, above the surface of which it stands about 136 feet.

Copenhagen, the third station, is built upon the southern coast of the Categate, or entrance of the Baltic. The Royal Observatory, at which the register was kept, stands about 136 feet above the mean level of the sea; its latitude is 55° 41′ N., and its longitude 12° 40′ E. It is distant about 300 miles to the south-west of the preceding station.

Gottingen, the next in succession, is situated almost exactly upon the meridian from which we set out; its longitude being 9° 53′ E., and its latitude 51° 52′. It is about 290 miles distant from the North Sea, above the level of which it stands about 450 feet. The river Leine flows near it, and it is surrounded by moderate hills. The Hartz Mountains rise in the N. E., at a distance of not more than 15 miles.

Sagan, the fifth station, has been selected as corresponding with the preceding in latitude, *viz.*, 51° 42′ N., but being far removed in longitude, *viz.* 15° 27′ E. It is situated upon the Bober, and surrounded on all sides by extensive plains. The place of observation was raised about 60 feet above the level of the river.

Erfurt, the sixth station, again approaches the first meridian, but advances us to the south. Its longitude is 11° 23′ E., and its latitude 51° N. The surrounding country is open.

Brussels, the seventh station, is the most western point of our present comparison, and particularly remarkable, as we shall hereafter have occasion to observe, on account of its being the nearest to the Western Sea. It is situated in a very open country, upon the banks of the small river Senne. The latitude is 50° 51′ N., the longitude 4° 28′ E. The barometer was placed about 175 feet above the level of the river.

Prague, the eighth station, carries us again more than 10° to the east; its longitude being 14° 50′ E., and its latitude 50° 5′ N. It is situated in a hilly country upon the banks of the Moldaw.

Mannheim, the ninth station, was the head quarters of the Meteorological Society; and here the observations, under the immediate superintendence of the Secretary Hemmer, were more varied and more complete than in any other place: on this account, as well as on that of its central situation, it furnishes the best standard of comparison for all the other observatories. All the particulars of its situation are most accurately described in the Transactions. It is placed in a vast plain, and nearly surrounded by the waters of the Necker and the Rhine. Its latitude is 49° 26′ N., and its longitude 8° 31′ E.; not very far removed from the meridian from which we set out. The barometer was placed about 51 feet above the mean level of the Rhine.

Ratisbon, the tenth station, is placed upon the Danube, in latitude 48° 56′ N., and longitude 12° 5′.

Munich, the eleventh station, is situated about 62 miles to the south of the preceding. It is seated in a plain on the river Iser, in latitude 48° 10′ N., and longitude 11° 36′ E. The barometer was placed about 48 feet above the ground.

Peissenberg, or Hohenpeissenberg, the twelfth station, is a mountain of Upper Bavaria, distant not more than 3 or 4 miles from the mountains of the Tyrol. The place of observation was its very summit, about 1300 feet above the level of the River Amber, and 1100 above the Leike. Its latitude is 47° 47′ E., and its longitude 10° 59′ E. It is remarked in the Transactions, as a place peculiarly adapted by nature for meteorological observations. Its horizon extends on all sides, but the south, to a distance of above 12 miles; but in the south the

Tyrolese Mountains overtop it considerably. It is surrounded by marshy land; and there are no less than three considerable lakes within two miles of it, and several rivers. The northern side of the mountain is covered with wood, and there are large forests in its vicinity. The barometer was placed about 30 feet above the ground.

Buda, the thirteenth station, is the most eastern point of the present comparison. Its longitude being 18° 22′ E., and its latitude 47° 29′ N. It is situated upon the side of a hill upon the banks of the Danube, and surrounded on all sides by hills. The barometer was placed in the Royal Observatory, about 290 feet above the mean height of the river.

Geneva, the fourteenth station, is situated upon the extensive lake to which it gives its name, in latitude 46° 12′, and longitude 6° 5′. The Rhone takes its course through the city, and it is surrounded on all sides by lofty mountains. The level of the lake is about 1350 feet above that of the sea.

St. Gothard, the fifteenth station, was the Hospice of the Capuchin Monks, almost upon the summit of the mountain. It is situated 6800 feet above the level of the Mediterranean Sea. It is surrounded on all sides by lofty rocks, some of which rise to the height of 2000 feet above it. The situation is most open to the north and the south, but it is closely hemmed in on every other quarter. There is a small lake close by the dwelling; it is considerably raised above the forests in its neighbourhood.

Marseilles is the sixteenth station, and stands upon the shores of the Mediterranean Sea. Its latitude is 43° 18′ N., and its longitude 5° 27′. The ground upon which it stands is uneven, and it is surrounded on the land side by mountains, some of which are not less than 2500 feet high. The observatory is built upon one of the highest points of the town, and the height of the barometer above the sea was 153 feet.

Rome, the seventeenth station, is the most southern point to which the observations extend. It is not very far removed from the Norwegian meridian from which we set out, and agrees almost exactly with that of Copenhagen, thus extending the comparison to upwards of 1200 miles in a straight line from north to south. The exact latitude is 41° 54′ N., and the longitude 12° 55′ E. The city is situated upon the river Tiber, which runs through a part of it, and at no great distance from the Mediterranean on the south, and the Adriatic on the north. The barometer was fixed at a height of about 90 feet above the level of the sea.

Padua, the last station of this comparison, is a few degrees more to the north, being in latitude 45° 23′ N., and longitude 12° 1′ E. It is seated at the top of the Adriatic, in a fine plain, at the confluence of the rivers Brenta and Bachiglione. The surrounding country is distinguished by its beauty and fertility. The barometer stood about 60 feet above the level of the sea.

I regret very much the not being able to include London amongst the stations of this interesting survey. The Royal Society, as might be supposed was one of the first scientific bodies to which the Meteorological Society of the Palatinate addressed themselves for co-operation in the great and truly scientific work which they had undertaken; and it is very remarkable, and, to an Englishman, very mortifying, to remark, that the answer of the Royal Society to the invitation is the only one amongst a vast number which does not appear in the *Transactions*. By some unfortunate coincidence, the years which are included in the Ephemerides are precisely those during which no Meteorological Register was published in the *Philosophical Transactions*: so that the comparison fails at a point which, for many reasons, is one of the utmost interest and importance; but particularly on account of the situation of London being on the extreme west of Europe, and of its being surrounded by the waters of the Atlantic Ocean.

The names of the stations are inserted on one side of the plates, and their longitudes and latitudes on the other: in conjunction with the former, I have placed the mean heights of the barometer for the year, by which a judgment may at once be formed of their relative elevations above the level of the sea.

I may here remark that I have laid down several more curves, at intermediate places, which are not included in the plates, for fear of rendering them confused. In selecting the present series, I have been guided in my choice by such circumstances as might be supposed to produce the greatest difference between them. Those which I have omitted all concur in the general result.

The principal fact disclosed by the comparison of observations at all these various points must at once strike the eye of the most careless observer—namely, the near coincidence of all the curves. From the shores of the Baltic to those of the Mediterranean; from the level of the sea to the height of nearly 7000 feet above it; in every variety of country, from the Alps to the sandy plains of Germany; in every season, in every change of weather, the continual movements of the barometer correspond in a most wonderful manner. The general law which governs these effects, within these limits, is constant in its operation, although subject to modifications which it will be highly instructive to trace and appreciate.

In the first place, the oscillations of the mercurial column decrease in proceeding from north to south; and in this way some of the minor movements are obliterated before they reach the extreme southern point. Hence, partly, it is that the three Mediterranean curves are not only flatter than the three Baltic, but less serrated and uneven. If the extreme northern and southern lines were placed in juxta-position, the resemblance would be but faint and imperfect; but they pass so gradually into one another, through the whole series, that their connexion is very manifest. These observations apply most accurately to those places which are nearest situated to the same meridian.

Secondly. There are other modifications which depend upon the relative distances of the places in longitude. By comparing together the most eastern and most western curves which nearly agree in latitude, it may be remarked, that their several points would better correspond, if the former, that is, the eastern, were moved to the left hand about the space of a day or a half. This is more striking upon the larger scale, upon which they were originally laid down, but still may be satisfactorily established by an attentive examination of the plates. It is particularly obvious in the curves of Gottingen and Sagan, and in those of Marseilles and Rome. It is not dependent upon the difference of time, which is consequent upon the difference of longitude, for it is in the opposite direction; and, by taking the latter into consideration, the amount of the deviation is increased.

Thirdly. Besides this regular difference, there are occasional greater discrepancies as we change the longitude, traces of which are generally preserved throughout the meridian upon which they occur. For example, about the 4th of January, at Sagan, a very remarkable elevation of the line occurs, which differs very much from those at the same period at Gottingen and Erfurt, but which corresponds exactly with the curves of Prague and Buda. Again, about the 18th of March, a much bolder depression of the line takes place at Stockholm than at the same time at Spidberg and Copenhagen; and the same excess of effect may be traced down the more eastern meridians at Sagan, Prague, and Buda, by comparing them with their western neighbours. This difference, dependent upon longitude, will, however, be more distinctly marked, when we come to the explanation of the next plate.

Fourthly. The effect of the difference of the meridians becomes greater as we approach the Western Sea. It will at once be evident that the Brussels curve agrees the least with any of the others. Although it accords in the general outline, almost all the remarkable elevations and depressions are strongly modified. The Marseilles curve, on the contrary, which is only one degree more to the east, agrees very closely with that of Rome. It will be remarked that the whole width of France interposes between this station and the Atlantic, while the former is at a comparatively small distance from the North Sea.

Fifthly. There is another modification which is dependent upon height. The mountain curve of St. Gothard is manifestly flatter, and its inequalities more rounded, than that of Geneva; and it is also worthy of remark how the relative distance of the two decreases as the temperature of the months increases, and augments with their decrease. In the summer the space between them is not half as great as in the winter. This is obviously caused by the expansion of the atmospheric columns in the former season; the difference of course manifesting itself in the increased weight of the upper section.

Sixthly. The abrupt and angular changes of the winter portion of the curves is strongly contrasted with the more gentle and rounded undulations of the summer months, and even the stormy portions of each with those of the more settled weather. These changes extend through the whole series; but it cannot but be remarked that the southern members of the group generally partake more of the latter, and the northern of the former, character.

Seventhly. It may be observed how very generally the corresponding angles of ascent and descent agree. There are certainly many exceptions to the rule, but it is almost universally true, that when the barometer falls very abruptly, it rises to the same amount as suddenly, and *vice versâ.* This evidently points to the equality of action and reaction, and the cause of the exceptions themselves would form an interesting object of research.

Amongst the instructive relations of these comparisons, will be found the theory and practice of the mensuration of heights by the barometer. I shall not now attempt to trace out this connexion in detail, but shall probably recur to it again upon some future occasion.

Before I turn from the consideration of the third and fourth plates, I must remark, that the observations, which were all registered in French inches, &c., have been laid down with the greatest fidelity, except in one or two instances, where the general accordance of the curves manifestly pointed out an error of *a whole inch.* Thus, for example, on the 5th of January, in the Spidberg curve, I have shown, by a dotted line, the course of such a mistake, which would have caused a wide departure from the character of the contemporaneous movements at other stations. I have corrected two other similar misprints; and there are a few other analogous discrepancies in different parts of the synoptic views, which may probably be referrible to errors of the same description, but with which I have not ventured to interfere. Such, I am almost tempted to believe, is the sudden rise in the Stockholm line on the 31st January, and the anomalous fall in that of Spidberg on the 12th February. One of the uses to which, in future, this method of laying down observations may be applied, is, to check, within certain limits, the accuracy of different observers. This is the test to which I have occasionally brought the late observations of the Royal Society; and I have in this way discovered, on several occasions, much wider differences between them and the observations of Mr. Howard at Tottenham, than between any two, even of the most remote stations in Europe, which we have just been comparing.

The next most interesting objects of inquiry are the limits of this correspondence, and the commencement of the reflux which must necessarily accompany these extensive undulations; for the laws of equilibrium require that a fall of the mercurial column throughout the greater part of Europe, should be accompanied by a corresponding rise in other regions. Unfortunately the observations of the Ephemerides, extended as they were, are not sufficiently comprehensive to fix these important points with all the precision which we could desire. In Plate V., however, I have collected two groups which throw considerable light upon this part of the subject; and I shall now proceed to describe the stations from which they have been selected.

Pyschminsk, the nineteenth station, was the office of the mines in the Ural Mountains of Siberia, situated in the government of Perm. Its latitude is 57° N., and longitude 41° 4′ E. It is the most eastern point to which the registers of the Society extended. Its height was upwards of 2000 feet above the level of the sea.

Petersburg, the twentieth station, is situated on the river Neva, close to the Gulf of Finland. The ground in the neighbourhood is flat and low, and was formerly a vast morass. Its latitude is 59° 56′ N., and its longitude 30° 25′ E.

Moscow, the twenty-first station, is placed about 460 miles to the S.E. of the last. The river Moskwa runs through the city, and in the open country around are some small lakes, which give rise to the Neglina. Its longitude is 37° 31′ E., and its latitude 55° 45′ N.

Mannheim has been here introduced as the type of the western portion of the Continent, in which we have already traced the movements of the atmosphere. From the comparisons afforded by this group we perceive,—

First, that there is neither agreement nor regular opposition in the course of the barometric curves, by extending the observations in the direction of the longitude. This is rendered quite obvious, by comparing together the lines of Pyschminsk, Moscow, and Mannheim, which are at nearly equal distances of 30° apart. I may further add that, in laying down the curve of contemporaneous observations at Cambridge, in the United States, contained, in the Register, in longitude 70° 45′ W., and latitude 42° 25′ N., the same want of correspondence is observable.

Secondly, that the accordance is still maintained upon this remote meridian, in the direction of the latitude, as appears from the curves of Petersburg and Moscow, which agree together very eaxctly, notwithstanding they are not situated so nearly north and south as might be wished for the comparison. The quiet state of the atmosphere at Pyschminsk in the summer months of July and August, is another very interesting feature of its curve. This is probably the most inland station at which a register of the barometer has ever been kept, being almost exactly in the centre of that immense extent of land which constitutes the continents of Europe and Asia.

Gothab, the twenty-second station, is situated upon the west coast of Greenland, in lat. 64° 10′ N., and in long. 51° 18′ W. It is the most northern point of the Mannheim Transactions. It stands on the shores of Davis' Strait, at the foot of some very lofty mountains, which rise to the height of nearly 7000 feet. The barometer was placed about 15 feet above the level of high-water.

Edsberg, the twenty-third station, is a small town in Norway, not very far removed from the station at Spidberg, to which the observations, for some reason which is not explained, appear to have been transferred. Its lat. is about 59° N., and its long. 9° E.

Copenhagen and Mannheim are again brought in, to connect the comparison with our previous remarks. I was unable to confine all the observations, as I should have preferred doing, to the same year, because there was no one year that included all the stations that were necessary to my purpose. This series of curves was not laid down like all the others, from two observations in each day, but from the mean of three; which gives them rather a rounder character, but does not materially alter their configuration. In this second group of the third plate, we can discover, I think, with sufficient distinctness, traces of that regular opposition of the curves, which could not but be anticipated from theory. Gothab is situated considerably to the north of all the other stations, and with the interposition of the Atlantic Ocean. It is unfortunately more removed to the westward of the European meridians than could have been wished; but nevertheless the opposition of the movements of the barometer at the end of January, the whole of March, and parts of April and May, is almost perfect. The accordance of the three European curves of this group is sufficient to show that they are under the influence of the same law as in the former year, which we have more minutely traced. I have been unable in any way to follow up this interesting part of the subject, for the observations of Gothab have been only published for the six months which I have laid down.

It is in vain, I fear, to point out how very instructive a good series of observations would be from the east coast of Greenland; but we may possibly look forward to a no less interesting set from Spitzbergen, from the expedition which has lately left our shores.

Such being the state of the facts, we are now to inquire whether they are consistent with the general theory which I have proposed. In a system of balancing currents, as I have before observed, any cause which equally affects the velocity of two antagonist streams will change the weight of any perpendicular column, comprehending sections of the two; but, if either of them should be retarded or accelerated in its course without the other, or should they be unequally affected, their compound pressure must alter. We must conclude further, that such unequal change taking place at any particular point of the course of two antagonist currents, must manifest itself throughout their own line by opposite effects on each side of such point. Let us imagine two such currents: the lower flowing from the poles to the equator, and the upper from the equator to the poles, with such regular velocities that they exactly balance each other, and their perpendicular pressure never alters. Suppose a sudden local check to be given at the centre point of the lower current: the upper will still move forward with its original velocity from its *vis inertiæ;* while on one side of the intervening obstacle its expenditure is fed by a reduced supply, and on the other it meets with the retarded course of the lower current, and must accumulate upon it. The barometer, therefore, which measures the combined pressure of the two currents must rise on one side and fall on the other. This rise and fall will also be sensible to the two extremities of the current, but in a decreasing ratio from the original point of disturbance, on account of the particles of the upper gradually losing their original impulse, and adapting themselves to a new arrangement and a new balance of velocities. Now the order of the phenomèna which we have been contemplating appears to me exactly to correspond with these theoretical conclusions. That part of the Continent of Europe over which the observations extend lies about midway in the course of two antagonist currents of the atmosphere; which, from the distribution of the temperature of the globe, must, in an undisturbed state of a gaseous fluid, as I have elsewhere shown, flow between the pole and the equator. The whole line of its northern extremity, however, is bounded by the ocean, the evaporation from which must constantly disturb that regular progression of temperature upon which the balance of the two streams depends. The mode of this operation I have described at length in my first essay: it will be sufficient here to recall to remembrance, that it is by the heat evolved from the condensation of the rising vapour in the upper regions, and by the cold communicated by its re-evaporation, that such inequalities are produced. The actions and consequent re actions produced by these irregularities we find constantly communicated from the north, where they originate, in a decreasing degree, to the south. We also find that the counter-action of the barometer, which must necessarily exist somewhere, is not to be traced to the east or to the west; and although observations are wanting at the exact points where theory would teach us to seek it,—namely, to the north of the sea which bounds the northern coasts of Europe,—yet we find evidence of its existence upon the coast of Greenland; which, however, is too remote from the meridian of the chief observations, to present more than a slight correspondence.

The minor circumstances attending the principal phenomena accord no less with the theory. The modification to which the curves are obviously subject upon the shores of the Mediterranean are referrible to the evaporation of that sea. The northern ocean is exactly situated in the best position with regard to the Continent of Europe for producing powerful effects: by its means the warm waters of the Atlantic are brought in contact with the cold winds and ice of the Arctic Regions. In such a combination of circumstances, evaporation and condensation must reach the most violent extremes, particularly in the autumn and spring of the year. The vast undulations of the aërial ocean dependent upon these circumstances are sometimes broken, but not destroyed, by the gentler influence of the southern waters *.

It is evident that the closest accordance of the curves is to be found amongst those which are

* While speaking of the circumstances which are best calculated to produce the greatest disturbance in the aërial currents, and which theory suggests to be the nearest approach of hot water to ice and cold, I may be permitted to observe, that I anticipated that the greatest oscillations of the barometer in this hemisphere would be found about the point where the Florida stream makes its nearest approach to the north. This conjecture I have had an opportunity of verifying by the kindness of Lieutenant Bullock, of his Majesty's ship the Snap, who, while employed by Government in making a survey of the coasts of Newfoundland, kept a very accurate register of the barometer. These observations I have laid down upon the scale; and, although they only extend over the summer months, and consequently those least calculated to exhibit the effect, their curve is decidedly more bold than that of any other situation I have traced at the same season of the year in any other part of the world.

situated the nearest to the same meridians; and the curves of the different meridians differ more from one another as they approach the Western Ocean. This is well exhibited by the Brussels line, which varies very considerably from those of Erfurt and Prague; while the curve of Marseilles, which is situated nearly as much to the west, but far removed from the western waters, agrees much more closely with those of Rome and Padua. The influence of the approach of water is also exhibited by the curves of Spidberg and Stockholm, which, in their differences from that of Copenhagen, are doubtless modified by the intervention of the Baltic Sea.

The precession of the western curves before the eastern may also be explained by the situation of the immense reservoir of vapour which is continually rising from the Atlantic Ocean: this, when wafted by the currents to the northern shores of Europe, can only progressively exert its influence along the successive meridians. We cannot but lament in this comparison the want of a corresponding station upon the north coast of Africa, from which, amongst many other interesting questions, the influence of the Mediterranean Sea would, probably, have been more clearly determined. In indulging, however, a wish upon this subject, it is difficult to restrain it within these limits, and not to extend it to the revival of a society similar to that of the Palatinate, and the extension of its operations in all directions.

It may possibly be objected to the theory, that these undulations, extending over such a vast tract of the globe, cannot be supposed to depend upon any regular current of the air; because the wind is blowing at the same time in all directions with every modification of force. But it must be remembered that the winds, which are sensible to us, are influenced by local circumstances upon the surface of the earth, and are mostly parts of minor systems of compensation, compared with the grand movement of the atmospheric ocean. It would be as reasonable to expect that the current which flows out of the upper part of a heated room, and is balanced by a counter-current at the lower part, should affect the barometer, as that the direction which the wind assumes in a particular valley, or the retardation which it may experience against any particular mountain, should influence the general movement of the mercurial column. The currents to which the theory refers overtop the Alps, and sweep along uninfluenced by anything but changes in the elasticity of the medium, of which they form a part;—and thus it is that we find the same curve described upon the summit of St. Gothard and at the level of the sea. I do not mean to deny that these accidents of situation have a local and circumscribed influence; for even the sudden shutting of the door of a heated room, by which we affect the balance of its little system of currents, will cause a delicate barometer to oscillate; but they are lost in the grand outline which we have been considering.

Having traced the general accordance of the barometric curves in the situations which I have pointed out, it would doubtless be a very interesting and instructive task to enter into a further comparison of them, with a view of tracing the causes of their minuter differences; to inquire how these may be referrible to peculiarities of local situation, and to what extent they are connected with changes of weather. To do this to advantage, however, it would be necessary that the comparison should extend through a number of years. The materials are not wanting in the Mannheim Transactions; and indeed I have already laid down a far more extended number of observations for the purpose; but the expense of engraving them is very considerable, and the attention at present given to the study of Meteorology is not sufficient to induce a publisher to undertake their execution with the hope of any advantage. The discovery, however, that some general law governs the oscillations of the barometer, which extends with unerring constancy over large tracts of the earth's surface, seems to me to give a new interest to the pursuits of meteorology. It removes, in a great measure, the charge of uncertainty from its conclusions, and it opens a view of practical utility which is more likely to excite inquiry than the speculations of abstract science. That we may, hereafter, from observations properly arranged, be able to judge, at any moment of time, of the force of the winds at distant points, will not, I think, appear chimerical to those who will attentively consider the phenomena which have just been developed; and it is not difficult to perceive that the interests of navigation may be deeply concerned in a knowledge of the laws of such fluxes and refluxes of the aërial ocean as those which we have been contemplating.

ON THE GRADUAL DETERIORATION OF BAROMETERS

AND THE

MEANS OF PREVENTING THE SAME.

IN my previous remarks upon the construction of barometers, I have stated my reasons for differing from the high authority of the President of the Royal Society, upon the *cause of the existence of elastic matter in barometer tubes,* suggested by him in a paper upon "the electrical phenomena exhibited in vacuo," and published in the *Philosophical Transactions* for the year 1822. Sign. Bellani also arrived at the same conclusion as myself, from a series of experiments which he undertook, expressly to determine whether the air or vapour, the last portions of which are found to remain so obstinately in barometers and thermometers, is introduced with the mercury, or is a portion of that which originally occupied the tube before the introduction of the metal. The conclusion he comes to is, that it is always a portion of that which previously adhered to the glass, and *that mercury is utterly incapable of absorbing either air or moisture.* One of his experiments is so simple, and at the same time so conclusive, that I cannot refrain from giving a short account of it. He filled a barometer tube, and boiled it very carefully, and then prepared a funnel made of a small capillary tube, which reached through the mercury in the barometer tube to the closed end, and was enlarged at the top. When introduced, it had been recently made, and perfectly dry. Some mercury was then prepared by agitating it in a bottle with water and air, and dried by means of filtering paper, and afterwards passing it through paper cones, three or four times, into dry vessels. A little of this mercury was poured into the funnel-tube, and the air extracted by means of a fine wire, so that the column was continuous. So much of this prepared mercury was then poured in as fully to displace the mercury which had been boiled in the tube. The barometer was found to stand exactly at the same height as before in the same circumstances; and, when the mercury was heated, none of those bubbles appeared which arose on the first boiling *.

Still further to illustrate this subject, which I thought of the highest importance, and to ascertain the difference of capillary action in boiled and unboiled barometer-tubes, I undertook the following experiments. The apparatus which I made use of consisted of an upright pillar of brass, standing upon a mahogany foot, upon which two horizontal arms of unequal lengths were made to slide; at the extremity of each of these a steel needle, with a fine point, was fixed perpendicularly downwards.

* *Giornale de Fisica.* Vol. vi. p. 20, or see *Royal Institution Journal,* Vol. xv. p. 371.

These points could be adjusted to the same plane, or their relative distance be measured, by means of a nut and screw upon the pillar, which carried a *nonius;* and the slightest contact of these points with the clean surface of a basin of mercury was instantly perceptible. I satisfied myself, by repeated trials, that the adjustment might be depended upon to the one-thousandth part of an inch. I made a contrivance to hold a glass tube perpendicularly immersed in a basin filled with mercury; and when one of the steel points was made to touch the surface of the fluid in the tube, and the other the surface in the basin, the depression of the former was accurately measured by the *nonius.* In this manner I determined the capillary action of several tubes, varying in their diameters from one to six-tenths of an inch. The results agreed as nearly as possible with Dr. Young's table, calculated from the experiments of Mr. Cavendish. The end of the tubes, opposite to those at which the measures had been taken, were then hermetically sealed, in such a manner as to be readily re-opened under mercury: they were immediately filled with mercury, and carefully boiled. I expected to be able to ascertain the differences of depression by opening them in the basin of mercury, and proceeding as before. The experiment was performed as soon after the operation of boiling, as the mercury in the tube had cooled down to the temperature of that in the basin. At first the attraction between the mercury and the glass appeared to be perfect, and no depression could be perceived. When, however, the tubes were left some time exposed, either before or after they were opened, the air and moisture insinuated themselves between the metal and the glass, and an immediate depression was the consequence. This depression increased gradually, till at length it became fixed at the exact point of that of the unboiled tube. The progress of this effect was easily perceptible with a magnifying glass, and was rendered still more visible by heating the tube, when air-bubbles were immediately detached. This is obviously the same effect as that described by Sir H. Davy in his paper before alluded to, in which he says that, "on keeping the stop-cock of one of the tubes, used in the experiment on the mercurial vacuum, open for some hours, it was found that the lower stratum of mercury had imbibed air, for when heated *in vacuo* it emitted it distinctly from a space of a quarter of an inch of the column; smaller quantities were disengaged from the next part of the column, and its production ceased at about an inch high in the tube." Now I conceive that the fact above related absolutely proves that the air had insinuated itself between the mercury and the tube, and shows that there is no "reason to believe that this air existed in mercury in the same invisible state as in water, that is, distributed through its pores." For, if the latter had been the case, the mercury, which contained no air after being boiled, would, from its greater density, have sunk, in the tube, when surrounded by mercury which had not been boiled, and would have risen gradually as it became saturated with air. I am justified in drawing the conclusion from the contrary effect, that the air had insinuated itself between the metal and the tube; for the capillary depression is known to be in inverse proportion to the affinity of the fluid for the containing tube, and nothing could have affected that affinity in the case before us, but the gradual interposition of a thin stratum of air and moisture.

Having thus traced and measured the progress of the air down the sides of small tubes filled with mercury, and boiled with the greatest care, I was naturally led to suspect that the same action might take place in barometers, to their gradual deterioration. I soon saw reasons to conclude that such a process actually was going on in the most carefully constructed instruments, and that, in time, air would thus insinuate itself into the best Torricellian vacuum. In my paper upon the construction of the barometer, I have given all the particulars of the making of two barometers, in which every precaution was used to dispel every particle of air. One of these, it will be recollected, was of very large dimensions, and was fixed up in the apartments of the Royal Society, under the superintendence of the Meteorological Committee; the other was of the mountain construction, and intended for my own use: the agreement between these two instruments, when all corrections were made for the differences in their sizes and forms, was very perfect, and proved that the care which had been bestowed upon them had not been thrown away. In the latter, however, I lately remarked that a small quantity of air had ascended into the *vacuum*. I could not discover any way in which this could have obtained admission; but, attributing it to accident, I laid it aside, and thought no more of it till the present experiments recalled it to my recollection. By a singular coincidence I was, about this time, informed that the barometer of the Royal Society had assumed a very remarkable appearance, and that the mercurial column, which was originally perfectly bright and compact, now seemed dull and speckled. I immediately proceeded to examine it carefully, and I at once perceived that it was copiously studded with minute air-bubbles. As far as the mercury was exposed to view, the specks could be traced decreasing in size, from the upper to the lower part. The manner in which this instrument is fixed rendered it impossible to suspect that this air could have obtained admission by any accident; for, unlike the mountain-barometer, the column of metal is exposed to no oscillations but such as arise from differences of atmospheric pressure. I was myself quite satisfied, and those who have read the account of the precautions taken in filling the tube will also, I think, be satisfied, that this air was not left at its original construction; and the manner of its intrusion is I think pointed out by the experiments which I have detailed.

While I was occupied with these considerations, and sufficiently vexed to find that all my care had been thrown away to prevent my adopting that opinion without very strong grounds, it occurred to me that I had, in the course of my experience, observed a phenomenon, which was calculated to throw some light upon the present question; namely, that gases were more readily preserved from mixture with atmospheric air over water than over mercury. I was unable to refer to any notes of experiments to confirm this suspicion, but I proposed the question to Mr. Faraday, who, I made no doubt, from his great accuracy and experience, must have made the observation, if it were founded in fact. Without at the time having any knowledge of the ulterior object which I had in view, he at once answered me, that mercury, he believed, would not confine gases for a long period so well as water; and he thought that, by referring to his note-book, he could furnish me with the particulars of a case in point. He accordingly did me the favour to extract the following particulars:—

In June, 1823, he made a mixture of one volume of oxygen and two volumes of hydrogen; with this he filled five *dry* bottles over mercury, and also four bottles over water. He left the glasses inverted over mercury and water, placing three mercury and two water bottles in the windows, so as to receive the sun's rays and day-light; and two mercury and two water bottles he placed in a dark place. In July, 1824, he examined the bottles; the water bottle in the light contained hydrogen and some common air, and there was no alteration of volume; the mercury bottle in light contained common air only. The water bottles in the dark place showed no alteration of volume, and the air contained in them proved to be the *original mixture;* the mercury bottles in the same situation contained *nothing but common air.*

Hence it appears that a fluid which has attraction enough for glass, to enable it to wet its surface, effectually prevents the passage of gases into or out of vessels, of that substance; while a fluid which does not wet the surface permits their slow penetration. The case of the confined air is exactly analogous to that of the barometer; for its escape and the admission of the atmospheric air can only be in virtue of the law discovered by Mr. Dalton, that the gases are as *vacua* to one another. The inference is also pretty strong, that the infiltration takes place along the surface of the glass, and not through the pores of the fluid.

It has been attempted to contravert this conclusion, by the observation that gases have been preserved a considerable time by mercury; but when it is considered that the slightest film of moisture, or any foulness of the mercury, will form a connexion between the metal and the glass, the objection can be of no avail, unless these circumstances have been attended to. To ensure the maximum of the effect which I have been describing, it is necessary that both glass and mercury should be in the driest and cleanest possible state; that is to say, exactly in the state in which they exist in a well made barometer.

In consequence of doubts which were thrown over the fact, in the course of some discussions which arose upon the subject, Mr. Faraday was induced to repeat the experiment in the most careful manner, and the following are the results as recorded by himself.

"Two volumes of hydrogen gas were mixed with one volume of oxygen gas, in a jar over the mercurial trough, and fused chloride of lime introduced, for the purpose of removing hygrometric water. Three glass-bottles, of about three ounces capacity each, were selected for the accuracy with which their glass stoppers had been ground into them; they were well cleaned and dried, no grease being allowed upon the stopper. The mixture of gases was transferred into these bottles over the mercurial trough, until they were about four-fifths full, the rest of the space being occupied by the mercury. The stoppers were then replaced as tightly as could be, the bottles put into glasses in an inverted position, and mercury poured round the stoppers and necks, until it rose considerably above them, though not quite so high as the level of the mercury within. Thus arranged, they were put into a cupboard which happened to be dark, and were sealed up. This was done on June 28, 1825, and on September the 15th, 1826, after a lapse of fifteen months, they were examined. The seals were unbroken, and the bottles found exactly as they were left; the mercury still being higher on the inside than the outside. One of them was taken to the mercurial trough, and part of its gaseous contents transferred; upon examination it proved to be common air, no traces of the original mixture of oxygen and hydrogen remaining in the bottle. A second was examined in the same manner; it proved to contain an explosive mixture. A portion of the gas introduced into a tube, with a piece of spongy platina, caused dull ignition of the platina; no explosion took place, but a diminution to rather less than one-half. The residue supported combustion a little better than common air. It would appear, therefore, that nearly a half of the mixture of oxygen and hydrogen had escaped from it, and been replaced by common air. The third bottle, examined in a similar manner, yielded also an explosive mixture, and upon trial was found to contain nearly two-fifths of a mixture of oxygen and hydrogen, the rest being a very little better in oxygen than common air.

"There is no good reason for supposing that this capability of escape between glass and mercury is confined to the mixture here experimented with; probably every other gas, having no action on the mercury or the glass, would have made its way out in the same manner. There is every reason for believing that a small quantity of grease round the stoppers would have made them perfectly tight." *Journal of the Royal Institution*, vol. xxii. p. 220.

I also repeated the experiment with the following variations. I inclosed a portion of pure hydrogen in a glass jar, standing over mercury, carefully preserving the same level both on the inside and the out. I then passed a lump of spongy platinum into the jar, which floated upon the surface of the metal, and it was carefully put by in a dark cupboard. At the expiration of thirteen months, it was examined, and the mercury had risen in the jar an inch above the level of that without. The atmospheric air had evidently insinuated itself between the glass and the metal; and the oxygen, combining with the hydrogen, from the action of the platinum, had created a partial vacuum, and caused the mercury to rise.

I was no sooner convinced that the most carefully constructed barometers were liable to a slow and gradual deterioration, in the manner which I have indicated, than I endeavoured to find a remedy to the evil; without which, it is clear, that some of the most interesting problems of meteorology must be for ever left in a state of vagueness and uncertainty. For a long time I despaired of success; but I was fortunate enough at length to discover an effectual method of preserving the Torricellian *vacuum*.

I soon perceived that the only possibility of effecting the object which I had in view, consisted in finding out some method of making the mercury *wet* (if I may be allowed the term) the tube in which it is contained. I was fearful, at first, that all the substances to which its attraction is sufficiently strong for this purpose, would be so much acted upon as to become disintegrated or dissolved. I, however, fortunately recollected that, in some experiments in which I was formerly engaged, platinum, immersed in boiling mercury, became completely coated by it, and afterwards retained its coating for a long time. I repeated the experiment with some platinum foil, and found it to succeed perfectly. The mercury adhered strongly to the foil, and the latter, after a long immersion, was found to have lost none of its tenacity. I availed myself of this property in the following way:—I caused a small thin piece of platinum tube to be made about the third of an inch in length, and of the diameter of the glass tube; this was carefully welded to its open end, so that the barometer tube terminated in a ring of platinum. The tube was filled and boiled as usual, and the infiltration of air was completely prevented by the adhesion of the mercury, both to the interior and exterior surface of the platinum guard. I have no doubt that a mere ring of wire welded, or even cemented upon the exterior surface of the glass, which would be a much easier and less expensive operation, would be a sufficient protection, as the slightest line of perfect contact must effectually arrest the passage of the air: but in the first attempt I was desirous that the experiment should be tried in the most perfect manner. When a piece of glass, armed either with a ring or tube, is immersed in mercury, the effect is easily perceived; instead of any depression being visible around it, the mercury may be lifted by it considerably above its proper level. Time, of course, alone would fully confirm the efficacy of the guard; but as far as the experiments have gone, they have been completely satisfactory.

Hydrogen gas introduced into a glass jar with spongy platinum, exactly in the same manner as before described, except that the edge of the jar was protected by a rim of platinum wetted with mercury, was completely cut off from any mixture of the atmosphere, and the mercury maintained its level without change.

It was long before I could find evidence of the deterioration of barometers in the numerous registers that are kept of their oscillations: few have been continued for a sufficient length of time, with the same instruments, to answer this purpose satisfactorily. Instances abound of observers having taken the pains to re-boil their barometers, from air having obtained admission, in some unknown way which has always been attributed to accident;

but the fact of their gradual deterioration cannot, in this way, be established, by modern observations. so completely as might have been supposed.

The register of the Royal Observatory at Paris has only been published since the year 1816, in the *Annales de Chimie;* a period which is not sufficient so far to neutralize the annual oscillation as to afford the means of a satisfactory comparison. Mr. Howard, however, in his work upon the climate of London, states the mean height of the Royal Society's barometer for ten years, from 1797 to 1806, to be 29.882 inches, while for the ten succeeding years it is only 29.849 inches, which gives a depression of .033 inches in that interval.

The observations of the following ten years will not, I fear, be available in the same comparison, from the carelessness with which they have been made.

The difference in the height of the old and new barometer, which have now been placed side by side, was, in the latter part of the year 1824, .07 inch, upon a mean of twenty observations; the new barometer standing upon the average so much higher than the old one. Whether this be wholly owing to deterioration, it is not possible to say; for the old barometer does not appear to have been boiled: but from the well-known accuracy of Mr. Cavendish, under whose superintendence it was constructed, it is impossible not to ascribe a great portion of it to this cause. The mercury of this instrument is now thickly studded with air-bubbles of much larger size than those of the new barometer; and when I last examined it, some of them were just upon the point of making their escape.

In the *Ephemerides* of the Meteorological Society of the Palatinate, however, of which I have given an account in a previous essay, I found what I had long sought amongst modern observations. The first astronomers of the continent did not at that time think the science of meteorology beneath their notice, and themselves attended to the irksome labour of registering the indications of the instruments, and calculating the mean results. Whoever will take the trouble of examining these faithful and laborious records, must come to the conclusion, that this branch of natural knowledge, not only has made no progress since the unfortunate dissolution of this society, but has seriously retrograded; both as to the accuracy of the instruments of research, and the proper method of pursuing the investigation.

From the immense repository of these volumes I have selected* eight registers, in which the same instruments, all carefully compared together, were used during the greatest length of time; and from them I have extracted the following mean annual heights of the different barometers. The observations were taken three times in the day, and the means are calculated from all the observations:—

The FIRST series is that of Mannheim, which consists of 12 years, from 1781 to 1792 inclusive: this I have divided, in the following table, into two periods of six years each. The height of the barometer is registered in French inches, lines, and tenths:—

* These were the only instances which were calculated to throw any light upon the subject, from the length of time during which they were continued.

Year.	Ins.	Ls.	Tenths.	Year.	Ins.	Ls.	Tenths.
1781 ..	27	9	9	1787 ..	27	9	8
1782 ..	47	9	3	1788 ..	27	9	6
1783 ..	27	9	6	1789 ..	27	8	3
1784 ..	27	9	1	1790 ..	27	9	2
1785 ..	27	9	9	1791 ..	27	8	9
1786 ..	27	9	4	1792 ..	27	7	5
Mean ..	27	9	5	Mean ..	27	8	8

From this it appears that the mean of the last six years is .7 of a line, or .062 in. English lower than that of the first six. The SECOND series is that of Padua for the same twelve years, divided into similar periods.

Year.	Ins.	Ls.	Tenths.	Year.	Ins.	Ls.	Tenths.
1781 ..	28	0	84	1787 ..	28	2	1
1782 ..	28	1	05	1788 ..	28	1	5
1783 ..	28	1	65	1789 ..	28	1	46
1784 ..	28	1	2	1790 ..	28	2	7
1785 ..	28	1	68	1791 ..	27	11	2
1786 ..	28	1	7	1792 ..	27	10	1
Mean	28	1	3	Mean	28	0	8

The result of this comparison is, that the mean of the last six years is lower than the first six .5 of a line, or .044 English inch.

The THIRD series is that of Rome, in which, however, the first year is deficient, the observations not having been commenced till the year 1782.

Year.	Ins.	Ls.	Tenths.	Year.	Ins.	Ls.	Tenths.
1781 ..				1787 ..	28	0	6
1782 ..	27	10	49	1788 ..	28	0	3
1783 ..	27	10	71	1789 ..	27	9	3
1784 ..	27	11	7	1790 ..	27	9	0
1785 ..	28	0	2	1791 ..	27	10	2
1786 ..	28	0	5	1792 ..	27	8	3
Mean	27	11	5	Mean	27	10	2

The average of the last six years is here lower than that of the first five 1 line .3, or .114 English inch.

The FOURTH series is that of Buda, which likewise wants the first year.

Year.	Ins.	Ls.	Tenths.	Year.	Ins.	Ls.	Tenths.
1781 ..				1787 ..	27	5	98
1782 ..	27	5	76	1788 ..	27	6	2
1783 ..	27	6	09	1789 ..	27	3	5
1784 ..	27	5	89	1790 ..	27	6	5
1785 ..	27	5	90	1791 ..	27	5	9
1786 ..	27	5	85	1792 ..	27	4	6
Mean	27	5	8	Mean	27	5	4

The difference is here .4 line, or .035 English inch.

The FIFTH series is that of Brussels, which, however, consists of only eight years, wanting the four first.

Year.	Ins.	Ls.	Tenths.	Year.	Ins.	Ls.	Tenths.
1785 ..	27	10	72	1789 ..	27	8	7
1786 ..	27	9	98	1790 ..	28	0	9
1787 ..	27	10	0	1791 ..	27	9	9
1788 ..	28	0	5	1792 ..	27	9	9
Mean	27	10	8	Mean	27	10	3

The difference between the means of the first and second four years is .5 line, or .044 English inch.

The SIXTH series is taken from a higher station; viz. Munich. The first six years are complete, but the eighth is wanting.

Year.	Ins.	Ls.	Tenths.	Year.	Ins.	Ls.	Tenths.
1781 ..	26	5	69	1787 ..	26	6	3
1782 ..	26	5	01	1788			
1783 ..	26	5	35	1789 ..	26	3	8
1784 ..	26	5	50	1790 ..	26	6	9
1785 ..	26	4	99	1791 ..	26	4	8
1786 ..	26	4	88	1792 ..	26	2	8
Mean	26	5	2	Mean	26	4	9

The mean of the last five years is lower than that of the first six .3 line, or .026 English inch.

The SEVENTH series is from the summit of Peissenberg, a mountain in Bavaria.

Year.	Ins.	Ls.	Tenths.	Year.	Ins.	Ls.	Tenths.
1781 ..	25	0	14	1787 ..	24	11	89
1782 ..	24	11	27	1788 ..	24	11	73
1783 ..	24	11	42	1789 ..	24	11	09
1784 ..	24	11	03	1790 ..	25	0	1
1785 ..	24	11	36	1791 ..	24	11	28
1786 ..	24	11	07	1792 ..	24	8	9
Mean	24	11	3	Mean	24	11	0

The number of years is complete, and the mean of the first six is .3 line higher than that of the last six, or .026 English inch.

The EIGHTH and last series is taken from the summit of Mount St. Gothard. The first year only is deficient.

Year.	Ins.	Ls.	Tenths.	Year.	Ins.	Ls.	Tenths.
1781 ..				1787 ..	21	10	2
1782 ..	21	8	91	1788 ..	21	9	0
1783 ..	21	10	0	1789 ..	21	9	9
1784 ..	21	9	3	1790 ..	21	10	3
1785 ..	21	9	7	1791 ..	21	8	0
1786 ..	21	9	24	1792 ..	21	7	4
Mean	21	9	4	Mean	21	9	1

The mean of the last six years is lower than that of the first five .3 line, or .026 English inch.

All these examples clearly concur in establishing the supposition of the gradual depression of the mercurial column by the infiltration of the air. There is also another conclusion, derivable from the same facts, which might have been anticipated from theory; namely, that the amount of the effect depends, in some degree, upon the elasticity of the atmosphere in which it takes place. The five series of observations whose mean pressure is 29 .235 in. English, show an average depression of .059 inches in twelve years; while the three series, whose mean pressure is 25 .977, exhibit a depression of only .026 inches in the same interval.

From the same valuable record of facts I have also derived a strong confirmation of my opinion of the manner in which the air gains access to the vacuum of the barometer; that is to say, that it is by means of the glass, and not of the mercury. While I quote the observation in support of my ideas upon this subject, it will at the same time serve for a specimen of the skill and exactness with which all the proceedings of the Society were regulated. The observation occurs in the directions given by the Secretary Hemmer for boiling the mercury in thermometers. He thus expresses himself, "Notatu perquam dignum hoc in labore est, nihil ingentis illius vis bullarum aërearum conspici, quæ, ubi mercurius sine igne immissus fuit, inter hunc et vitrum in coctione apparere solent, *manifesto indicio, eas non tam a mercurio quam a vitro provenire,* cujus parietibus adhærebant, quemadmodum in universum omnes corporum superficies densiore aëris lamina stipatas esse et ratio et experientia evincunt. Hanc aëris massam jam tum a cylindro expuleram, cum exiguam hydrargyri portionem initio immissam fortius ebullire facerem unde quæ in secunda coctione in conspectum veniebant bullæ admodum raræ erant*."

I have lately been directed by my friend Mr. R. Phillips, to another very singular confirmation of my ideas upon this interesting subject. The authority upon which this confirmation rests, is that of Dr. Priestley, whose acuteness of observation few, I imagine, will presume to doubt. The following extract is taken from the third volume of his Observations on Air (p. 236, and sequel), published in 1786:—

"In the course of these experiments with the sun, I observed a remarkable source of fallacy with respect to the increase of the quantity of air confined by mercury, when there is so much moisture in the inside as to be subject to sudden dilations

* There is a defect which may often be observed in old looking-glasses, which may probably be referred to the same cause as the deterioration of barometers. I allude to a dulness which takes place in large spots over their surface, and which generally seems to radiate from a centre. I have frequently remarked this in the very old mirrors in some of the palaces upon the continent. I imagine that this arises from the slow insinuation of air by the edges, or some accidental crack in the metal at the back of the glass. It is also, I understand, well known to the dealers in mirrors, that when placed against a damp wall, looking-glasses are particularly liable to become cloudy; and it is most likely that moisture greatly facilitates the action to which I have been referring.

and compressions. For a considerable quantity of common air would get into the inside of the vessel when there was the depth of an inch of mercury on the outside of it, and of two or three inches within. In these circumstances I have seen more than an ounce-measure of the external air gain admission in less than one minute. This must have been occasioned by the mercury never being in perfect contact with glass; so that when the mercury was in a state of undulation, the air that was confined between it and the glass was continually protruded, and more air from the atmosphere was forced into its place, by the same pressure which supported the column of mercury within the glass. This effect I prevented by having a quantity of water upon the mercury on the outside of the vessel. For this would be in perfect contact with the glass; and in this case I never found either air or water to get into the vessel to disturb my experiment."

From these few facts thus briefly, but clearly, described, the whole of my conclusions with regard to the barometer might have been deduced with as much justness as from the more extended and varied observations upon which I have hitherto rested them. If mercury, in Dr. Priestley's experiments, could not be brought in perfect contact with glass, neither in the common construction of a barometer can it so be brought. And as, when the mercury in his jars was in a state of undulation, the air that was confined between the two was continually protruded; so, in the barometer, will it ascend by the continual oscillations and vibrations to which it is exposed.

Again—Dr. Priestley argued, that a fluid which would be in perfect contact with the glass would effectually interrupt this action; and he accordingly found that when he put a quantity of water upon the mercury, on the outside of the vessel, neither air nor water got in to disturb his experiments. It follows, therefore, that if perfect contact between the mercury and any complete circle of the barometer tube can be produced, the air will be effectually prevented from ascending into the vacuum. I have already described an easy method of producing this contact.

The time which has elapsed since my first proposition and application of the guard to barometers enables me now to add the testimony of experience to its sufficiency and permanency. At the expiration of nearly three years from its adoption, the first instrument to which it was applied was lately examined and opened in the presence of several scientific gentlemen. The mercurial column from top to bottom was found perfectly bright and free from specks or air-bubbles, and the platinum ring as sound as on the first day when it was fixed and thoroughly *wet* with the mercury. Upon wiping the fluid off, the platinum exhibited the original marks of the hammer upon it; so perfectly free was it from any signs of corrosion or disintegration! Thus has time removed the only remaining doubt about the ultimate solution of the platinum by the mercury, and confirmed the conclusion which I originally drew from the experiment with the thin platinum foil.

ADDENDA AND NOTES.

ADDENDA AND NOTES.

ON THE CONSTITUTION OF THE ATMOSPHERE.

Page 13.

THE illustrative tables of this essay were not intended as accurate representations of the different states of the atmospheric columns, but mere rough approximations; and their use is to assist the mind in following the train of reasoning in the same way that rudely-sketched diagrams assist the mathematician in solving the problems of Euclid. The principle upon which they were constructed is this: I assumed the mean temperature of the latitude for which I wished to calculate the atmospheric column as the temperature of an homogeneous atmosphere; and I thence derived the pressures at different altitudes from the surface, and from them again the regular decrease of temperature for the density. It is clear that, for accurate purposes, both the pressures and temperatures so obtained require correction, and that the tables include an error which should be divided between the two. The labour of applying these corrections would have been very considerable; and the tables which, even in their present state, cost much pains, would not have better answered the purposes of illustration.

Page 96.

It has been objected to my theory of the fluctuations of the barometer, " that it rests upon the sandy foundation of assumed partial changes of temperature in the higher regions of the atmosphere, of the existence of which we have very insufficient evidence:" but I certainly conceive that no meteorological fact rests upon better authority than these variations—I appeal to every ascent of every mountain, and to every aërostatic voyage, during which the thermometer has been consulted; for they all appear to me to agree in the same result: amongst the latter, more particularly, there are abundant instances in which, not only the temperature of the atmosphere has not followed the progression due to the density, but warm strata have been found interposed between cold, and cold between warm. The contemporaneous registers, moreover, kept at the summit of St. Bernard and at Geneva, and which are published every month in the Bibliothéque Universelle, place the fact beyond dispute. The difference of temperature between the stations is perpetually varying; and although the changes oscillate round the point of equilibrium, the general law of the decrease for the altitude is only developed from a mean of many observations.

Page 102.

Mr. De Buch, in his remarks upon the climate of the Canary Islands, observes that the existence of a current flowing in a contrary direction to that of the trade winds, was nothing but a conjecture till of late years, but that every day this counter-current may be observed in the Canaries. The Peak of Teneriffe rises into it even in the middle of summer. There is scarcely an account of a journey to the summit of the mountain which does not mention the force of the western wind. M. de Humboldt ascended to the top in the middle of June; and when he had arrived at the edge of the crater, the wind from the west was so furious that he could hardly stand upon his feet.—*Ann. de Chimie*, t. xxii. p. 288.

For a more accurate account of the direction of the trade winds, see Captain Basil Hall's essay in the present volume.

Page 103.

The currents of a heated room in some measure exemplify the

great currents of the atmosphere. If the door be opened, the flame of a candle held to the upper part will show by its inclination a current flowing outwards: but, if held near the bottom, it will be directed inwards. If the door be closed suddenly from without, it moves with the in-coming current and against the out-going, and a condensation of air takes place in the room, which is shown by the rattling of the windows, and the bursting open of any other door in the room, if slightly closed. If the door close from within, it moves against the in-coming current and with the out-going, and a rarefaction of the air in the room takes place, which is evidenced by rattling of the windows, and the bursting open of another door in the contrary direction.

In mines, the following means are employed to produce a current of air. In an adit in which a stream flows there is placed, at a small elevation on the latter, a partition of boards: below this partition the air follows the course of the stream; above, it takes the contrary direction.

Page 110.

It is a singular fact that the same kind of uncertain squalls are frequent in the interval between the two trade winds, which is called *the Variables*. In this region rain is common and very heavy; and Captain Basil Hall has informed me that he has frequently observed a ship, at a distance of five or six miles, struck by a squall which was some hours in reaching his own vessel; although, when it arrived, the wind blew with a velocity of twenty or thirty miles per hour, and in the direction towards him.

Page 118.

I have been favoured with a most interesting illustration of the fact that the vapour of the atmosphere is arranged in beds of equal density, by Captain Basil Hall, in the following account of his ascent of the Peak of Teneriffe:—

"On the 24th of August we left Oratara to ascend the Peak. The day was the worst possible for our purpose, as it rained hard, and was so very foggy that we could not see the Peak, or indeed any object beyond one hundred yards distant.

"After riding slowly up a rugged path for four hours, it became extremely cold, and as the rain never ceased for an instant, we were by this time drenched to the skin, and looked with no very agreeable feelings to the prospect of passing the night in wet clothes. At length the night began to close in, and the guides talked of the improbability of reaching the English station before night. It was still raining hard; but we dismounted and took our dinner as cheerfully as possible, and hoping for clearer weather next day. On remounting, we soon discovered that the road was no longer so steep as it had been heretofore, and the surface was comparatively smooth; we discovered, in short, that we had reached a sort of table-land, along which we rode with ease. Presently we thought the fog less dense, and the drops of rain not so large, and the air less chilling. In about half an hour we got an occasional glimpse of the blue sky; and as we ascended, for our road, though comparatively level, was still upon the rise, these symptoms became more manifest. The moon was at the full, and her light now became distinct and we could see the stars in the zenith. By this time we had reached what is called the Llano de los Rememos, or Retamos Plain, which is many thousand feet above the sea; and we could distinctly see that during the day we had merely been in a cloud, above which having now ascended, the upper surface lay beneath us like a country covered with snow. It was evident, on looking round, that no rain had fallen on the pumice gravel over which we were travelling. The mules were much fatigued, and we got off to walk. In a few minutes our stockings and shoes were completely dried; and in less than half an hour all our clothes were thoroughly dried. The air was sharp and clear, like that of a cold frosty morning in England, and, though the extreme dryness, and the consequent rapid evaporation, caused considerable cold, we were enabled by quick exercise to keep ourselves comfortable. I had various instruments with me, but no regular hygrometer; accident, however, furnished me with one sufficiently indicative of the dry state of the air. My gloves, which I kept on while mounted, were completely soaked with the rain; and I took them off during this walk, and, without considering what was likely to happen, rolled them up and carried them in my hand. When at the end of an hour, or somewhat less, we came to remount our mules, I found the gloves as thoroughly dried and shrivelled up as if they had been placed in an oven, and I could not manage to get them on again.

"During all the time we were at the Peak itself on the 26th, the sky was clear, the air quite dry, and we could distinguish, several thousand feet below us, the upper and level surface of the stratum of clouds through which we had passed the day before, and into which we again entered on going down, and found precisely in the same state as when we started."

ON THE CONSTRUCTION AND USES OF A NEW HYGROMETER.

Page 157.

I HAVE been favoured by Mr. Galbraith of Edinburgh with the following Table of the Elastic Force of Aqueous Vapour, calculated from the experiments of Dr. Ure, by the formula which Mr. Ivory has lately proposed:—

TABLE of the Elastic Force of Aqueous Vapour, calculated by Mr. Galbraith, from the Experiments of Dr. URE, by the Formula of Mr. IVORY.

Temp. Fah.	Force. Inches of Mercury.	Temp. Fah.	Force. Inches of Mercury.	Temp. Fah.	Force. Inches of Mercury.	Temp. Fah.	Force. Inches of Mercury.
0°	0.0511	32°	0.1857	63	0.5705	94°	1.5617
1	0 0533	33	0.1929	64	0.5904	95	1.6104
2	0.0556	34	0.2004	65	0.6109	96	1.6604
3	0.0580	35	0.2082	66	0.6320	97	1.7117
4	0.0605	36	0.2162	67	0.6537	98	1.7645
5	0.0631	37	0.2244	68	0.6761	99	1.8187
6	0.0658	38	0.2329	69	0.6993	100	1.8743
7	0.0686	39	0.2417	70	0.7231	101	1.9315
8	0.0715	40	0.2509	71	0.7477	102	1.9902
9	0.0745	41	0.2604	72	0.7730	103	2.0504
10	0.0776	42	0.2702	73	0.7991	104	2.1122
11	0.0809	43	0.2804	74	0.8259	105	2.1756
12	0.0843	44	0.2909	75	0.8536	106	2.2408
13	0.0879	45	0.3017	76	0.8821	107	2.3076
14	0.0916	46	0.3129	77	0.9114	108	2.3762
15	0.0954	47	0.3244	78	0.9416	109	2.4466
16	0.0993	48	0.3363	79	0.9727	110	2.5189
17	0.1033	49	0.3487	80	1.0047	111	2.5931
18	0.1075	50	0.3615	81	1.0377	112	2.6689
19	0.1118	51	0.3747	82	1.0716	113	2.7466
20	0.1163	52	0.3884	83	1.1064	114	2.8263
21	0.1210	53	0.4025	84	1.1422	115	2.9082
22	0.1259	54	0.4170	85	1.1791	116	2.9920
23	0.1309	55	0.4320	86	1.2170	117	3.0780
24	0.1362	56	0.4475	87	1.2560	118	3.1663
25	0.1417	57	0.4635	88	1.2962	119	3.2568
26	0.1473	58	0.4800	89	1.3374	120	3.3494
27	0.1531	59	0.4970	90	1.3799	121	3.4444
28	0.1591	60	0.5145	91	1.4235	122	3.5417
29	0.1654	61	0.5325	92	1.4683	123	3.6412
30	0.1720	62	0.5512	93	1.5144	124	3.7434
31	0.1787						

Page 201.

In the course of a tour which I made in Germany in the early part of the year 1825, an instrument was shown to me by some of the ingenious artists of that country, which was intended for a simplification of my hygrometer. It consisted of a large-bulbed thermometer, the stem of which was bent twice at a right angle. On the top of the bulb a small cup was formed, in which a piece of cotton was placed and moistened with ether; by the evaporation of which the mercury was cooled till dew appeared upon the surface of the glass. It was supposed that the temperature of the dew-point would be indicated at the moment this precipitation took place by the scale of the thermometer. Experience soon proved that this indication could not be depended upon, and the instrument was consequently rejected. I purchased one of these hygrometers at Berlin, and another was presented to me by the Grand Duke of Weimar, and both are now in my possession.

Under these circumstances, it was with a considerable degree of astonishment that I read in the second part of the *Philosophical Transactions* for the year 1826, a description of this same instrument, under the title of an *Improved Hygrometer by Mr. Thomas Jones. Communicated by Captain Henry Kater, F.R.S.*

The sole difference between Mr. Jones's instrument and those which I brought from Germany, consists in a piece of muslin being wrapped round the lower part of the bulb of the former, instead of the piece of cotton placed upon the top of the latter; a variation which, although sufficiently insignificant, is, upon the whole, as I shall presently show, detrimental to the intended observation. Mr. Jones, indeed, himself hardly asserts that his invention is new; for he says, at the conclusion of his paper, "I ought also, perhaps, to mention that an instrument somewhat similar in principle, has been used in Vienna, and was mentioned by Professor Baumgarten, of that capital, to a friend, who communicated the fact to myself." The instrument, however, under either form, is utterly incapable of answering its intended purpose.

This, I should have supposed, would have been so extremely obvious to any person in the least acquainted with the time which necessarily must elapse before a thermometer, containing a large quantity of mercury, will mark the temperature of a fluid in which it is even wholly immersed, that I could only have been induced to have expended any words in pointing it out by the magnitude of the authority by which it has been brought forward.

In the first place, the thermometer, being graduated by the total immersion of its bulb into a fluid of known temperature, cannot for a moment be supposed to indicate the temperature of an evaporating fluid, into which it is afterwards only half immersed? Or, one half of the inch-long bulb being cooled by evaporating ether, and the other half being exposed to the temperature of the surrounding air, by what process of induction can it be established that the temperature denoted by its scale is that of one-half of its surface only?

In the second place, mercury, being a fluid, is an extremely bad conductor of heat, and when heat is applied to any mass of it unequally, an equal distribution is only brought about by a circulation of its different parts. This circulation will be more or less rapid, according as the differently heated portions are more or less favourably arranged for the operation. Thus, in the original German contrivance, the cooling power being placed above, the circulation will be more quick than in that of Mr. Jones, where the refrigerating ether is applied below. In the former case, the heavy cold particles will rapidly fall and give place to the lighter warm particles; while, in the latter, the cold particles being below, will but slowly be driven from their station by the lighter warm particles above. Mr. Jones, moreover, contracts the diameter of his bulb below, and thus creates another obstacle to the interchange of particles, upon which so much depends.

In the third place, an equal distribution of temperature must take place in very unequal times, according as the partial cooling influence is more or less rapid in its effects; and as the evaporation of the ether will vary with every breath of wind, the result must always be uncertain. That these very simple propositions were not wholly unknown to Mr. Jones, appears from the following observation extracted from his paper:—"Should it be objected against the principle of the instrument here proposed, that the indications do not exhibit the true temperature of the upper surface of the bulb on which the deposition of dew takes place, but that of the lower part to which the ether is applied; it may be answered, that by inclining the whole instrument, so as to render the axis of the bulb horizontal, and establish thereby a free circulation of the mercury in every part, this objection may be obviated." To this it may be replied, that the indications do not exhibit the true temperature of either the upper surface on which the deposition of dew takes place, or of that of the lower part to which the ether is applied, but a perpetually varying combination of both; and with regard to inclining the axis of the instrument, it does not at all appear how this can establish a free circulation in every part; and indeed, Mr. Jones goes on to observe with perfect sincerity, "but on repeated trials I have not found this to produce any difference in the results."

I shall not now stop to inquire whether a stream of the vapour of ether rising so immediately under the condensing surface of the glass, and issuing with some force into the very portion of the atmosphere whose dew-point it is proposed to ascertain, may not interfere with the result, although I think that it would not be difficult to show that such interference must follow; but I shall detail the particulars of an experiment which will, probably, work conviction upon those who have overlooked the preceding arguments.

I caused a glass bulb to be made of the exact form and dimensions of that of Mr. Jones's hygrometer, and having inserted one of the very delicate thermometers used by Mr. Newman in the construction of my hygrometers into the upper part, and another into the lower, I filled it with mercury: I then covered one-half with muslin, by wetting which with ether I readily produced a difference of 16° or 17° between the two small thermometers! This difference was never constant, but varied with every inclination of the bulb, and every variation of the cooling process. After this, it is scarcely necessary to say that I have inquired of those who, with much perseverance, have endeavoured to make observations with Mr. Jones's hygrometer, and that they report that its indications are inconsistent and not to be depended upon.

Page 204.

The previous account of the applications of the Hygrometer would, I consider, be incomplete were I to omit to subjoin the details of the observations made with it during Captain Parry's third voyage to the arctic regions. The few opportunities which will, probably, ever occur of repeating such observations under circumstances of so much interest render them particularly valuable, and they admirably illustrate the powers of the instrument in one extreme of climate. The performance of the experiments was rendered particularly difficult by the negligence of the instrument-maker who provided the supply of ether for the voyage. This was of such very bad quality as scarcely to produce a depression of two or three degrees in the instrument. I have heard similar complaints from two other quarters of the same instrument-maker, who is nevertheless greatly distinguished for ingenuity and ability; and in one case a most valuable series of observations has been lost by his carelessness. On the present occasion the deficiency was ingeniously but laboriously supplied by a mixture of snow and salt or muriate of lime.

"*H. M. S.* Hecla, *at Port Bowen;*
December 1824.

"Two experiments to ascertain if any moisture existed in the atmosphere were made in the course of this month with Mr. Daniell's hygrometer; but none could be detected. On the 21st, the wind being light from the northward, with a perfectly clear sky, the instrument was exposed till both thermometers indicated the temperature of the atmosphere, which was $-30°$; and the freezing mixture (muriate of lime and snow) being then applied to the covered ball, the ether soon became frozen, and the thermometer immersed in it indicated $-46°$, without the slightest appearance of deposit. Mr. Foster repeated this experiment on the 25th, with very similar results, the temperature of the atmosphere being then $-25°.5$ with calm and clear weather.

"*January* 1825.—Mr. Daniell's hygrometer was twice tried during this month: on the 3d, the temperature being $-30°$, and the instrument subjected to the same process as before, the ether froze without producing any deposit. The wind at this time was light from the eastward; the sky perfectly clear, except to the westward, when a dense haze indicated the vapour arising from open water in that direction.

"On the 24th, the temperature of the atmosphere was $-35°$, the sky clear, with the exception of a few thin clouds near the horizon to the eastward, and the wind light from the north,—the experiment was repeated, and when the ether became frozen, the thermometer indicated $-50°$, without the slightest appearance of deposit on the coloured ball.

"*February* 1825.—The hygrometer twice tried at the temperature of $-39°$ and $-28°.5$, and the ether froze without any visible deposit.

"*April* 1825.—Twice in this month Mr. Foster succeeded in obtaining a deposit on the coloured ball of Mr. Daniell's hygrometer. On the 21st, the temperature of the atmosphere being $+15°$, the sky partially clear, with large well-defined clouds to the westward, a broad white belt of frozen vapour appeared on the instrument, coincident with the surface of the ether, on the temperature being reduced to $-4°$. On the 25th, the temperature of the atmosphere being $+6°$, and the sky densely overcast, a similar deposit took place on the coloured ball on the ether being reduced to the temperature of $-1°.5$.

"Mr. Daniell's hygrometer was tried on several occasions, in different parts of the ship. The following examples will show how great a degree of dryness was maintained below*:

January 9. $11^h 30^m$ A.M.

Temp. of external air.	Middle of lower Deck.	Dew Point.	Remarks.
−22	+67.5	+53.5	All the people had been on the lower deck for an hour and a half previously, but were off the deck at the time.

April, $11^h 30^m$.

External air.			D. Point.	Remarks.
−20	Captain's cabin	+64	+48	A few people below, the copper boiling, and meat taking out.
	Gun-room .	+64.2	+50	
	Middle of lower deck . .	+63.5	+55	
11th, 9.30 P.M.	ditto,	+66	+55	The ship's company in bed.

* Parry's Third Voyage, p. 46.

ON THE RADIATION OF HEAT IN THE ATMOSPHERE.

Page 226.

My observations upon this subject have been misrepresented* —perhaps misunderstood. I cannot forget that the objector is one of the first philosophers of the age; but in attempting to meet his arguments, I shall endeavour to forget the spirit in which they have been conveyed.

By fixing upon one expression in my paper, it has been made to appear, that I have advanced certain most untenable conclusions with *confidence* and *presumption*. I trust, however, that I have taken more than ordinary pains to guard against any such imputation. I have expressly stated, that I was sensible that "my observations were in a very imperfect state; and that I only ventured to bring them forward in the hope of exciting some attention to a subject which appeared to me to be well worthy of elucidation, and to suggest some experiments, which, to render them beneficial, require much perseverance and extensive co-operation." To those who have done me the honour to read the previous essay, I may confidently appeal whether I have not, throughout, maintained that tone of diffidence which the sense of incompleteness required.

From eighteen months' observations of my own, in this country, compared with twelve months' observations of Captain Sabine, at different intervals, and at different places, between the tropics, I thought I saw reason to conclude that the force of radiation from the sun was greater in the former than in the latter situation. I was aware (and I have so expressed myself) that the instruments made use of were not sufficient to determine the

* Annales de Chimie et de Phisique, tom. xxvi. p. 29.

question with any degree of nicety; but I thought that the irregularities to which they were subject would be in some measure counterbalanced by the number of the observations. It was from the *entire* number of observations in this country compared with the *entire* number between the tropics, that I conceived myself entitled to argue. My argument is this: Captain Sabine undertook at my suggestion to measure the force of solar radiation between the tropics, by observing its effects upon a thermometer prepared to receive its greatest impression, placed in the most unexceptionable manner that circumstances would allow, and by comparing them with another screened from its influence, marking as nearly as possible the mean temperature of the air. To accomplish this object, he selected his opportunities at different stations: but *only once*, at Bahia, did he obtain a result which even equalled the mean power of the sun for two years, in this country, in the month of June; *all the cloudy days* being included in the average; and which fell short, by one-third, of the maximum effect which often occurred in clear weather, measured by the same means. It has been objected that the thermometers were not always placed at equal distances from the ground and from the vegetation on it, and that they were not equally secured from currents of air, &c. &c.; but it must be admitted that there is ample room for allowances of this kind, and yet to save the conclusion, which is only general, that the power of solar radiation is less between the tropics than in higher latitudes. These results were moreover confirmed by others obtained with instruments of more delicate construction; in which the thermometers were placed *in vacuo*, one being armed with a case of polished silver to repel the rays, and the other with a blackened surface to absorb them. I repeat, that if Captain Sabine *but once* succeeded, at any *one station* between the tropics, in obtaining the full impression of the sun upon a blackened thermometer, or even approached the full impression within one-third, that there is ground for the hypothesis.

Let us now turn to the unpremeditated observations of those who, having no object in view, and being unprepared with even the rough apparatus which has been described, had their attention called to the prodigious power of the sun's rays in high northern latitudes. The bulb of the instruments in these instances was not covered with black wool, or even blackened superficially; but then it has been objected, " Qui ne sait combien la force réfléchissante de la neige est considérable ? Il aurait fallu faire, par le calcul ou par l'expérience, la part de cette réflexion, avant de comparer les observations de Londres avec celles du Capitaine Parry." Let the calculation of the effects of this reflection on one side, and of the blackened bulb on the other, be fairly made, and I do not think that my reasoning will be shaken by the result.

To assist in forming a right conclusion upon the subject, let the following additional fact, extracted from Captain Lyon's interesting Journal, be taken into consideration; the place of observation and the date are, Igloolik, 16th February:—" I observed, even while the temperature in the shade was 35° below zero, that fine powder of snow melted under the influence of the sun when sprinkled on a stick covered with soot; thus making a difference of temperature existing at the same time as great as 67° and upwards."—*Lyon's Journal*, p. 389.

Here the coating of soot renders the experiment very closely comparable with a thermometer covered with black wool, with which the utmost effect I ever obtained in the month of February, in this country, was 36°. The difference, therefore, of the power of the sun in the two situations was 31°; from which let any reasonable deduction be made for reflection, and the remainder will be amply sufficient to support my conclusion.

The confirmation of my argument, upon which I place the greatest reliance, is that of M. de Humboldt, who " *often* endeavoured to measure *the power of the sun* between the tropics, by two thermometers of mercury perfectly equal, one of which remained exposed to the sun, while the other was placed in the shade. The difference resulting from the absorption of the rays in the ball of the instrument *never* exceeded 3°.7 (6°.6 Fahr.); *sometimes* it did not even rise higher than one or two degrees."

Captain Sabine tried the very same experiment with a naked thermometer at Jamaica, and obtained the same result; namely, 3.1 centigrade degrees between the sun and the shade.

Will it be said, that we have no analogous experiments in these northern latitudes to compare with those of the uncoated thermometers? I refer to the Ephemerides of the Meteorological Society of the Palatinate, published in 1783 and following years; in which will be found a register of the power of the sun at Mannheim, measured by equal and carefully-adjusted thermometers, with naked bulbs, nearly every day in the year for several years. It will there be seen that a difference of from 5 to 7 octogesimal degrees (6.3 to 8.7 centig.) is often recorded.

I will take the opportunity of introducing another argument in favour of my hypothesis, derived from a very different source, which I have lately met with, and which has afforded me much satisfaction. Mr. Andrew Knight, the President of the Horticultural Society, well known for his admirable and practical remarks upon the physiology of the vegetable kingdom, has observed that pine-apples, ripened in the house during the winter, have proved of great excellence. He suggests that this fruit will ripen better early in the spring than in the summer months: "for," he says, "this species of plant, though extremely patient of a high temperature, is not by any means so patient of the action of very continued bright light as many other plants, and much less so than the fig and orange tree: possibly, having been formed by nature for intertropical climates, its powers of life may become fatigued and exhausted by the length of a bright English summer's day in high temperature."—*Hort. Trans.* vol. iv. p. 548.

But to conclude the whole, I must again refer to the last part which has appeared, of that admirable work to which I have so often been indebted,—the Personal Narrative of the

Baron de Humboldt, from which I have made the following extract. Taken by itself, it almost proves the general proposition for which alone I have ever contended.

"The thermometer was placed in the shade, in an airy spot, far from the reflection of the soil, at the Faubourg of the Guayqueries Indians. Cumana being regarded as one of the hottest, dryest, and healthiest places of the low regions of equinoctial America, it is important to make known these partial observations. I take them by chance out of 1600 I possess. They will serve, above all, to certify that the climate of the tropics is much more characterized by the *duration of the heat*, than by its *intensity*, that is, by the *maxima* of temperature which the thermometer attains on certain days. I never saw that instrument at Cumana below 20°.8, *nor above* 32°.8 *cent.;* and I found on the registers of M. Orta, whose thermometers were compared by mine with those of the observatory at Paris, that at Vera Cruz the maximum of heat in thirteen years *had only three times attained* 32° *cent., and once* 35°.7 ; *while we have seen the centesimal thermometer at Paris at* 38°. 4*."

* Humboldt's Personal Narrative, translated by H. M. Williams, vol. vi. part ii. p. 779.

ON THE HORARY OSCILLATIONS OF THE BAROMETER.

Page 262.

SINCE the preceding hasty and imperfect sketch of the horary fluctuations of the barometer was written, Baron de Humboldt, with his characteristic zeal for the promotion of science, has presented to the public a connected view of the observations made to verify the progress of this interesting phenomenon, from the level of the sea to the ridge of the Cordillera of the Andes. None but those who have examined this very interesting memoir can have an idea of the immense labour of such a work; and science is indeed greatly indebted to him, who not only made so valuable a compilation, but himself contributed so large a portion of the most accurate and important observations. Those who are at all interested in this subject will, of course, deeply study this work, where they will not only find all the known facts connected with it, but a candid, historical account of the progress of the investigation.

I have extracted from this work the following Table, which contains the result of the observations made between the parallels of latitude 25° S. and 55° N., from the level of the ocean to the elevation of 1400 toises.

RESULT OF THE OBSERVATIONS OF THE HORARY VARIATIONS MADE BETWEEN THE PARALLELS OF LATITUDE 25° SOUTH AND 55° NORTH, FROM THE LEVEL OF THE OCEAN TO THE ELEVATION OF 1400 TOISES.

ZONES.	Names of the observers.	LIMIT-HOURS. Minima after midnight.	Maxima of the morning.	Minima afternoon.	Maxima of the evening.	Mean extent of the oscillations of the barometer (in hundredths of millim.)	PLACES OF OBSERVATION.
EQUATOR. North and South Torrid Zone.	Lamanon & Monges ...	-4^h	$+10^h$	-4^h	$+10^h$	......	Equatorial and Atlantic Ocean.
	Humboldt & Bonpland	$-4\frac{1}{2}^h$	$+9\frac{1}{4}^h$	$-4\frac{1}{2}^h$	$+11^h$	2.55	Equatorial America, from 23° N. lat. to 12° S. lat. between 0° and 1500 toises of elevation.
	Duperrey......	-3^h	$+9^h$	$-3\frac{1}{2}^h$	$+11\frac{1}{4}^h$	3.40	Payta (on the coast of Peru), lat. 5° 6′ S.
	Boussingault & Rivero......	......	$+9\frac{1}{2}^h$	$-3\frac{1}{2}^h$	$+10^h$	2.44	La Guayra, lat. 10° 36′ N.
		-4^h	$+9^h$	-4^h	$+10^h$	2.29	Santa Fé de Bogota (lat. 4° 35′ N.) height 1366 t.
	Horsburgh ...	-4^h	$+8\frac{1}{2}^h$	-4^h	$+11^h$	......	Indian and African Sea, lat. 10° N., 25° S.
	Langsdorff & Horner..	$-3\frac{1}{2}^h$	$+9\frac{3}{4}^h$	-4^h	$+10\frac{1}{2}^h$	......	Equatorial Pacific Ocean.
	Sabine.........	-5^h	$+9\frac{1}{2}^h$	$-3\frac{3}{4}^h$	$+10^h$	......	Sierra Leone, lat. 8° 30′ N.
	Kater	-5^h	$+10\frac{1}{2}^h$	-4^h	$+10\frac{1}{2}^h$	......	Table-land of Mysore, (lat. 14° 11′ N., height 400 t.) Rainy season.
	Simonoff	$-3\frac{1}{2}^h$	$+9\frac{1}{2}^h$	$-3\frac{1}{2}^h$	$+9\frac{3}{4}^h$	......	Pacific Ocean, from lat. 24° 30′ N. to 25° 0′ S.
	Richelet	-5^h	$+9^h$	-5^h	$+10^h$	......	Macao, lat. 22° 12′ N.
	Balfour	-6^h	$+9\frac{1}{4}^h$	-6^h	$+10^h$	......	Calcutta, lat. 22° 34′ N.
	Dorta, Freycinet, Eschwege. .	-3^h	$+9\frac{1}{4}^h$	-4^h	$+11^h$	2.34	Equinoctial Brazil, at Rio Janeiro (lat. 22° 54′ S.), and at the Missions of the Coroatos Indians.
	Hamilton......	......	......	......	......	......	Table-land of Katmandoo (in India) lat. 27° 48′ N.

RESULT OF THE OBSERVATIONS OF THE HORARY VARIATIONS MADE BETWEEN THE PARALLELS OF LATITUDE 25° SOUTH, AND 55° NORTH, FROM THE LEVEL OF THE OCEAN TO THE HEIGHT OF 1400 TOISES. (*Continuation.*)

ZONES.	Names of the observers.	LIMIT-HOURS.				Mean extent of the oscillations of the barometer (in hundredths of millim.)	PLACES OF OBSERVATION.
		Minima after midnight.	Maxima of the morning.	Minima afternoon.	Maxima of the evening.		
TROPIC.	Leopold de Buch	......	$+10^h$	-4^h	$+11^h$	1.10	Las Palmas, in the Island Grand Canaria, lat. 28° 8′ N.
Temperate Zone.	Coutelle	-5^h	$+10^h$	-5^h	$+10\frac{1}{2}^h$	1.75	Cairo, lat. 30° 3′.
	Marque-Victor	summer. winter.	$+8\frac{1}{2}^h$ $+10^h$	$-5\frac{1}{2}^h$ $-2\frac{1}{2}^h$	$+11^h$	1.20	Toulouse, lat. 43° 34′. (Mean of five years.)
	Billiet	summer. winter.	$+7\frac{1}{2}^h$ $+10^h$	-3^h -2^h	.....	1.00	Chambery, lat. 45° 34′ (height 137 t.)
	Ramond	summer. winter.	$+8^h$ $+9^h$	-4^h -3^h	$+10^h$ $+9^h$	0.94	Clermont-Ferrand, lat. 45° 46′ (height 210 t.)
	Herren-schneider	-5^h	$+8\frac{1}{2}^h$	$-3\frac{1}{2}^h$	$+9\frac{1}{2}^h$	0.80	Strasbourg, lat. 48° 34′. (Mean of six years.)
	Arago	......	$+9^h$	-3^h	......	0.72	Paris, lat. 48° 50′. (Nine years of the most precise observations.)
	Nell de Breautté	......	$+9^h$	-3^h	......	0.36	La Chapelle, near Dieppe, lat. 49° 55′.
	Sommer & Bessel	......	$+8\frac{1}{2}^h$	$-2\frac{1}{2}^h$	$+10^h$	0.20	Kœnigsberg, lat. 54° 42′. (Eight years.)

The astonishing regularity in the periods of this oscillation over so large a portion of the surface of the globe, and its gradual decrease as we proceed from the equator to high latitudes, are thus ascertained in the most unexceptionable manner.

With regard to the hypothesis, which I have ventured in the preceding pages to propose, for the explanation of these fluctuations of the atmospheric ocean, M. de Humboldt observes:— "M. Daniell a cru reconnoître dans des observations faites pendant les derniers voyages aux régions polaires, surtout à l'île de Melville, et aux Montagnes Rocheuses, que le baromètre monte par les 74° de latitude, lorsqu'il baisse par les 41°. Ce savant physicien paroît attribuer ce phénomène à des courans atmosphériques dont l'existence n'est pas facile à constater."

No one can wonder less than myself at this hesitation to admit an explanation which, while I indicated an unerring test of its correctness, I have constantly referred to future observations for confirmation. I shall now proceed to show that such further experience has fully justified my reasoning, and to offer such additional arguments and illustrations as maturer consideration enables me to propose.

Upon the return of Capt. Parry's second expedition from the northern coast of America, I was extremely anxious again to bring my hypothesis to the test of experience, and for this purpose was favoured, upon application, with the loan of Captain Lyon's Meteorological Journal. This, as well as all other nautical registers which I have had an opportunity of examining, has been kept with the utmost precision and neatness; and it is highly gratifying to find so much attention to the interests of science amongst our naval officers, who have such opportunities of enlarging our acquaintance with the different climates of the globe. The periods of the day were almost as favourable as possible to the comparison, but the latitudes were not as far removed as that of Melville Island from the influence of variations of daily temperature. The following table presents the monthly means of the observations for two years, during which the Hecla was confined between the latitudes 66° and 70°.

TABLE—Showing the Mean Heights of the Barometer and Thermometer at four different Hours of the Day on board H. M. S. HECLA, between the Latitudes 66 and 70.

Date.	A. M. 4.		A. M. 8.	P. M. 4.		P. M 8.
1821.	BAR.	THER.	BAR.	BAR.	THER.	BAR.
August . .	29.835	33.5	29.846	29.848	39.9	29.825
September .	29.958	29.8	29.974	29.973	34.3	29.977
October . .	29.881	8.9	29.876	29.889	17.6	29.898
November .	30.166	2.7	30.156	30.165	12.6	30.159
December .	29.904	—19.2	29.898	29.914	−11.5	29.918
1822.						
January . .	29.921	−26.9	29.924	29.933	−20	29.952
February. .	29.762	−27.5	29.746	29.753	−18.5	29.761
March . . .	29.849	−17	29.854	29.864	− 3.8	29.852
April . . .	29.895	− 0.2	29.893	29.907	+13.9	29.918
May	29.985	+13.5	29.957	29.973	+31.5	29.978
June	29.886	26.9	29.877	29.897	38.2	29.868
July	29.682	32.7	29.693	29.694	40.6	29.702
August . .	29.643	31.5	29.636	29.661	[illegible]	29.667
September .	29.883	22.2	29.883	29.895	28.3	29.894
October . .	29.967	10.7	29.981	29.981	18.1	29.985
November .	29.875	−22.6	29.876	29.884	−13.4	29.882
December .	29.756	−32.5	29.739	29.741	−25.4	29.726
1823.						
January . .	29.877	−20.2	29.902	29.898	−10.6	29.893
February .	29.904	−24.9	29.906	29.905	−13.4	29.907
March . . .	30.050	−24.1	30.055	30.050	−12	30.061
April . . .	29.957	− 9.	29.955	29.957	+ 7.5	29.954
May	29.929	+16.9	29.916	29.920	33.3	29.921
June	29.922	23.4	29.910	29.909	41.2	29.909
July	29 507	33.2	29.499	29.509	43.8	29.508
Mean . . .	29.874		29.872	29.880		29.879
Difference	−.005		−.002	+.008		−.001

It appears from this table that the rise in the mercurial column from 8 A. M. to 4 P. M. was nearly constant; and upon further examination it will be found that in the only two exceptions of any amount, namely, the months of January and March 1823, some unusual influence prevailed in the atmosphere. The first was distinguished by an unusually high mean temperature, and frequent storms of wind. Captain Parry remarks in his Journal, " from the morning of the 24th till midnight on the 26th, the mercury in the barometer was never below 30.32 inches, and at noon on the latter day had reached 30.52 inches, which was the highest we had yet observed it in the course of this voyage. This unusual indication of the barometer was followed by hard gales on the 27th and 28th, first from the south-west, and afterwards from the north-west, the mercury falling from 30.51 inches at 8 P.M. on the 26th, to 30.25, about 5 P.M. on the 27th, or about 0.26 of an inch in nine hours before the breeze came on. At midnight on the 27th, it had reached 29.30; and on the following night 29.05, which was its minimum indication during the gale. These high winds were accompanied by a rise in the thermometer very unusual at this season of the year, the temperature continuing above 0° for several hours, and very near this point of the scale for the whole two days.

The month of March, on the contrary, was as much below the mean temperature, as January was above it; and the observation renders it probable that the usual course of the season was modified by some extraneous cause.

I am aware that it may be objected, that these observations were not made with all the precision that the accurate determination of such small quantities requires, and particularly that the heights of the barometer were not corrected for the variations of temperature. The objection, to some extent, is certainly valid, and it is much to be lamented that the advantages of the utmost attainable degree of precision in these observations had not been duly appreciated: but when it is recollected that the instrument made use of was placed in the cabin of the ship, where considerable pains were taken to maintain an equal temperature, it will be found that less importance attaches to the omission in this particular instance than might at first be supposed. In this voyage, more especially, the precautions which were adopted to secure this important end were eminently successful. It appears, for instance, by Captain Parry's register, that in the months of October and November, the mean temperature of the external air varied 32°, while that of the air of the lower deck only varied 5°; so that the changes in the course of the twenty-four hours could have been scarcely appreciable.

The return, however, of Captain Parry's last expedition has left nothing to be desired in the way of accuracy. The instruments which were provided upon this occasion were of the first excellence; and the barometers were independently graduated and compared with one another, as well as with the standard of the Royal Society. The proper corrections were applied as the observations were made; and the meteorological journal, kept by Lieut. Foster, is a complete model of all that is desirable in such a register. As opportunities will not probably soon again occur for making similar observations in such high latitudes, it is well that it should be known that this series, and this alone, is to be depended upon for accurate purposes; and they will, doubtless, hereafter furnish important data in resolving the question of the mean height of the barometer at the level of the sea, in different latitudes; which the carelessness of observers and instrument-makers does not yet permit to be determined in more accessible regions.

While upon this subject, it may be as well to mention, that the thermometric observations are entitled to equal confidence. The thermometers, at the lowest temperatures, agreed within three degrees with one another, while, upon former occasions, there was a difference of many degrees.

At my particular request, the register of the barometer was

kept in such a way as to elicit most advantageously the phenomena of the horary oscillations. The following observations and tables of results are extracted from Captain Parry's interesting Journal:—" The most rigid attention to the observation and correction of the column during several months, discovered an oscillation amounting only to ten thousandth parts of an inch; the times of the maximum and minimum altitude appear, however, decidedly to lean to four and ten o'clock, and to follow a law directly the reverse, as to time, of that found to obtain in temperate climates, the column being *highest at four*, and *lowest at ten o'clock*, both A. M. and P. M. The whole of the observations, being comprised in the ' Meteorological Abstracts,' with the general results stated at the bottom of each, can be consulted with great convenience; and the following Table will afford one comprehensive view of six months' observations on this interesting subject."

ABSTRACT.

" THE mean result of six months' observations at Port Bowen, (N. lat. 73° 48′, long. 88° 54′,) in which the barometer was registered at the hours of 3, 4, 9, and 10, are here collected into one table; and in a second table is given a comparative view of three months' observations, in which it was registered at the additional hours of 5 and 11. On reference to these tables, it will be seen that the general tendency seems to indicate high barometer at four o'clock, and low at ten in the morning. The evening tide, though less regular, is also highest at four, but lowest at eleven o'clock. The changes, however, are in themselves so extremely minute (amounting to only the hundredth part of an inch), that a sudden alteration in the atmosphere causing the barometer to rise or fall rapidly on any one day, is sufficient to introduce an anomaly sensibly affecting the mean result of a whole month.

COMPARATIVE VIEW of the MEAN PRESSURE of the ATMOSPHERE at the Hours of 3, 4, 9, and 10 A.M. and P.M., during six successive Months, 1824, 25.

DURING THE MONTH OF	MEAN PRESSURE of the ATMOSPHERE at							
	3 A.M.	4 A.M.	9 A.M.	10 A.M.	3 P.M.	4 P.M.	9 P.M.	10 P.M.
1824.	Inches.	Inches.	Inches.	Inches.	Inches.	Inches.	Inches.	Inches.
November . .	29.8943	29.8990	29.8986	29.8971	29.9067	29.9037	29.8971	29.8909
December . .	29.8726	29.8767	29.8729	29.8687	29.8693	29.8695	29.8663	29.8551
1825.								
January . . .	29.7610	29.7668	29.7580	29.7541	29.7614	29.7599	29.7677	29.7617
February . .	29.8921	29.8938	29.8889	29.8788	29.8864	29.8890	29.8865	29.8835
March . . .	30.1064	30.1103	30.1101	30.1041	30.1095	30.1105	30.1080	30.1049
April	30.0639	30.0697	30.0653	30.0594	30.0698	30.0681	30.0739	30.0665
Means . . .	29.9317	29.9359	29.9323	29.9270	29.9338	29.9334	29.9332	29.9271

COMPARATIVE VIEW of the MEAN PRESSURE of the ATMOSPHERE at the Hours of 3, 4, 5, 9, 10, and 11 A.M. and P.M., during three successive Months, 1825.

DURING THE MONTH OF	MEAN PRESSURE of the ATMOSPHERE as observed at											
	3 A.M.	4 A.M.	5 A.M.	9 A.M.	10 A.M.	11 A.M.	3 P.M.	4 P.M.	5 P.M.	9 P.M.	10 P.M.	11 P.M.
	Inches.	Inches.	Inches.	Inches.	Inches.	Inches.	Inches.	Inches.	Inches.	Inches.	Inches.	Inches.
Feb.	29.8921	29.8938	29.8890	29.8889	29.8788	29.8811	29.8864	29.8890	29.8899	29.8865	29.8835	29.8812
Mar.	30.1064	30.1103	30.1088	30.1101	30.1041	30.1028	30.1095	30.1105	30.1079	30.1080	30.1049	30.1025
April	30.0639	30.0697	30.0670	30.0653	30.0594	30.0610	30.0698	30.0681	30.0712	30.0739	30.0665	30.0649
Means	30.0208	30.0246	30.0216	30.0214	30.0141	30.0150	30.0219	30.0225	30.0230	30.0228	30.0183	'30.0162

Thus, I think that I am entitled to say, that the test which I had proposed, to ascertain whether my hypothesis of the cause of the horary oscillations of the barometer were founded upon the laws of nature, has confirmed its truth, and furnished results which I was enabled to anticipate by its conclusions, and its conclusions alone. Further consideration has suggested to me other unforeseen conditions of the problem which the proposed solution appears to me quite sufficient to satisfy.

The hypothesis supposes, that two horizontal currents of air flowing in opposite directions, one above the other, balance each other so exactly, that, as long as no disturbing influence deranges the adjustment, their perpendicular pressure, as measured by the barometer at the surface of the earth, is equal in every part of their course, and never varies: every particle of these two currents is carried forward with a given momentum, which would continue its motion even after the cause which gave the original impulse had ceased to act. Now, if we suppose the course of the lower current to be checked by some extraneous influence, the upper current will still move forward from its momentum with its original velocity; and at the commencement of its course will create a deficiency, and a diminished pressure of the perpendicular column comprising both the currents. At the same moment, however, at its other extremity, it must accumulate upon itself to the same degree; and perpendicular columns of the atmosphere, weighed at equal intervals between the two extremities, will afford a series of gradually increasing pressures, exceeding the mean on one side, and deficient on the other.

The momentum, however, which caused this wave being expended, what must be the course of the phenomena? The increased pressure, at the point of greatest accumulation, acts against the natural flow of the upper current, but impels the lower in its proper course. The latter, in its turn, now rushes forward with undue velocity, and, continuing its course even after the impulse has been expended, accumulates a wave at the other extremity of the system. A series of increasing pressures, in the opposite direction, will hence result, and the same neutral point will separate the deficient and redundant halves of the progression.

Now, it has been proved, from the constitution of the atmosphere, that a system of compensating currents must prevail between either pole and the equator; whose general tendency upon the surface of the globe must be from the colder regions to the hotter, but in the upper stratum from the hotter to the colder. These systems are subjected to the alternate contracting and expanding forces produced by the changes of temperature from day to night; which changes, before they can equally diffuse themselves through the system, affect the regular adjustment of the currents: those, which are in contact with the surface from whence the changes are derived, first feel their influence.

The course of the phenomena in the earth's atmosphere, as indicated by the horary oscillations of the barometer, would appear to be as follows:—The rising heat of the day, communicated to the air from the surface of the globe, checks the lower current in its progress from the poles to the equator, by causing the warm particles to ascend, and deflecting them from their horizontal course. This heating process is not instantly, or perhaps equally, communicated to the upper current, which meanwhile flows on with its acquired momentum, and the aërial fluid is thus drawn from the equatorial regions and heaped upon the polar; passing by a neutral point, where the balance is maintained by an accumulation exactly equal to the excess of velocity: the maxima of these effects occur about one hour and a half or two hours after the earth has acquired its greatest temperature. By this time some of the momentum of the upper current is expended, and the accumulation at the poles commences its re-action, which presses the lower current forward and checks the upper. It is the former which now advances with a velocity greater than is due to an exact balance of the two, and carried forward, in its turn, by its acquired momentum, draws the atmosphere from the poles, and heaps it upon the equator: the same neutral point is, however, maintained by the same means as before. While this action proceeds, the surface of the earth returns to its medium state of temperature, and the maximum effect of this revulsion occurs about two hours after. The cooling process, however, proceeds beyond this point; and the effect of radiation falling primarily upon the lower current, checks its progress by an analogous, though contrary, effect to that by which it was before retarded. The particles of air are now drawn downwards, and the stratum contracts within itself. The upper current meanwhile having again acquired momentum from the accumulation at the equator, now acting with it and against the lower, flows forwards past the point of equipoise, and again accumulates upon the poles the fluid which is drawn from the equator. The maximum of this effect occurs not far from the period at which the surface of the earth is furthest removed from its mean state of temperature in the opposite extreme to that which we before considered. Another reaction follows, while the earth recovers the mean heat of the twenty-four hours; and, as the influence of the sun carries it beyond this term, the course of the phenomena is again renewed.

I am well aware that my previous statement of this hypothesis has been very incomplete, and that from want of sufficient explanation it has been misunderstood. I have here endeavoured to re-state it in other words, with more distinctness, but I am not quite sure that I have been more successful; and I frankly own that I have great difficulty in conveying my ideas of this succession of intricate phenomena in a manner satisfactory even to myself. I can only hope that those who will devote sufficient attention to the remarks, may perceive the clue which may guide them to a further elucidation of a problem, which yet requires the assistance of many well-combined observations to enable us to solve it in a satisfactory manner.

The grand test of the correctness of an hypothesis is its capability of explaining phenomena which were either unknown or overlooked at the time of its original construction. I have shown that one consequence necessarily flowing from my explanation of the horary oscillations of the barometer, and which was wholly unknown at the time, has been confirmed by subsequent observations; and I shall now proceed to show that another uncontemplated circumstance is not inconsistent with it.

Baron Humboldt observes, that, " in the torrid zone the limit hours (that is, the instants when the oscillations attain the *maximum* and *minimum*) are the same at the level of the sea, and on table-lands, at the elevation of from 1300 to 1400 toises. It is asserted, that this isochronism is not manifested in some parts of the temperate zone, and that at the convent of the Great St. Bernard, for instance, the barometer lowers at the same hours when it rises at Geneva."—(Personal Narrat. vol. vi. p. 758.)

This assertion is confirmed by the registers of the Palatine Society, in which it will be found that the barometer upon the summit of St. Gothard rose from morning to afternoon, and fell from afternoon to night, with almost as much regularity as it followed the contrary course at the level of the sea. It is true that two o'clock, the time of the afternoon observation, is not the hour the best calculated to exhibit the full amount of the oscillation, but it sufficiently establishes its general tendency.

Now, in what way does the hypothesis meet this exception to the general law? At page 16 of my essay upon the constitution of the atmosphere, I have shown, that the weight of a column of air at its base will not be affected by any alteration of temperature, but that the weight of any of its superior sections will be affected by the contraction and expansion. As the column expands, the barometer, which measures the pressure of any upper section, will rise in proportion to its height above the base, on account of the different distribution which takes place of the ponderable matter.

The summit of an insulated mountain, or the narrow ridge of

a chain of mountains, may be considered to approximate in their situations, with regard to the atmosphere, to a section of an atmospheric column of an equal height; for, although it is true that the column of the atmosphere immediately above such summit or ridge, rests upon it as a base, yet this base, presenting as it were but a point, or a line, with regard to the ambient mass, the incumbent air must necessarily be subject to all the fluctuations of the elastic ocean, by which it is hemmed in. The great body of the atmosphere expanding with the rising heat of the day flows over the insulated station, and obliterates the effect of that regular drain which would be due to it, at a lower level, from the cause which we have been investigating.

The same reasoning does not apply to extensive table-lands, however lofty: the extent of the basis upon which the aërial fluid rests, in such situations, constitutes it an independent atmosphere; in which we find the same effects following the same causes, as in equally extended plains at the level of the sea.

Trusting to the sufficiency of this explanation, and the general correctness of the hypothesis, I will venture to anticipate, that although, in the torrid zone, the horary oscillations of the barometer are found upon *table-lands* at the height of 1300 or 1400 toises, to be isochronous with those at the level of the sea, if ever a sufficient number of observations should be obtained at equal heights upon *insulated peaks*, in the same latitudes, they will indicate that the regular law is masked by the action which I have endeavoured to explain; while, on the other hand, with Baron Humboldt, I do not doubt that, notwithstanding the contrary result upon the summits of St. Gothard and St. Bernard, " in the elevated plains of La Mancha in Spain, at 320 toises we should see the barometer ascend at the same hours as at Valencia or Cadiz."

CLIMATE CONSIDERED WITH REGARD TO HORTICULTURE.

Page 534.

During my stay at Berlin, I was informed that one of the hot-houses of the Botanical Garden in the neighbourhood of that city was constructed with double glass. I was prevented by several unforeseen disappointments from seeing the construction, and making such inquiries as might have explained the steps by which experience had been led to the adoption of means which I, being totally unaware of the fact, had recommended from theoretical considerations alone. Upon my return, I found that the existence of such houses was totally unknown in this country, and Mr. Lindley kindly undertook to ascertain all the particulars from the correspondents of the Horticultural Society. The following is an extract of a letter from Mr. Otto, which he has lately received, in answer to his inquiries :—

" In the year 1804, a hot-house with double windows was erected in the Botanic Garden at Berlin, which completely answered the purpose expected from it. Since that time most of the hot-houses have been built in a similar way, and provided with double windows. The main objects in the application of these windows are the following: viz. to save the very laborious task of covering and uncovering the windows during rainy or cold weather, by means of wooden shutters or mats ; to admit, uninterruptedly, the light to the plants—which cannot be the case if the windows are covered; protection against cold and wet, whereby the dropping of rain or snowy weather is totally avoided. These would be the chief advantages of houses with double windows. On the other hand, it cannot be denied, that a great deal of the sun s light and warmth is withheld from the plants by the obstruction of the sun's rays; and on this account the double windows are applied only to larger houses, in which stout and full-grown plants are cultivated. In smaller ones, where young plants only are kept, as also in forcing-houses for fruit, &c., I do not find them equally applicable, because much light and warmth of the sun is lost by them. We generally put on the double windows here in September, on the tropical houses; but on the conservatories, later in November. Those standing in the front are generally put on later, and again removed earlier, which is entirely regulated by the state of the weather. The tropical, as well as all New Holland and Cape plants, keep extremely well in these houses, and I shall probably never follow any other method. Even the snow never lies long on the upper windows, but melts and runs down on them.

"17 *March*, 1827."

INDEX.

The reader is requested to correct the following errata—the only two which materially affect the sense:—

Page 36, line 5, *for* 1,000, *read* 10,000.
—— 389, — 11, — sixth, — sixtieth.

LONDON:
Printed by WILLIAM CLOWES,
Stamford-street.

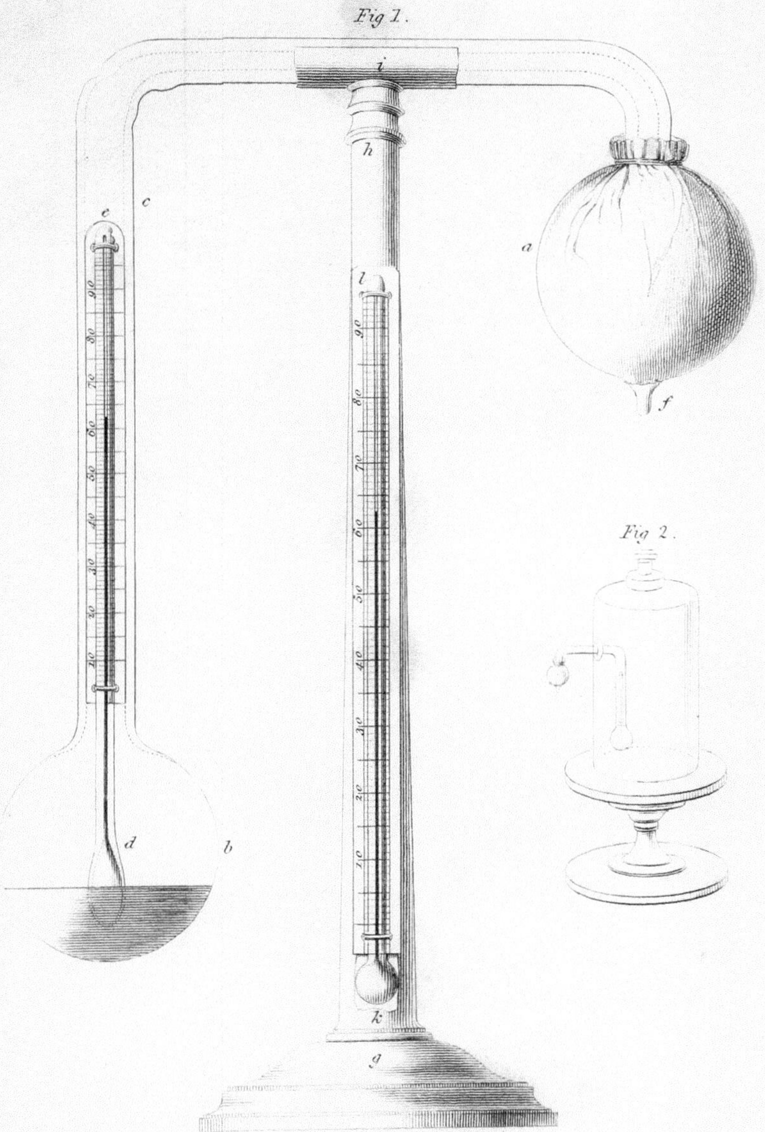

Mr Daniell's Hygrometer

Wm. Tite delt. J. Basire sculpt.

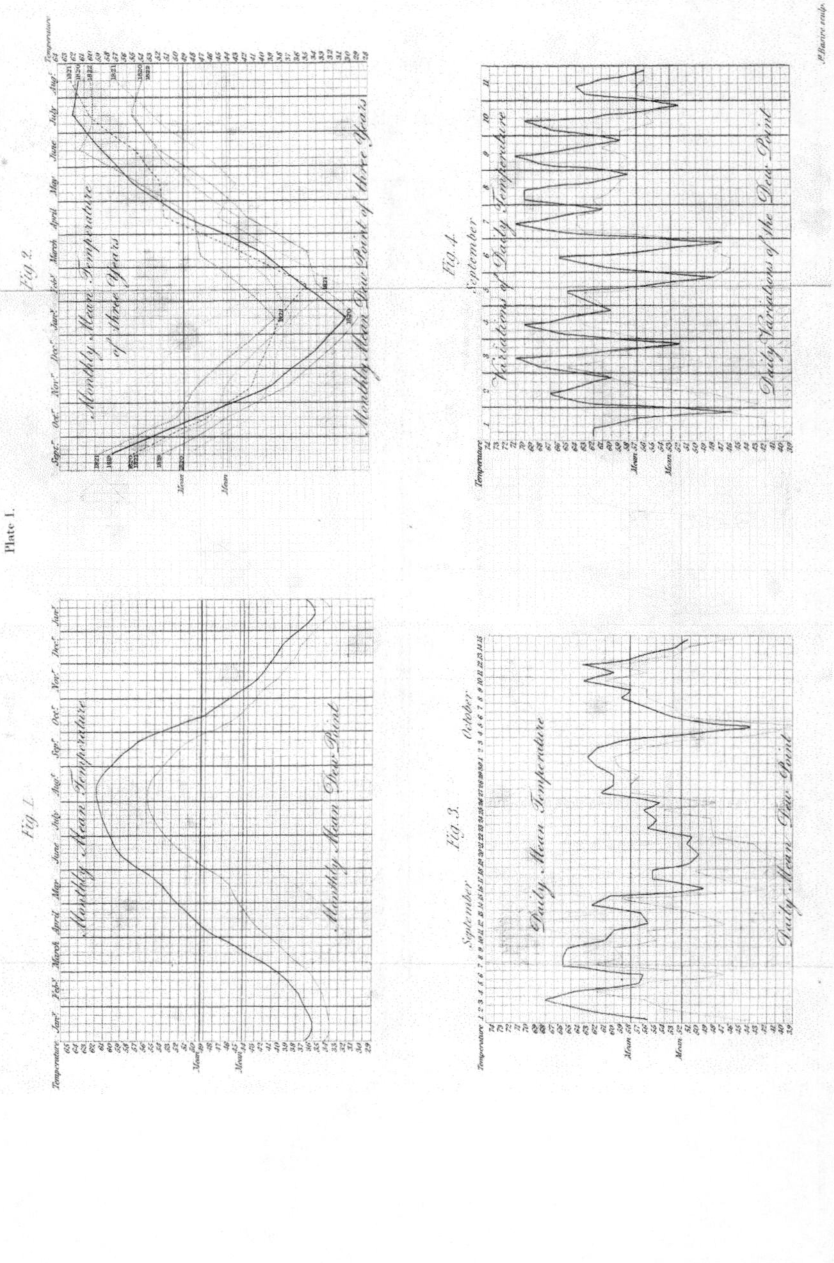

The material originally positioned here is too large for reproduction in this reissue. A PDF can be downloaded from the web address given on page iv of this book, by clicking on 'Resources Available'.

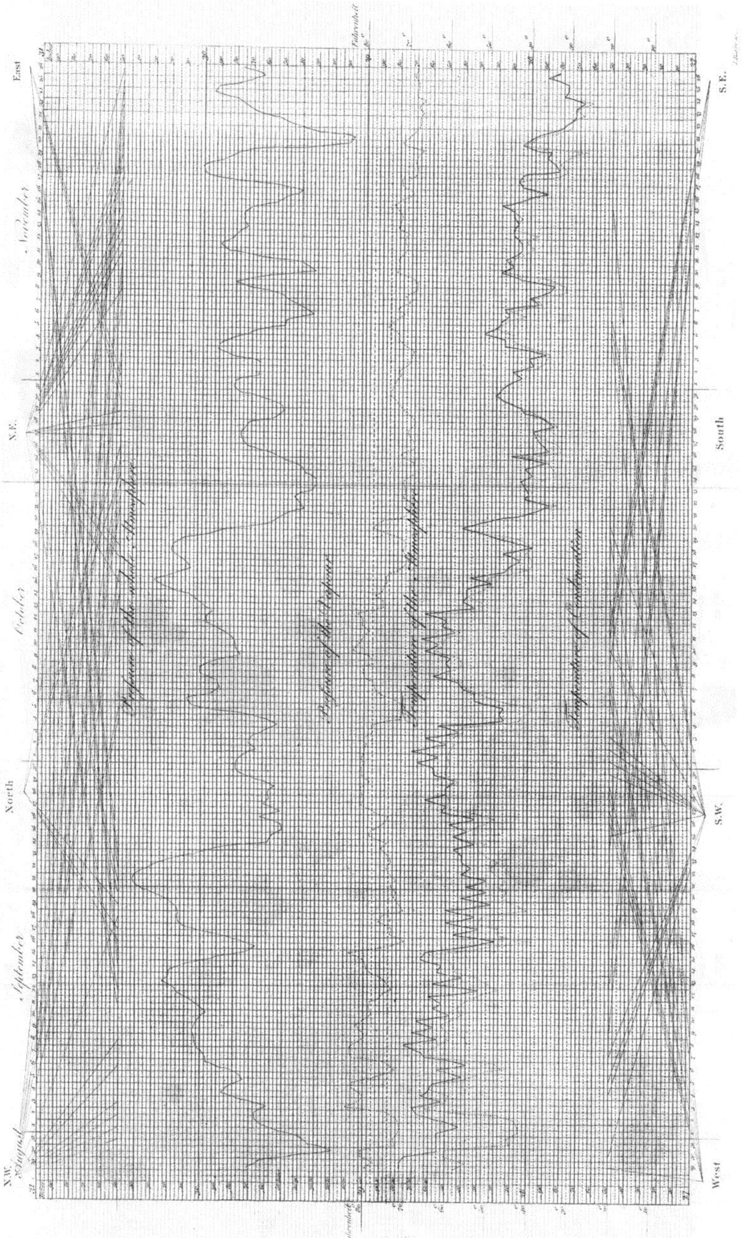

The material originally positioned here is too large for reproduction in this reissue. A PDF can be downloaded from the web address given on page iv of this book, by clicking on 'Resources Available'.

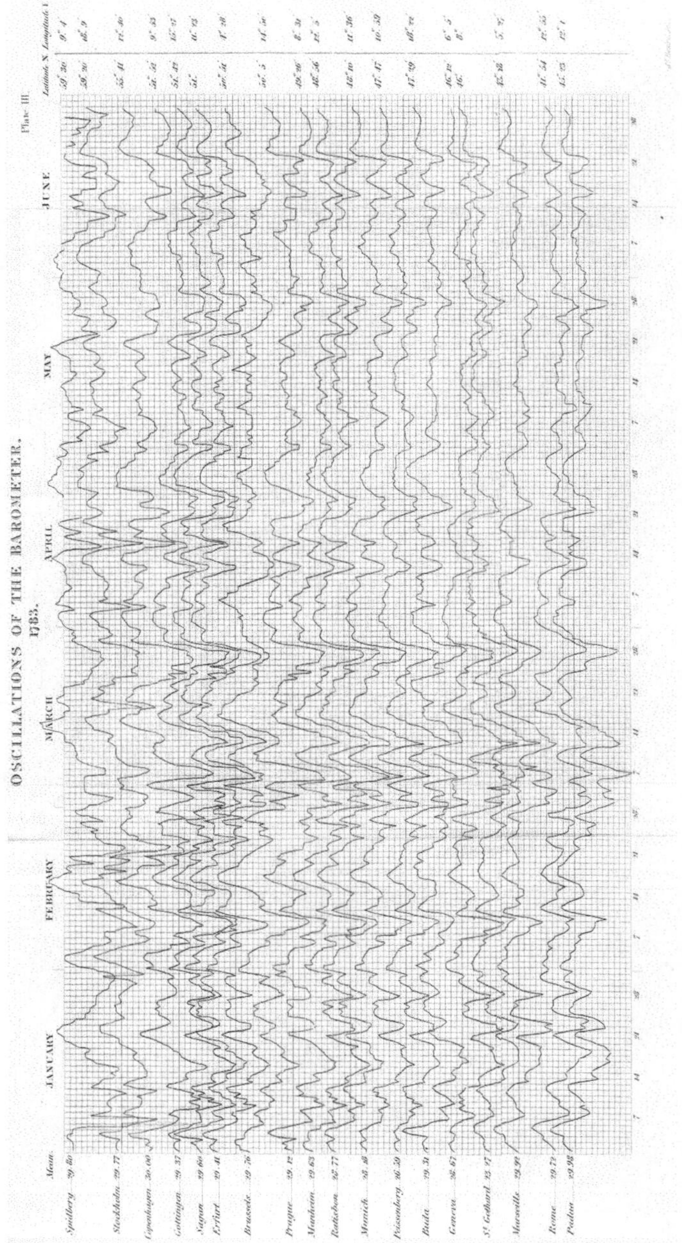

The material originally positioned here is too large for reproduction in this reissue. A PDF can be downloaded from the web address given on page iv of this book, by clicking on 'Resources Available'.

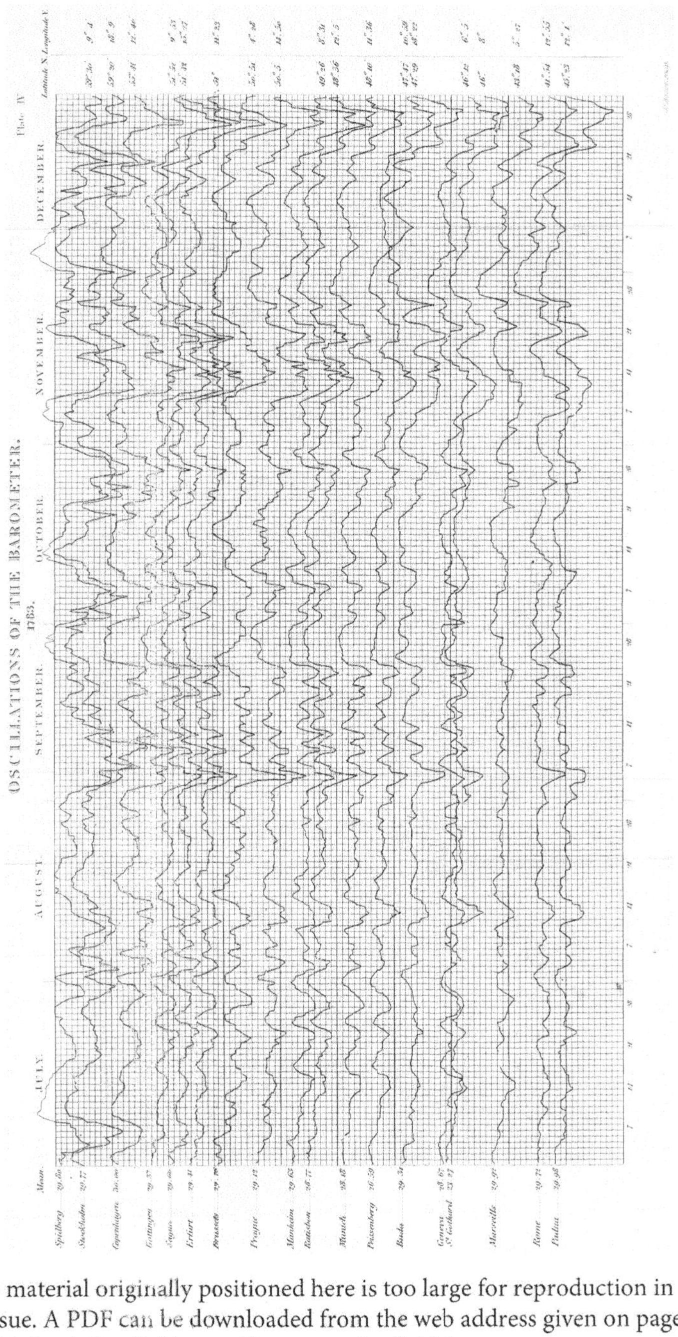

The material originally positioned here is too large for reproduction in this reissue. A PDF can be downloaded from the web address given on page iv of this book, by clicking on 'Resources Available'.

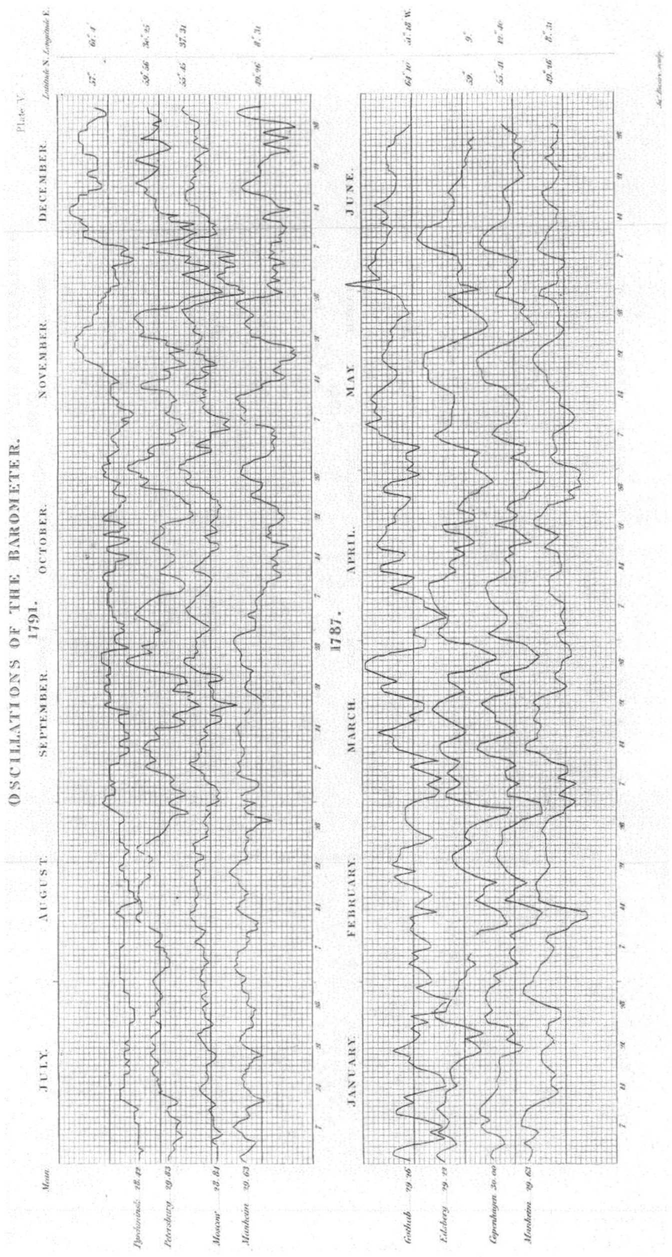

The material originally positioned here is too large for reproduction in this reissue. A PDF can be downloaded from the web address given on page iv of this book, by clicking on 'Resources Available'.